BIOMATERIALS SCIENCE AND ENGINEERING

BIOMATERIALS SCIENCE AND ENGINEERING

JOON BU PARK

College of Engineering
University of Iowa
Iowa City, Iowa

PLENUM PRESS • NEW YORK AND LONDON

Library of Congress Cataloging in Publication Data

Park, Joon Bu.
  Biomaterials science and engineering.

  Bibliography: p.
  Includes index.
  1. Biomedical materials. 2. Biomedical materials — Physiological effect. 3. Biomedical engi-
neering. 4. Prosthesis. I. Title.
R857.M3P38 1984   610'.28   84-16016
ISBN 0-306-41689-1

First Printing — October 1984
Second Printing — July 1987

© 1984 Plenum Press, New York
A Division of Plenum Publishing Corporation
233 Spring Street, New York, N.Y. 10013

Dedicated to

my family

Bea Young
Mi Sun
Yoon Ho
Yoon Il

my brothers

Si Joon
Hyun Joon

my former teachers

Jae Gee Cha (*Korea*)
K. L. DeVries (*University of Utah*)
A. S. Hoffman (*University of Washington*)
W. O. Statton (*Hawaii*)
D. R. Uhlmann (*MIT*)

PREFACE

This book is written for those who would like to advance their knowledge beyond an introductory level of biomaterials or materials science and engineering. This requires one to understand more fully the science of materials, which is, of course, the foundation of biomaterials.

The subject matter of this book may be divided into three parts: (1) fundamental structure–property relationships of man-made materials (Chapters 2–5) and natural biological materials, including biocompatibility (Chapters 6 and 7); (2) metallic, ceramic, and polymeric implant materials (Chapters 8–10); and (3) actual prostheses (Chapters 11 and 12).

This manuscript was initially organized at Clemson University as classnotes for an introductory graduate course on biomaterials. Since then it has been revised and corrected many times based on experience with graduate students at Clemson and at Tulane University, where I taught for two years, 1981–1983, before joining the University of Iowa. I would like to thank the many people who helped me to finish this book; my son Yoon Ho, who typed all of the manuscript into the Apple Pie word processor; my former graduate students, M. Ackley Loony, W. Barb, D. N. Bingham, D. R. Clarke, J. P. Davies, M. F. DeMane, B. J. Kelly, K. W. Markgraf, N. N. Salman, W. J. Whatley, and S. O. Young; and my colleagues, Drs. G. H. Kenner (University of Utah), F. W. Cooke, D. D. Moyle (Clemson University), W. C. Van Buskirk (Tulane University), and Y. King Liu (University of Iowa). Special thanks are extended to Mr. L. S. Marchand of Plenum Publishing Corporation, who encouraged me to finish this manuscript and gave me the opportunity to publish an earlier introductory book on biomaterials.

Every attempt has been made to make the book short and readable. Most illustrations were drawn by me in order to make them more presentable. Any mistakes are mine and I hope the reader will bring them to my attention.

Joon Bu Park

Iowa City, Iowa

CONTENTS

1

INTRODUCTION

1.1. DEFINITION OF BIOMATERIALS

The word *biomaterials* can be defined in two ways: as commonplace biological materials such as tissues and woods or as any materials that replace the function of the living tissues or organs. In legal terms (Clemson Advisory Board for Biomaterials "Definition of the word 'biomaterials,'" The 6th Annual International Biomaterial Symposium, April 20–24, 1974) "a biomaterial is a systemically, pharmacologically inert substance designed for implantation within or incorporation with a living system." This definition clearly emphasizes biomaterials as implant materials although the conventional usage of the prefix *bio* is somewhat violated; for example, biochemistry and biophysics refer to the study of biological materials rather than man-made materials. In order to avoid confusion, *biomaterials* will refer to implants replacing and restoring living tissues and their functions. From this definition, (implantable) *biomaterials* includes anything that is intermittently or continuously exposed to body fluids although they may actually be located outside of the body proper. Included in this category are most dental materials although traditionally they have been treated as separate entities. Devices such as external artificial limbs, hearing aids, and external facial "prostheses" are not implants.

Because the ultimate goal of using biomaterials is to restore function of natural living tissues and organs in the body, it is essential to understand relationships among properties, functions, and structures of biological materials. Thus, three aspects of study on the subject of biomaterials can be envisioned: biological materials, implant materials, and interaction between the two in the body. This is a very difficult task to master unless one possesses a fundamental knowledge of the whole system under study.

1

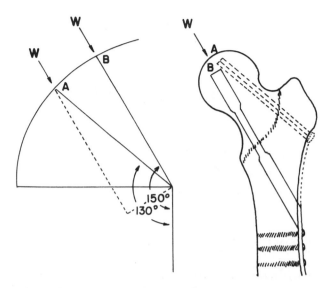

Figure 1-1. A biomechanical analysis of femoral neck fracture fixation. Note that if the implant is positioned at 130° rather than 150°, there will be a force component which will generate a bending moment at the nail–plate junction. The 150° implant is harder to insert and therefore not preferred by surgeons. (From Ref. 1.)

Another important area of study is that of the mechanics and dynamics of tissues and the resultant interactions between them. Generally, this study, known as *biomechanics*, is incorporated in the design and insertion of implants as shown in Figure 1-1.

The performance of an implant after insertion can be considered in terms of reliability. For example, there are four major factors contributing to the failure of hip joint replacements. These are fracture, wear, infection, and loosening of the implants as shown in Figure 1-2. If the probability of failure of a given system is assumed to be f, then the reliability, r, can be expressed as

$$r = 1 - f \qquad (1\text{-}1)$$

The total reliability r_t can be expressed in terms of the reliabilities of each contributing factor for failure:

$$r_t = r_1 \cdot r_2 \ldots r_n \qquad (1\text{-}2)$$

where $r_1 = 1 - f_1$, $r_2 = 1 - f_2$, and so on.

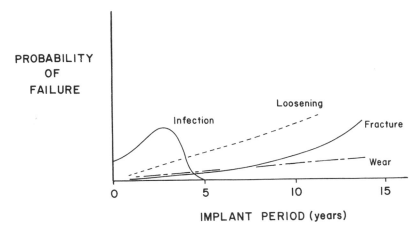

Figure 1-2. A schematic illustration of probability of failure versus implant period for hip joint replacements. (From Ref. 2.)

Equation (1-2) implies that even though the implant is a perfect one (i.e., $r = 1$), if an infection occurs every time it is implanted, then the total reliability of the operation is zero.

The study of structure—property relationships of biological materials is as important as that of biomaterials but traditionally this subject has not been treated fully in biologically oriented disciplines. This is due to the fact that these studies pertain to biochemical function rather than viability or nonviability of "materials". In many cases one can study biological materials while ignoring the fact that they are made of and made from living cells. In other cases the functionality of the tissues or organs is so vital that it is meaningless to replace them with biomaterials, e.g., the spinal cord or the brain. In this case the subject will be treated as if it does not contain living cells unless this is a contributing factor to the properties or function of the materials in the body.

The success of a biomaterial or an implant is highly dependent on three major factors: properties and biocompatibility of the implant, health condition of the recipient, and competency of the surgeon who implants and monitors progress of the implant. It is easy to understand the requirements for an implant by examining the required characteristics of a bone plate for stabilizing a fractured femur after an accident. These are:

1. Acceptance of the plate to the tissue surface, i.e., biocompatibility (this is a broad term and includes points 2 and 3)
2. Nontoxic and noncarcinogenic

Table 1-1. Materials for Implantation

Materials	Advantages	Disadvantages	Examples
Polymers Silastic® rubber Teflon® Dacron® Nylon	Resilience; easy to fabricate; low density	Low mechanical strength; time-dependent degradation	Sutures, arteries, veins; maxillofacial: nose, ear, maxilla, mandible, teeth; cement, artificial tendon
Metals 316, 316L S.S. Vitallium® Titanium alloys	High-impact tensile strength; high resistance to wear; ductile adsorption of high strain energy	Low biocompatibility; corrosion in physiological environment; mismatch of mechanical properties with soft connective tissues; high density	Orthopedic fixation: screws, pins, plates, wires; intermedullary rods, staples, nails; dental implants
Ceramics Aluminum oxides Calcium aluminates Titanium oxides Carbons	Good biocompatibility; corrosion resistance; inert; high compression resistance	Low-impact tensile strength; difficult to fabricate; low mechanical reliability; lack of resilience; high density	Hip prosthesis; ceramic teeth; transcutaneous device
Composites Ceramic-coated metal Carbon-coated material	Good biocompatibility; inert; corrosion resistance; high tensile strength	Lack of consistency of material fabrication	Artificial heart valve (pyrolytic carbon on graphite); knee joint implants (carbon fiber-reinforced high-density polyethylene)

3. Chemically inert and stable (no time-dependent degradation)
4. Adequate mechanical strength
5. Adequate fatigue life
6. Sound engineering design
7. Proper weight and density
8. Relatively inexpensive, reproducible, and easy to fabricate and process for large-scale production

Most of the requirements are pertinent to the material properties that are the subjects of this book. The list in Table 1-1 illustrates some of the advantages, disadvantages, and applications for four groups of synthetic (man-made) materials used for implantation. Reconstituted (natural) materials such as collagen have been used for replacements such as arterial wall, heart valve, and skin.

An alternative to the artificial implant is transplantation, such as kidney and heart, but this effort has been hindered due to social, ethical, and immunological problems. However, in the case of kidney failure the patient's hope lies in transplantation because an artificial kidney has many disadvantages, such as high cost, immobility, and constant care and maintenance of the dialyzer.

The surgical uses of implant materials are given in Table 1-2.

Table 1-2. Surgical Uses of Biomaterials

Permanent implants
    Muscular skeleton system—joints in upper (shoulder, elbow, wrist, finger) and lower (hip, knee, ankle, toe) extremities, permanently attached artificial limb
    Cardiovascular system—heart (valve, wall, pacemaker, entire heart), arteries, veins
    Respiratory system—larynx, trachea, and bronchus, chest wall, diaphragm, lungs, thoracic plombage
    Digestive system—esophagus, bile ducts, liver
    Genitourinary system—kidney, ureter, urethra, bladder
    Nervous system—dura, hydrocephalus shunt
    Special senses—corneal and lens prosthesis, ear, carotid pacemaker
    Other soft tissues—hernia, tendons, visceral adhesion
    Cosmetic implants—maxillofacial (nose, ear, maxilla, mandible, teeth), breast, eye, testes, penis, etc.
Transient implants
    Extracorporeal assumption of organ function—heart, lung, kidney, liver
    Decompressive—drainage of hollow viscera-spaces, gastrointestinal (biliary), genitourinary, thoracic, peritoneal lavage, cardiac catheterization
    External dressings and partial implants—temporary artificial skin, immersion fluids
    Aids to diagnosis—catheters, probes
    Orthopedic fixation devices—general (screws, hip pins, traction), bone plates (long bone, spinal, osteotomy), intertrochanteric (hip nail, nail–plate combination, threaded or unthreaded wires and pins), intramedullary (rods and pins), staples
    Sutures and surgical adhesives

1.2. BRIEF HISTORICAL BACKGROUND

Historically speaking, until Dr. J. Lister's aseptic surgical technique was developed in the 1860s, various metal devices such as wires and pins constructed of iron, gold, silver, platinum, etc. were not largely successful mainly due to infection after implantation. Most of the modern implant developments have centered around repairing long bones and joints. Lane of England designed a fracture plate in the early 1900s using steel as shown in Figure 1-3a. Sherman of Pittsburgh modified the Lane plate to reduce stress concentration by eliminating sharp corners (Figure 1-3b). He used vanadium alloy steel for its toughness and ductility. Subsequently, Stellite® (Co–Cr-based alloy) was found to be the most inert material for implantation by Zierold in 1924. Soon 18-8 (18 w/o Cr, 8 w/o Ni), 18-8sMo (2–4 w/o Mo) stainless steels were introduced for their corrosion resistance, with 18-8sMo being especially resistant in saline solution. Later, another stainless steel (19 w/o Cr, 9 w/o Ni) named Vitallium® was introduced into medicine. Another noble metal, tantalum, was introduced in 1939 but its poor mechanical properties and difficulties of processing it from the ore made it unpopular in orthopedics, yet it found wide use in neurological and plastic surgery. During the post-Lister period, all the various designs and materials could not be related specifically to the success of an implant and it became customary to remove any metal implants as soon as possible after its initial function was served.

Fracture repair of the femoral neck was not initiated until 1926 when Hey-Groves used carpenter's screws. Later, Smith-Petersen (1931) designed the first nail with trifins to prevent rotation of the femoral head. He used stainless steel but soon changed to Vitallium®. Thornton (1937) attached a metal plate to the distal end of the Smith-Petersen nail and secured it with screws for better support. During this time Smith-Petersen (1939) used an

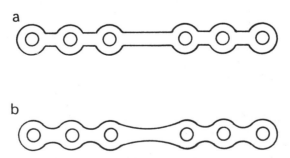

Figure 1-3. Early design of bone fracture plate: (a) Lane; (b) Sherman.

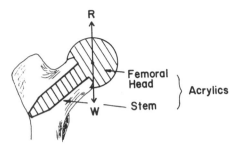

Figure 1-4. The Judet prosthesis for hip surface arthroplasty. (From Ref. 3.)

artificial cup over the femoral head in order to create new surfaces for the diseased joints. He used glass, Pyrex®, Bakelite®, and Vitallium®. The latter was found more biologically compatible and 30–40% of the patients gained usable joints. Similar mold arthroplastic surgeries were performed successfully by the Judet brothers of France who used the first biomechanically designed prosthesis made of an acrylic (methylmethacrylate) polymer (Figure 1-4). The same type of acrylic polymer was also used for corneal replacement in the 1940s and 1950s due to its excellent properties of transparency and biocompatibility.

Figure 1-5. An early model of the Starr–Edwards heart valve made of a silicone rubber ball and metal cage. (Courtesy of Starr–Edwards Co.)

Table 1-3. Notable Developments Relating to Implants[a]

Year	Investigator	Development
Late 18th–19th century		Various metal devices to fix fractures; wires and pins from Fe, Au, Ag, and Pt
1860–1870	J. Lister	Aseptic surgical techniques developed
1886	H. Hansmann	Ni-plated steel fracture plate
1893–1912	W. A. Lane	Steel screws and plates for fracture fixation
1909	A. Lambotte	Brass, Al, Ag, and Cu plate
1912	W. D. Sherman	Vanadium steel plate, first alloy developed exclusively for medical use; lesser stress concentration and corrosion
1924	A. A. Zierold	Stellite® (CoCrMo alloy), a better material than Cu, Zn, steels, Mg, Fe, Ag, Au, and Al alloy
1926	M. Z. Large	18-8sMo (2–4% Mo) stainless steel for greater corrosion resistance than 18-8 stainless steel
1936	C. S. Venable, W. G. Stuck	Vitallium® (developed in 1929; 19 w/o Cr–9 w/o Ni stainless steel)
1939	J. C. Burch, H. M. Carney	Ta
1926	E. W. Hey-Groves	Used carpenter's screw for femoral neck fracture
1931	M. N. Smith-Petersen	Designed first femoral neck fracture fixation nail made originally from stainless steel, later changed to Vitallium®
1938	P. Wiles	First total hip replacement
1946	J. and R. Judet	First biomechanically designed hip prosthesis. First plastics used in joint replacement
1947	J. Cotton	Ti and its alloys
1940s	M. J. Dorzee, A. Franceschetti	Acrylics for corneal replacement
1952	A. B. Voorhees, A. Jaretzta, A. H. Blackmore	First blood vessel replacement made of cloth
1958	S. Furman, G. Robinson	First successful direct stimulation of heart
1958	J. Charnley	First use of acrylic bone cement in total hip replacements
1960	A. Starr, M. L. Edwards	Heart valve
1970s	W. J. Kolff	Total heart replacement

[a] From Refs. 3 and 4.

Due to difficulty of surgical techniques and to material problems, cardiovascular implants were not attempted until the 1950s. Blood vessel implants were constructed with rigid tubes made of polyethylene, acrylic polymer, gold, silver, and aluminum. However, the major advancement was made by Voorhees, Jaretzta, and Blackmore (1952) when they used a cloth prosthesis made of Vinyon N copolymer (polyvinyl chloride and poly-acrylonitrile) and later experimented with nylon, Orlon®, Dacron®, Teflon®, and Ivalon®. Through the pores of the various cloths a pseudo- or neointima was formed by the tissue ingrowth which is compatible to blood and, in turn, prevented further blood coagulation. Heart valve implantation was made possible only after the development of open-heart surgery in the mid-1950s. Starr and Edwards (1960) made the first commercially available valve, consisting of a silicone rubber ball poppet in a metal strut (Figure 1-5). Concomitantly, artificial heart and heart assist devices have been developed. Table 1-3 gives a brief summary of the historical developments relating to implants.

PROBLEMS

1-1. Calculate the probability of failure of a hip arthroplasty after 1 and 10 years if the following probabilities are known:

a. Infection: after 1 year the probability of infection is about 5% but after 5 years it is negligible.
b. Probabilities of failure for wear, fracture, and loosening are given by the following equations:

$$f_w = 0.01 \exp(0.1t)$$

$$f_f = 0.01 \exp(0.12t)$$

$$f_l = 0.01 \exp(0.2t)$$

where w, f, l, and t are wear, fracture, loosening, and time (years), respectively.

1-2. List the advantages and disadvantages of kidney transplantation as compared to use of a dialysis machine.

1-3. Discuss the ethical and economic impact of artificial heart implants and heart transplants.

REFERENCES

1. W. K. Massie, Fractures of the hip, *J. Bone J. Surg. 46A*, 658–690, 1964.
2. J. H. Dumbleton, Elements of hip joint prosthesis reliability, *J. Med. Eng. Technol. 1*, 341–346, 1977.
3. D. F. Williams and R. Roaf, *Implants in Surgery*, Chapter 1, Saunders, Philadelphia, 1973.

4. D. C. Mears, *Materials and Orthopedic Surgery*, Chapter 1, Williams & Wilkins, Baltimore, 1979.

BIBLIOGRAPHY

J. Black, *Biological Performance of Materials*, Dekker, New York, 1980.

J. H. U. Brown, J. E. Jacobs, and L. Stark, *Biomedical Engineering*, Chapter 11, Davis, Philadelphia, 1971.

J. B. Park, *Biomaterials: An Introduction*, Plenum Press, New York, 1979.

L. Stark and G. Agarwal (ed.), *Biomaterials*, Plenum Press, New York, 1969.

S. A. Wesolowski, A. Martinez, and J. D. McMahon, *Use of Artificial Materials in Surgery*, Year Book Medical Publishers, Chicago, 1966.

D. F. Williams and R. Roaf, *Implants in Surgery*, Chapter 1, Saunders, Philadelphia, 1973.

CHARACTERIZATION OF MATERIALS

The physical characterization or properties required of a material for medical applications vary widely according to the particular application. Moreover, due to our imperfect understanding of tissue–material interactions, it is difficult if not impossible to translate the values of physicochemical properties of materials into *in vivo* performances. However, this should not prevent us from a thorough investigation of the characteristics of materials *in vitro* before using them as implants. On the contrary, the study of implant materials must start from a basic understanding of the behavior of materials under various conditions. In this chapter, we will try to limit our study by focusing attention on the basic elements of material characterizations.

2.1. MECHANICAL PROPERTIES

The fabrication and use of materials depend on such mechanical properties as strength, hardness, and ductility. The elastic and viscoelastic properties will be presented first before the static and dynamic characterizations.

2.1.1. Atomic Bonding and Elasticity (Hooke's Law)

The nature and strength of atomic bonding forces determine the stability of the solid under applied load (mechanical properties) or at varying temperatures (thermal properties). These binding forces can be described according to their attractive and repulsive forces as shown in

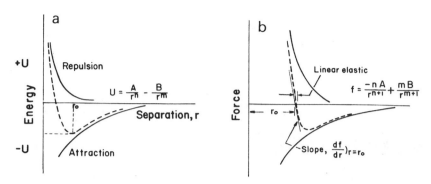

Figure 2-1. Schematic representation of atomic bonding energy (a) and force (b) versus distance in terms of attraction and repulsion of two atoms whose equilibrium distance is r_0.

Figure 2-1a. Generally, the repulsive forces act at a much shorter range than the attractive forces. Therefore, the equilibrium lattice spacing of a crystal structure is controlled largely by the repulsive forces. The interatomic energy or force of a pair of atoms separated by a distance r can be represented by

$$U = Ar^{-n} - Br^{-m} \qquad (2\text{-}1)$$

where Ar^{-n} is the attractive and Br^{-m} is the repulsive energy, A and B are constants, and $m > n$. Addition of the attractive and repulsive forces will give a minimum at some equilibrium distance, r_0, and the *cohesive* or binding energy of the bond is U_0.

When mechanical force f is applied to separate the atoms a distance dr along its direction of application, the interaction energy will be changed by the amount dU so that

$$dU = f dr \qquad (2\text{-}2)$$

Therefore,

$$f = dU/dr = -nAr^{-n-1} + mBr^{-m-1} \qquad (2\text{-}3)$$

Equation (2-3) will result in the same type of curve as in the energy versus distance curve (Figure 2-1b).

If we assume that U is a continuous function of r and, therefore, the displacement $x(=r-r_0)$ and $x \ll r_0$, then we can expand the energy function $U(x)$ as a *Taylor series* about $r_0(x=0)$:

$$U(x) = U_0 + U_0'x + 1/2U_0''x^2 + \text{higher terms} \qquad (2\text{-}4)$$

where U_0 means $U(x=0)$. Since $U_0 = 0$ at the minimum potential energy curve and for small strains the higher terms can be neglected, the force can be expressed as

$$f = dU(x)/dx = U_0''x^2 \qquad (2\text{-}5)$$

Since $U''[=d^2U(x)/dx]$ is constant at r_0, equation (2-5) is the statement of *Hooke's law*, that is, force is proportional to displacement. The proportionality constant is called the *elastic constant* and it has an indirect relationship to the binding energy, U_0. Equation (2-5) can be expressed in engineering terms as

$$\sigma = E\epsilon \qquad (2\text{-}6)$$

where σ is the stress, force per unit cross-sectional area (f/A), ϵ is the strain, change in length per original length $[(l-l_0)/l_0]$, and E is called the *elastic modulus* or *Young's modulus*. It can be shown further that Young's

Table 2-1. Moduli and Poisson's Ratios
for Some Materials[a]

	Elastic modulus E (GPa)	Shear modulus G (GPa)	Poisson's ratio
Cast iron	110.3	51.0	0.17
Mild steel	206.8	81.4	0.26
Aluminum	68.9	24.8	0.33
Copper	110.3	44.1	0.36
Titanium	106.9	—	—
Zirconium	93.8	35.8	0.31[b]
Lead	17.9	6.2	0.40
Soda lime glass	68.9	22.1	0.23
Nylon	2.8	1.0[b]	0.4
Hard rubber	2.8	0.8[b]	0.43
Rigid PVC[c]	3.5	1.3[b]	0.40

[a] From Ref. 1.
[b] Calculated value.
[c] Polyvinyl chloride.

modulus is related to the *shear* (G) and *bulk* (K) modulus for an isotropic material through *Poisson's ratio*, $(\nu = -\epsilon_x/\epsilon_z = -\epsilon_y/\epsilon_z)$ for cubic materials:

$$G = E/2(1+\nu) \qquad (2\text{-}7)$$

$$K = E/2(1-2\nu) \qquad (2\text{-}8)$$

Table 2-1 lists elastic moduli for various materials along Poisson's ratios.

2.1.2. Mechanical Property Measurements

The ability of a material to withstand *static* load can be determined by a standard tensile, compressive, and shear test. From a load–displacement curve a stress–strain diagram can be constructed by knowing its cross-sectional area and gauge length (Figure 2-2). The stress–strain curve of a solid can be demarcated by the *yield point* or *stress* (YS) into elastic and plastic

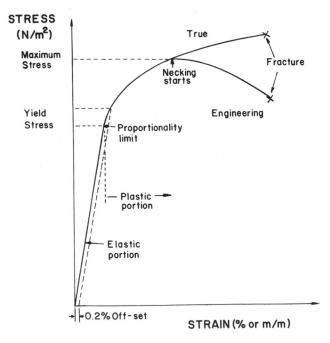

Figure 2-2. Schematic representation of a stress–strain curve for a ductile material.

regions. In the elastic region, the strain increases in direct proportion to the applied stress (Hooke's law) as given in equation (2-3).

In the plastic region, strain changes are no longer proportional to the applied stress. Further, when the applied stress is removed, the material will not return to its original shape but will be permanently deformed; this is termed *plastic deformation* (Figure 2-3).

The peak stress in Figure 2-2 is often followed by an apparent decrease until a point is reached where the material ruptures. The peak stress is known as the *tensile* or *ultimate tensile strength* (TS or UTS). The final stress where failure occurs is called the *failure* or *fracture strength* (FS).

In many materials such as stainless steel, definite yield points occur. This point is characterized by temporarily increasing strain without further increase in stress. Sometimes, when it is difficult to decipher the yield point for a given stress–strain curve, an offset yield point (usually 0.2%) is used in lieu of the original yield point.

Thus far, examples of "engineering" stress–strain curves have been examined. These differ from the true stress–strain curves in that the actual or true cross-sectional area is used to calculate the stress instead of the original cross-sectional area. The peak stress seen at the ultimate tensile point is due to the assumption of a constant cross-sectional area. When a specimen is loaded in tension, necking may occur (Figure 2-4), which changes the area over which the load is acting. If adjustments are made for the changes in the cross-sectional area, then a true stress–strain curve like that in Figure 2-2 can be obtained.

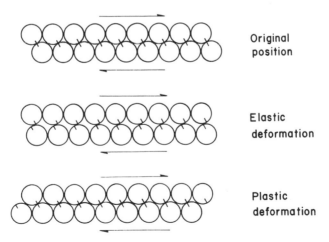

Figure 2-3. A two-dimensional atomic model showing elastic and plastic deformation.

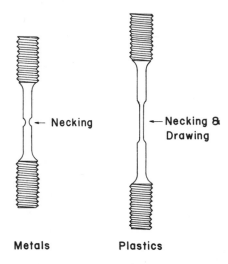

← Necking

← Necking & Drawing

Metals   **Plastics**

Figure 2-4. Deformation characteristics of metals and plastics under stress. Note that metals rupture without further elongation after necking occurs. In plastics the necked region undergoes further deformation called drawing, which is the basis of fiber formation.

Toughness is defined as the amount of energy required to produce failure and can be expressed in terms of stress and strain:

$$\text{Toughness (energy)} = \int_{\epsilon_0}^{\epsilon_f} \sigma\, d\epsilon = \int_{l_0}^{l_f} \sigma\, dl/l \qquad (2\text{-}9)$$

Expressed another way, toughness is the summation of (true) stress times the distance over which it acts (strain) taken in small increments. The area under the stress–strain curve provides a simple method of estimating toughness (Figure 2-5).

A material that can withstand high stresses and will undergo considerable plastic deformation (ductile-tough material) is tougher than one that resists high stresses but has no capacity for deformation (hard-brittle material) or one that has a high capacity for deformation but can only withstand relatively low stresses (ductile-soft or plastic material).

Two major characteristics of brittle fracture are that its fracture stress is far below the theoretical strength and that it is difficult to predict. The latter fact is the major reason why ceramic and glassy materials are not used extensively for implantation despite their excellent compatibility with tissues. The reason for the lower fracture stresses of brittle materials is discussed in Chapter 5.

Like toughness, *impact strength* is the amount of energy that can be absorbed by a material but the force is applied by impact. It can be measured by subjecting the specimen of known dimensions to a swinging pendulum (Figure 2-6). The amplitude change of the swing of the pendulum

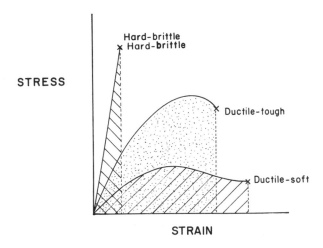

Figure 2-5. Stress–strain curves of materials exhibiting different characteristics under stress. The areas underneath the curves are the measure of toughness.

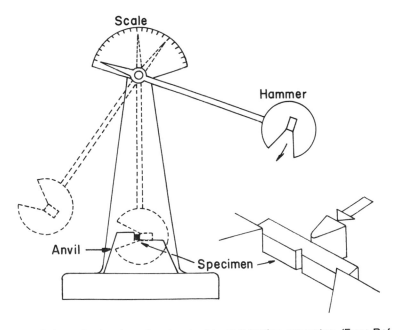

Figure 2-6. Schematic drawing of a standard impact testing apparatus. (From Ref. 2.)

is the measure of the energy absorbed by the specimen. From this the impact strength or energy can be calculated. Usually the impact testing requires a large number of samples because there is a large variation in results.

Hardness is a measure of plastic deformation and is defined as the force per unit area of indentation or penetration, and thus has the dimensions of stress. The stress figure obtained by the hardness test is 2.5–3 times higher than the tensile stress for a ductile material due to the impingement of the elastic material surrounding the plastic indentation which restricts free expansion of the indented region. A simplified view of the hardness test is shown in Figure 2-7.

Frequently used in hardness tests are Brinell, Vickers, Knoop, and Rockwell. They differ from each other mainly in their indenter material, its configuration, and the applied load (Table 2-2). For example, the Brinell hardness number (BHN) is obtained by using a 10-mm-diameter steel or tungsten carbide sphere that is pressed into the flat surface of a test specimen under a load of 300 or 3000 kg (29.42 kN) for 30s. The Vickers test utilizes a pyramid indenter made of diamond and low load (1.18 kN). The values of different hardness scales for various materials are given in Figure 2-8.

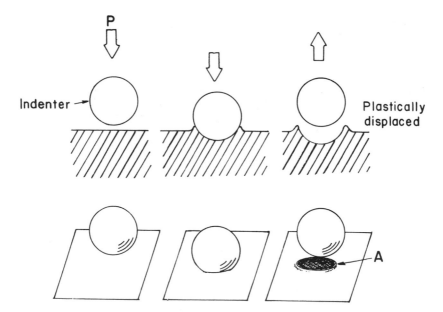

Figure 2-7. Schematic representation of a hardness test. The hardness is defined as *P/A*.

Table 2-2. Hardness Tests[a]

Test	Indenter	Shape of indentation — Side view	Shape of indentation — Top view	Load	Formula for hardness number
Brinell	10 mm sphere of steel or tungsten carbide	D, d	d	P	$\mathrm{BHN} = \dfrac{2P}{\pi D\left[D - \sqrt{D^2 - d^2}\right]}$
Vickers	Diamond pyramid	136°	d_1, d_1	P	$\mathrm{VHN} = 1.72\,P/d_1^2$
Knoop microhardness	Diamond pyramid	$l/b = 7.11$ $b/t = 4.00$	b, l	P	$\mathrm{KHN} = 14.2\,P/l^2$
Rockwell					
A / C / D	Diamond cone	120°		60 kg / 150 kg / 100 kg	$R_A =$, $R_C =$, $R_D =$ } 100–500t
B / F / G	$\frac{1}{16}$ in. diameter steel sphere			100 kg / 60 kg / 150 kg	$R_B =$, $R_F =$, $R_G =$
E	$\frac{1}{8}$ in. diameter steel sphere			100 kg	$R_E =$ 130–500t

[a] From Ref. 2.

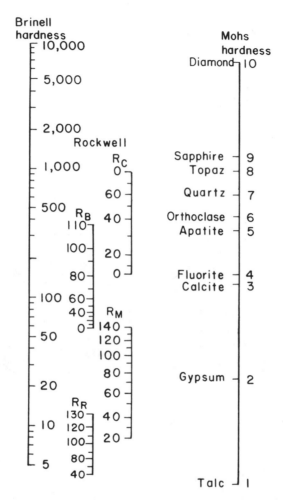

Figure 2-8. Comparison of hardness scales. (From Ref. 3.)

Creep properties of a material are measured by attaching a constant load or weight and monitoring the elongation with time (Figure 2-9). The initial elastic deformation by the constant applied load is followed by the primary, secondary (or steady-state), and tertiary *creep* before ultimate failure. Primary creep is partially recoverable; steady-state creep is irreversible and is commonly reported as creep rate, i.e., $d\epsilon/dt$, which is nearly constant in this range. Tertiary creep is accelerated elongation due to the increase in stress which, in turn, is due to the necking or internal stress cracking.

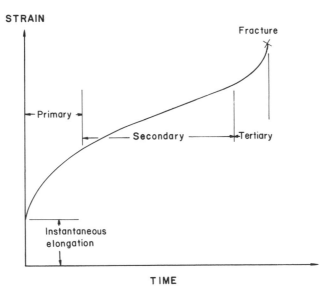

Figure 2-9. Schematic representation of a creep curve.

Another similar time-dependent behavior of a material is the *stress–relaxation* phenomenon in which a constant strain is applied and subsequent relaxation of the stress is measured. This phenomenon occurs through the irreversible deformation of a material although it contains the reversible, elastic deformation, which can be recovered completely upon relieving the applied strain. It is believed that a similar type of internal structural change occurs with the creep behavior. This type of stress–relaxation behavior can be understood better by the viscoelastic modeling as discussed in the next section.

Wear properties of an implant material are important especially for various joint replacements. Wear cannot be discussed without some understanding of friction between two materials. When two solid materials contact, they touch only at the tips of their highest asperities. Therefore, the real contact area is much smaller than the apparent surface area. It is found that the true area of contact increases with applied load (P) for ductile materials and for elastic materials like rubber and diamond. Ductile materials can be pressure welded due to the formation of plastic junctions as shown in Figure 2-10. The plastic junctions are the main source of an adhesive friction when two materials are sliding over with or without lubricating film. The resistance to the shear failure of the plastic junction results in a frictional force. Therefore, the sliding force F will be simply

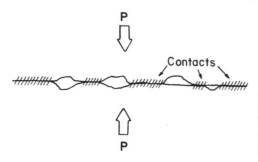

Figure 2-10. Schematic representation of two surfaces under pressure. Plastic junctions are formed when ductile materials are pressed together between asperities.

proportional to the shear yield strength k of the junctions and the contact area A:

$$F = Ak \qquad (2\text{-}10)$$

Since for ductile materials the area of contact increases with P,

$$P = HA \qquad (2\text{-}11)$$

where H is the penetration hardness or yield pressure. If we combine equations (2-10) and (2-11), the coefficient of sliding friction μ can be obtained:

$$\mu = F/P = k/H \qquad (2\text{-}12)$$

This equation implies that the friction coefficient is merely the ratio of two plastic strength parameters of the weaker material and is independent of the contact area, load, sliding speed including surface roughness, and geometry. Figure 2-11 shows an example of the effect of contact area changes on the friction coefficient. Note that there are no significant variations in friction coefficient with wide variations of the contact area.

Wear is obviously very important for the selection of implant materials especially for joint prostheses. Wear results from the removal and relocation of materials through the contact of two materials. There are several different types of wear.

Corrosive wear is due to chemical activity on one of the sliding materials. The sliding action removes the product of corrosion that would protect the surface from further attack, resulting in faster corrosion. *Surface*

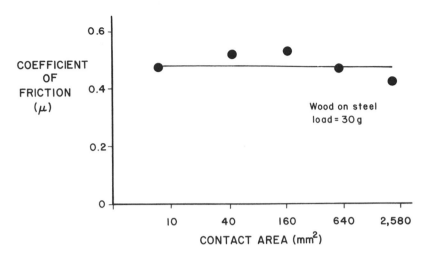

Figure 2-11. Friction coefficient versus sliding contact area, wood on steel.

fatigue wear is due to the formation of surface or subsurface cracks followed by breaking off of large chunks under repeated loading and sliding cycles.

Abrasive wear is a process in which particles are pulled off from one surface and adhere to the other during sliding. At a later time, the particles may be lost. This kind of wear can be minimized if the surfaces are smooth and hard particles are kept off the sliding surfaces (a small number of hard particles also come off from the hard surfaces). This type of wear is the most important process in implant materials and can be analyzed more easily than any other type of wear mentioned previously. It has been determined that the volume of wear ΔV is proportional to the applied load P across the two surfaces and to the sliding distances Δl and is inversely proportional to the hardness of the softer material, H; therefore,

$$\Delta V/\Delta l = KP/3H \qquad (2\text{-}13)$$

where K is the wear constant. Table 2-3 gives some wear constants for various sliding combinations.

When *lubrication* is present between two contacting surfaces, the friction and wear properties change drastically. Because in most implant applications there is some type of lubricant present, it is important to understand the lubrication process. Generally, there are two types of lubrication. *Fluid lubrication* can be achieved by a film of some liquid or gas thick enough to separate two solid surfaces completely. This type of

Table 2-3. Wear Constant of Various
Sliding Combinations[a]

Combination	K of first material[b]
Zinc on zinc	160×10^{-3}
Copper on copper	32×10^{-3}
Stainless steel on stainless steel	21×10^{-3}
Copper on low-carbon steel	1.5×10^{-3}
Low-carbon steel on copper	0.5×10^{-3}
Bakelite on Bakelite®	0.02×10^{-3}

[a] From Ref. 4.
[b] $K = (\Delta V/\Delta L)/(P/3H)$.

lubrication is mainly dependent on the properties of the lubricating fluids and is achieved more easily at high speed of sliding surfaces.

Boundary lubrication is achieved when a very thin film of lubricant or a softer material is introduced between two harder materials. The lubricant can reduce the direct contact of two sliding surfaces drastically, thus reducing friction and wear. This type of lubrication is more important for the artificial joints.

When a material is subjected to a constant or a repeated load below fracture stress, it can fail after some time (Figure 2-12). This is called static or dynamic (cyclic) *fatigue* and is usually plotted as stress versus time or log cycles (N), as shown in Figure 2-13. The time or number of cycles before failure depends on the magnitude and types of load, the test environment, and temperature.

Cyclic fatigue is characteristic of ductile or plastic materials, although the final fracture is rather rapid. The cause of cyclic fatigue is the inhomogeneity and anisotropy of materials. Imperfections, particularly those on the surface caused by machining and handling, can initiate cracks, and the growth of cracks leading to catastrophic failure can occur inside the material under cyclic loading. A material may undergo cyclic loading indefinitely below a certain stress level called the *endurance limit*.

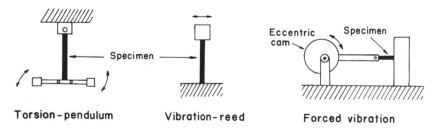

Torsion-pendulum   Vibration-reed   Forced vibration

Figure 2-12. Schematic representation of various dynamic testing machines.

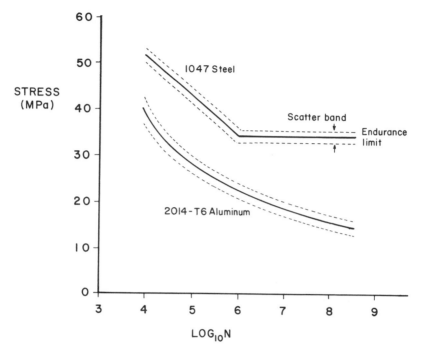

Figure 2-13. *S–N* curves for aluminum and low-carbon steel. (From Ref. 2.)

Fatigue tests in a simulated body environment will give a better evaluation of the material because the materials placed in the body undergo loading and unloading cycles. However, it is impossible to simulate the complicated loading and unloading conditions an implant undergoes *in vivo*. Nevertheless, the fatigue test is useful for comparing performances of various implants under given testing conditions.

2.2. VISCOELASTICITY

Although the simple equation (2-6) can describe the elastic behavior of many materials at low strains as shown in Figure 2-14, it cannot be used to characterize polymers and tissues that are a major concern of this book. The fluidlike behavior of a material (such as water and oil) can be described in terms of stress and strain as in the elastic solids, but the proportionality constant, *viscosity* (η), is derived from the following relationship:

$$\sigma = \eta \, d\epsilon / dt \qquad (2\text{-}14)$$

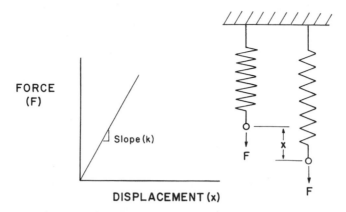

Figure 2-14. Force versus displacement of a spring.

It is noted that the stress and strain are shear rather than tensile or compressive although the same symbols are used to avoid complications.

A mechanical analog (*dashpot*) can be used to simulate the viscous behavior of equation (2-14) as shown in Figure 2-15. An automobile shock-absorbing cylinder has a similar construction, oil being the damping fluid. By examining equation (2-14) one can see that the stress is time dependent, that is, if the deformation is accomplished in a very short time ($dt \rightarrow 0$), then the stress becomes infinite. On the other hand, if the deformation is achieved slowly ($dt \rightarrow \infty$), the stress approaches zero regardless of the viscosity value.

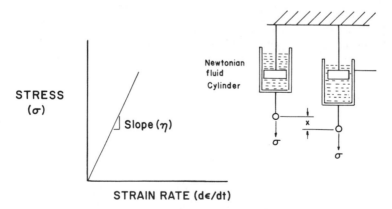

Figure 2-15. Stress versus strain rate of a dashpot. Note the stress and strain are in shear.

The simple equations (2-6) and (2-14) can describe, in principle, the *viscoelastic behavior* of a material when combined, as if the material is made of springs and dashpots. The stress–strain behavior of the spring and dashpot can be represented as shown in Figure 2-16. If the spring and dashpot are arranged in series and parallel, they are called *Maxwell* and *Voigt* (or *Kelvin*) models, respectively. Remember that equation (2-6) does not involve time, implying the spring acts instantaneously when stressed. When the Maxwell model is stressed suddenly, the spring reacts instantaneously while the dashpot cannot react since the piston of the dashpot cannot move due to the infinite stress required by the surrounding fluid. However, if we hold the Maxwell model after instantaneous deformation, the dashpot will react due to the restriction of the spring and this will take time (dt = finite). The foregoing description can be expressed concisely by a simple mathematical formulation. In general, the response to stress by the Maxwell model will result in cumulative strain, that is, total strain (ϵ_t) is a combination of the strain of spring (ϵ_s) and dashpot (ϵ_d):

$$\epsilon_t = \epsilon_s + \epsilon_d \qquad (2\text{-}15)$$

Differentiating both sides,

$$d\epsilon_t/dt = d\epsilon_s/dt + d\epsilon_d/dt \qquad (2\text{-}16)$$

and rewriting equation (2-6),

$$d\sigma_s/dt = E\, d\epsilon_s/dt \qquad (2\text{-}17)$$

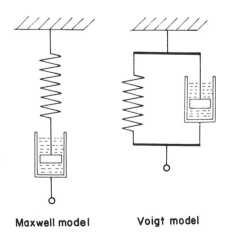

Maxwell model   Voigt model

Figure 2-16. Two-element viscoelastic models.

and substituting equations (2-17) and (2-14) into (2-16),

$$d\epsilon_t/dt = (1/E)d\sigma_s/dt + \sigma_d/\eta \qquad (2\text{-}18)$$

Also, one can see that the total stress is the same for the spring and the dashpot since each member has the same applied load internally (or else it breaks!). Thus, equation (2-18) becomes

$$d\epsilon/dt = (1/E)d\sigma/dt + \sigma/\eta \qquad (2\text{-}19)$$

where ϵ is the total strain (ϵ_t). Equation (2-19) can be applied easily for a simple mechanical test condition such as *stress relaxation* in which the specimen is strained (or stressed) instantaneously and the relaxation of the load is monitored while the specimen is held at a constant length (strain). Thus, the strain rate becomes zero ($d\epsilon/dt = 0$) and equation (2-19) can be rewritten as

$$(1/E)d\sigma/dt + \sigma/\eta = 0 \qquad (2\text{-}20)$$

Therefore,

$$d\sigma/\sigma = -E\,dt/\eta \qquad (2\text{-}21)$$

and by integrating and knowing $\sigma = \sigma_0$ at $t = 0$,

$$\sigma/\sigma_0 = \exp(-Et/\eta) \qquad (2\text{-}22)$$

The constant η/E can be substituted with another constant τ called *relaxation time* and equation (2-22) will become

$$\sigma = \sigma_0\exp(-t/\tau) = \sigma_0/\exp(t/\tau) \qquad (2\text{-}23)$$

Examining equation (2-23) one can see that if the relaxation time is short, then the stress σ decreases sharply. On the other hand, if the relaxation time is long, then the stress σ is the same as the original stress, σ_0.

Similar analysis can be made with the *Voigt* model. In this case the strain of spring and dashpot represent the total strain, that is,

$$\epsilon_t = \epsilon_s = \epsilon_d = \epsilon \qquad (2\text{-}24)$$

The total stress is a cumulative of spring and dashpot:

$$\sigma_t = \sigma_s + \sigma_d = \sigma \qquad (2\text{-}25)$$

Substituting equations (2-6) and (2-24) into (2-25),

$$\sigma = E\epsilon + \eta \, d\epsilon/dt \qquad (2\text{-}26)$$

If the applied stress is removed after a certain time, then

$$0 = E\epsilon + \eta(d\epsilon/dt) \qquad (2\text{-}27)$$

which is similar to equation (2-20) and can be solved likewise; hence,

$$\epsilon(t) = \epsilon_0 \exp(-Et/\eta) \qquad (2\text{-}28)$$

where ϵ_0 is the strain at the time of stress removal. The constant η/E is termed *retardation time* λ for this *creep recovery* process. Since the strain is being removed from the original strain ϵ_0,

$$\epsilon_{\text{recovery}} = \epsilon_0[1 - \exp(-t/\lambda)] \qquad (2\text{-}29)$$

2.3. VISCOELASTICITY IN DYNAMIC TESTS (ADVANCED TREATMENT, OPTIONAL)

Let us expand the study of viscoelasticity into the understanding of simple dynamic tests as shown in Figure 2-12. Assume that the Voigt model is subjected to a sinusoidal stress given by

$$\sigma = \sigma_0 \sin \omega t \qquad (2\text{-}30)$$

where σ_0 is an amplitude, ω is the angular frequency of the stress, and t is the time. Putting equation (2-30) into equation (2-26),

$$E\epsilon + \eta \, d\epsilon/dt = \sigma_0 \sin \omega t \qquad (2\text{-}31)$$

hence,

$$d\epsilon/dt + E\epsilon/\eta = (\sigma_0/\eta)\sin \omega t \qquad (2\text{-}32)$$

By assuming the solution to the equation (2-32) as follows

$$\epsilon = A \sin \omega t + B \cos \omega t \qquad (2\text{-}33)$$

one can get a solution,

$$\epsilon = \frac{\sigma_0}{E}\left(\frac{\sin \omega \tau}{1 + \omega^2 \tau^2} - \frac{\omega t \cos \omega t}{1 + \omega^2 \tau^2}\right) \qquad (2\text{-}34)$$

By examining equation (2-34) the strain also varies sinusoidally with time at the same frequency as the stress but is not in phase with it. The first term in the equation is in phase but the second term is 90° out of phase. This characteristic can be represented diagrammatically as shown in Figure 2-17. From Figure 2-17 the $0\epsilon_2$, representing the in-phase component of the strain, can be expressed as

$$0\epsilon_2 = 0\epsilon_0 \cos \phi = \sigma_0 / E(1 + \omega^2 \tau^2) \tag{2-35}$$

The length $0\epsilon_3$ is the projection of $0\epsilon_0$ representing the out-of-phase component of the strain:

$$0\epsilon_3 = \sigma\epsilon_0 \sin \phi = \sigma_0 \omega \tau / E(1 + \omega^2 \tau^2) \tag{2-36}$$

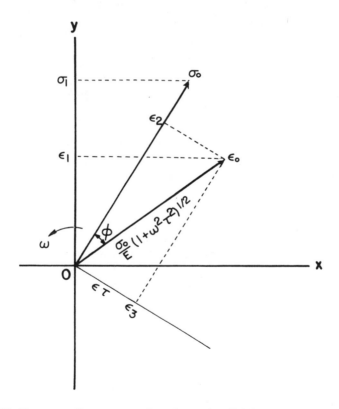

Figure 2-17. Diagrammatic representation of equation (2-34), where $0\sigma_0$ vector rotates counterclockwise with angular velocity ω. The stress is represented by the projection of $0\sigma_0$ vector on the $0y$ axis (σ_1). Likewise the strain is the projection of vector $0\sigma_0$ on the $0y$ axis (ϵ_1). $0\epsilon_0$ rotates at a constant angle ϕ behind $0\sigma_0$ and can be resolved into two components, ϵ_2 and ϵ_3.

The length $0\epsilon_0$ is called the *strain amplitude* and represents a maximum strain during a cycle and from equation (2-34),

$$0\epsilon_0 = \sigma_0/E(1 + \omega^2\tau^2)^{1/2} \qquad (2\text{-}37)$$

Let us now consider the Maxwell model to illustrate some further aspects of the dynamic behavior of viscoelastic materials. The strain of equation (2-33) can be expressed as a complex oscillatory function,

$$\epsilon = \epsilon^* \exp(-i\omega t) \qquad (2\text{-}38)$$

likewise for stress,

$$\sigma = \sigma^* \exp(i\omega t) \qquad (2\text{-}39)$$

and the complex dynamic modulus E^* is defined by

$$E^* = E' + iE'' = \sigma^*/\epsilon^* \qquad (2\text{-}40)$$

where E' is real and E'' is imaginary modulus. Also E' is in-phase with stress and hence is called *storage modulus*. The out-of-phase strain component causes a dissipation of energy as heat and is proportional to E''. The ratio E''/E' is called *loss factor* or *loss tangent* and can be expressed as

$$E''/E' = 1/\omega\tau = \tan\phi \qquad (2\text{-}41)$$

In addition, these E' and E'' can be expressed in terms of E for the Maxwell model as follows:

$$E' = \omega^2\tau^2 E/(1 + \omega^2\tau^2)$$
$$\qquad\qquad\qquad\qquad\qquad\qquad (2\text{-}42)$$
$$E'' = \omega\tau E/(1 + \omega^2\tau^2)$$

The dynamic modulus and damping of a Maxwell unit as a function of frequency are illustrated in Figure 2-18. In this example the modulus E of the spring is 10 GPa and the viscosity η of the dashpot is 10 GPa; therefore, the relaxation time is 1 s. As is evident at low frequencies the dynamic modulus E is very low since most of the deformation comes from the dashpot. At very high frequencies there is no time for the dashpot to adjust to the deformation cycles and thus the response is mostly from the spring. At the intermediate frequencies where the time for an oscillation is approximately equal to the relaxation time, motion of both spring and dashpot takes place.

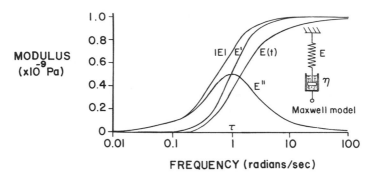

Figure 2-18. Dynamic behavior of a Maxwell unit as a function of frequency. The stress–relaxation modulus $E(t)$ of the same model is given for comparison. (From Ref. 5.)

The loss modulus E'' approaches zero at both low and high frequencies (Figure 2-18) since the dashpot cannot respond at high frequencies due to its energy dissipation and at low frequencies the motion is so slow that the rate of shear is small. It is also noted that the loss modulus goes through a maximum when $\omega = 1/\tau$.

2.4. ELASTICITY OF NON-HOOKEAN MATERIALS (ADVANCED TREATMENT, OPTIONAL)

When the deformation of the bonds becomes so large that the higher terms in a Taylor series of equation (2-4) are no longer negligible, then Hooke's law fails. Figure 2-19 shows tensile stress–strain curves for various materials in which the proportionality of the stress to strain extends only a little way from the origin. The nonapplicability of Hooke's law, however, is not in the linearity of elastic deformation but rather in the resistance of the material to other nonelastic processes of deformation. When a material is pulled beyond a certain limit (*elastic limit* or *yield stress*), it deforms plastically, causing atoms to move into new equilibrium positions with other neighbors (Figure 2-3). The elastic strains in the atomic bonds are still small and obey Hooke's law but they are overwhelmed by the much larger plastic strain. The two types of strains can be demonstrated easily by loading and unloading the specimen as shown in Figure 2-20. The plastic strain remains as a permanent set. Sometimes a small part of the plastic strain remains as a permanent set and a *hysteresis loop* is formed by unloading.

Rubbers are called elastomers and their elasticity is due to the thermal motion of the chain molecules which try to wriggle the chains up into

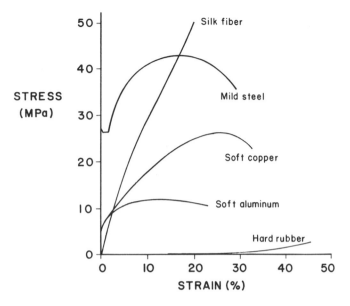

Figure 2-19. Tensile stress–strain curves of various materials.

random lengths in opposition to the applied force. The molecules return to their original position due to the cross-links which hold the individual chains together. Figure 2-21 illustrates a stress–strain curve of vulcanized natural rubber. The "initial" modulus of the material is several orders of magnitude smaller than the theoretical polymer chain strength. The elastic

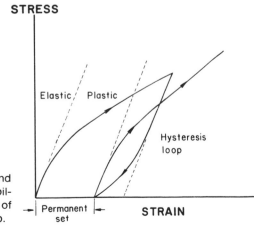

Figure 2-20. Effect of unloading and reloading which shows the reversibility of elastic strain, irreversibility of plastic strain, and a hysteresis loop.

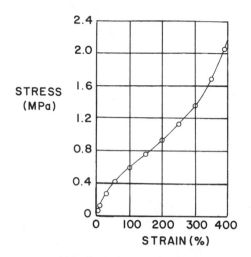

Figure 2-21. Tensile stress–strain curve for a lightly vulcanized natural rubber. (From Ref. 7.)

properties of this type of material can be satisfactorily explained by statistical thermodynamic considerations.[6]

If a material (rubber) undergoes a deformation without any exchange of mass with the surroundings (termed a *closed system*), and neglecting gravitational, kinetic, surface, and other effects, then the first law of thermodynamics (see Section 4.1.1) dictates that the internal energy change (dE) is equal to the change in enthalpy (dH) minus the applied work (dW):

$$dE = dH - dW \qquad (2\text{-}43)$$

For an *isothermal* (constant temperature) condition and under applied external force, f, equation (2-43) can be rewritten as

$$dE = T\,dS - P\,dV + f\,dl \qquad (2\text{-}44)$$

where dS are entropy changes by the *configurational* changes in chains, P is pressure, V is volume, and l is length of the specimen. Equation (2-44) can be related to the *Helmholtz free energy* changes (dF):

$$dF = dE - d(TS) \qquad (2\text{-}45)$$

By substituting equation (2-44) into (2-45), and at a constant temperature,

$$(dF)_{\mathrm{T}} = -P\,dV + f\,dl \qquad (2\text{-}46)$$

which yields the following at constant temperature and volume:

$$f = (\partial F/\partial l)_{T,V} = (\partial E/\partial l)_{T,V} - T(\partial S/\partial l)_{T,V} \qquad (2\text{-}47)$$

From equation (2-47) one can see that the force of retraction may be broken up into major contributions (Figure 2-22); one is the *internal energy* representing the changes in inter- and intramolecular energies on stretching and the other is the entropy changes in order or degree of randomness of the molecules. Note that the entropic contribution is much greater than the internal energy contribution as the strain becomes sufficiently high as shown in Figure 2-22. Indeed, for an "ideal" rubber the internal energy term $(\partial E/\partial l)$ becomes negligible, and therefore

$$f = -T(\partial S/\partial l)_{T,V} \qquad (2\text{-}48)$$

We can now apply the statistical theory of polymers to calculate the probabilities of stretched and unstretched configurations and obtain the changes in entropy from the Boltzmann relation [see equation (4-10)]:

$$S = k \ln W \qquad (2\text{-}49)$$

where k is the Boltzmann constant and W is the probability of a particular chain configuration. If we consider a freely jointed polymer chain with one

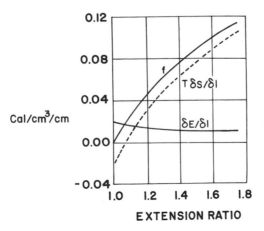

Figure 2-22. Changes in internal energy E and entropy S by stretching a rubber.

end fixed at the origin, then the problem becomes a random walk in three dimensions and the solution is the Gaussian distribution,

$$W(x, y, z) = (\beta/\pi^{1/2}) \exp(-\beta^2 r^2) \tag{2-50}$$

where $\beta^2 = 1.5nl^2$, n being the number of chain links with length l. By substituting equation (2-50) into (2-49),

$$S = \text{constant} - k\beta^2 r^2 \tag{2-51}$$

It follows that for a single polymer chain, the restrictive force f for an extension of magnitude dr is

$$f = -T(dS/dr) = 2kT\beta^2 r \tag{2-52}$$

The force is proportional to temperature which is in good agreement with elastomer behavior (if you stretch a rubber band suddenly and touch it with your lips you will sense warmness). This is contrary to the behavior of crystalline solids (the temperature rise in crystalline solids is due to internal friction) where the elastic constants decrease linearly as the temperature rises. It is also clear that equation (2-52) obeys Hooke's law since f is proportional to r. However, this is not true for the behavior of real elastomers as shown in Figure 2-21 and this simple model has rather severe limitations.

PROBLEMS

2-1. From the data obtained by using a standard tensile test specimen of 5.08-cm gauge length, 1.283-cm diameter:

a. Plot the engineering stress–strain curve.
b. Calculate the modulus of elasticity, 0.2% offset yield strength, and (ultimate) tensile strength.
c. Calculate the true tensile strength if the final diameter at fracture is 0.72 cm.

Load (kN)	Elongation ($\times 10^{-3}$ cm)
20	4.2
44	8.6
53	10.7
58	12.0
62	13.2
67	20.1
80	30.0
89	39.7
116	142.5
175 (max)	1200.0
142 (failed)	2000.0

2-2. A piece of polymer (polyethylene) is stretched 50% of its length. When the stress was released it recovered 50% of its strain after 1 h at room temperature.

a. What is the retardation time?
b. What is the amount of strain recovered after 5 h at room temperature?

2-3. The change in length of a rubber band with a cross-sectional area of 2.84 mm^2 was measured by putting weights at one end of it and yielded the following data:

Weight (g)	0	20	50	100	250
Length (cm)	13.0	13.8	15.7	19.7	26.4

a. Plot stress versus stretch ratio.
b. Calculate M_c by assuming ρ is 1 g/ml. and $M_c = \rho RT\,(\alpha - \alpha^{-2})/\sigma$.

2-4. Prove equation (2-34) by using equations (2-32) and (2-33).

2-5. Assume a Maxwell model which has $E_{spring} = 100$ MPa and $\eta = 10$ MPa-s.

a. Calculate E' and E''.
b. Plot E' and E'' versus frequency for the model on a graph.

2-6. Derive equation (2-42).

REFERENCES

1. Z. D. Jastrzebski, *The Nature and Properties of Engineering Materials*, 2nd ed., p. 205, Wiley, New York, 1976.
2. W. H. Hayden, W. G. Moffatt, and J. Wulff, *The Structure and Properties of Materials*, Volume 3, p. 13, Wiley, New York, 1965.
3. G. F. Kinney, *Engineering Properties and Application of Plastics*, p. 111, Wiley, New York, 1957.
4. F. A. McClintock and A. S. Argon (ed.), *Mechanical Behavior of Materials*, p. 668, Addison–Wesley, Reading, Mass., 1966.
5. L. E. Nielsen, *Mechanical Properties of Polymers*, p. 154, Reinhold, New York, 1962.
6. P. J. Flory, *Principles of Polymer Chemistry*, Cornell University Press, Ithaca, N.Y., 1953.
7. F. W. Billmeyer, Jr., *Textbook of Polymer Science*, 2nd ed., p. 195, Wiley, New York, 1971.

BIBLIOGRAPHY

A. H. Cottrell, *The Mechanical Properties of Matter*, Chapters 4 and 8, Wiley, New York, 1964.
H. W. Hayden, W. G. Moffatt, and J. Wulff, *The Structure and Properties of Materials*, Volume 3, Chapters 1 and 2, Wiley, New York, 1965.

F. A. McClintock and A. S. Argon (ed.), *Mechanical Behavior of Materials*, Chapters 1 and 2, Addison–Wesley, Reading, Mass., 1966.

L. H. Van Vlack, *Elements of Materials Science*, 3rd ed., Chapter 1, Addison–Wesley, Reading, Mass., 1975.

L. H. Van Vlack, *Materials Science for Engineers*, Chapters 1–5, Addison–Wesley, Reading, Mass., 1970.

L. H. Van Vlack, *Materials for Engineering: Concepts and Applications*, Addison–Wesley, Reading, Mass., 1982.

3

STRUCTURE OF SOLIDS

The structure of solids can be studied by using various instruments. The optical microscope was first used by Sorby in 1863 for studying the microstructure of carefully polished and acid-etched wrought iron.[1] The grains can be observed due to the imperfection of atomic packing at the grain boundaries (Figure 3-1).[2]

The atomic arrangement of materials could not be elucidated until the discovery of X radiation by Roentgen in 1895. Von Laue (1912) suggested that X rays can be diffracted from a regular array of atoms about 0.1 nm in diameter. This suggestion was proven by Bragg and Bragg in a paper published in 1913.[3] They described a method for measuring the spacing between planes of atoms as illustrated in Figure 3-2.

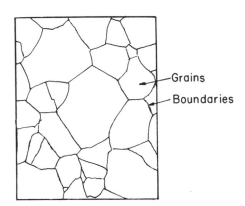

Grains

Boundaries

Figure 3-1. Grain boundaries of nickel, 170×. (From Ref. 2.)

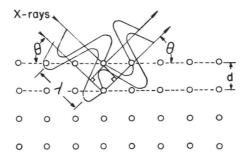

Figure 3-2. X-ray diffraction in an array of atoms is governed by the Bragg equation, $\lambda = 2d\sin\theta$. The interatomic spacing (d) can be calculated by knowing the wavelength λ and the angle θ of the X-ray beam.

Another important tool to study materials is the transmission electron microscope, which is very similar to the light microscope except that a beam of electrons replaces the light beam. In addition, carefully shaped magnetic and electrostatic fields replace the glass lenses. Although the first prototype was built in 1926 by Busch, the study of the internal structure of metals directly using an electron microscope was not possible until Heidenreich prepared a section thin enough for the electron beam to transmit.[4] In 1956, Hirsch, Horne, and Whelan observed dislocations in aluminum foil which elucidated the mechanical behavior of solids on an atomic scale for the first time.[5]

3.1. ATOMIC AND MOLECULAR BONDING

The nature of bonding determines the properties of a material. Metallic bonds allow high electrical and heat conductivity due to the free electrons which act as the medium. In addition, the nondiscriminate nature of the metal atoms for neighbors makes it easy to change their positions under load, resulting in yield point. On the other hand, ionic bonds can have a limited number of slip planes due to the repulsion of like-charged ions, making such materials very brittle. Ionic materials are also insulators of heat and electricity since their electrons are tightly held by the ions. Covalent bonds share valence electrons with neighboring atoms. Generally, they show directionality of bonding and poor electrical and thermal conductivity as for ionic bonding. The covalent bond is the strongest primary bond. Secondary bonding such as hydrogen and van der Waals bonding also play a major role in determining material properties.

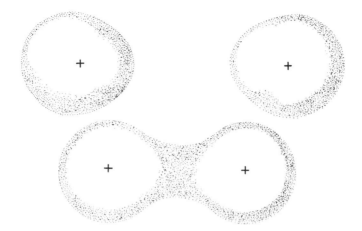

Figure 3-3. Bonding of two hydrogen atoms results in rearranged electrons to have opposite spins and their electronic orbital overlap to share the electrons (covalent bonding).

3.1.1. Primary Bonding

3.1.1a. Covalent Bonding

Sharing or exchanging of outer electrons between atoms is the basis for forming compounds. As two atoms such as hydrogen approach each other, each atom will experience electrostatic forces of attraction and repulsion. In order to minimize the interaction energy, the electrons will arrange themselves as to have the lowest energy state (1s level) and opposite spins (+ or −1/2). In this orbital state 1s atoms are stable and the electrons around the nuclei overlap and it appears that the nuclei share the electrons (Figure 3-3). This is called *covalent bonding* since it involves the valence (outermost shell) electrons (Figure 3-4, top). Obviously, this type of bond occurs between atoms that have partially filled valence electronic orbitals or energy levels so that the total energy of the system can be lowered by sharing of electrons between atoms. The greater the overlap of the valence shells, the stronger the bond becomes but it is limited by the electrostatic repulsive forces between nuclei.

In solids the bonds between atoms will be three-dimensional and an atom will have several neighbors to share valence electrons, as shown in Figure 3-5 for the carbon atom. The carbon atom (atomic number 12) can fill its second electronic orbitals with four neighboring electrons to minimize its energy state in the form of a tetrahedron.

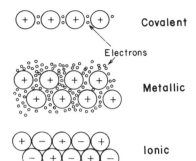

Covalent

Electrons

Metallic

Ionic

Figure 3-4. Schematic representation of covalent, metallic, and ionic bonding.

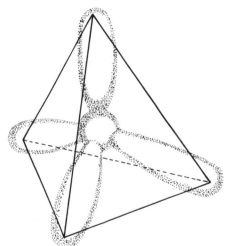

Figure 3-5. Electronic orbital state of a
carbon atom which shares its valence
electrons with four neighbors.

3.1.1b. Metallic Bonding

The metallic elements have a small number of valence electrons which
are weakly bound to the atomic system. This enables the valence electrons
to move around the atoms rather freely (Figure 3-4, middle). Metals,
therefore, can be considered as an array of metallic ions in a pool of free
valence electrons, all held together by electrostatic forces. These unbound
(free) electrons will move freely according to the electrical potentials of
metallic ions but are no longer bound to any individual atom. This is the
reason why metals exhibit high thermal and electrical conductivity and why
one type of metal can coexist with another metal in the form of a solid
solution, viz. alloy.

3.1.1c. Ionic Bonding

Electronegative (nonmetallic) and electropositive (metallic) atoms can form a strong primary ionic bonding by exchanging electrons to create closed-shell electronic structures (Figure 3-4, bottom). However, the valence electrons are much more likely to be found in the space around the negative ions than around the positive ions. Therefore, the bond is directional and the structure is limited by geometrical considerations and by the necessity of maintaining electrostatic neutrality in the crystal as a whole.

3.1.2. Secondary Bonding

3.1.2a. Hydrogen Bonding

The hydrogen bond is a secondary bond which is weaker than the primary metallic, covalent, and ionic bonds but plays an important role in determining material properties. Hydrogen bonding occurs when the atom is covalently bonded to an electronegative atom so that it becomes a positive ion which can then exert an attractive force. Since the hydrogen ion is very small, it can approach the negative ion very closely, resulting in an electrostatic force which can be substantial (Figure 3-6).

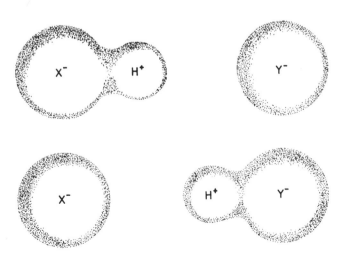

Figure 3-6. Principles of hydrogen bonding. The H^+ can be attracted to either negative atoms equally and appears to oscillate between them.

Table 3-1. Strengths of Different Chemical Bonds
as Reflected in Their Heat of Vaporization[a]

Bond type	Substance	Heat of vaporization (kJ/mol)
van der Waals	He	0.14
	N_2	13
Hydrogen	Phenol	31
	HF	47
Metallic	Na	180
	Fe	652
Ionic	NaCl	1062
	MgO	1880
Covalent	Diamond	1180
	SiO_2	2810

[a] From Ref. 6.

3.1.2b. van der Waals Bonding

The van der Waals attractive forces are attributed to the fluctuating electronic distribution between atoms. The most energetically favorable arrangement for neighboring atoms is to form dipoles which fluctuate in phase resulting in neighboring atoms being drawn together. This dipole–dipole bonding is not directional and can have effects over long distances since electron cloud overlap is necessary. This type of secondary bonding is much weaker than hydrogen bonding (Table 3-1).

3.2. CRYSTAL STRUCTURE OF SOLIDS

The arrangements of atoms can be represented by a *space lattice* which is defined as an infinite, three-dimensional array of (lattice) points, each of which has identical surroundings. The lattice space can be described conveniently by a repeating unit called a *unit cell*. Figure 3-7 shows a two-dimensional lattice in which any square can be repeated by moving one lattice spacing, "a," in any direction. If it is extended into three dimensions,

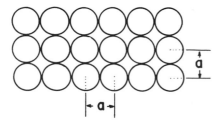

Figure 3-7. Stacking of spheres in simple cubic structure (a is lattice space).

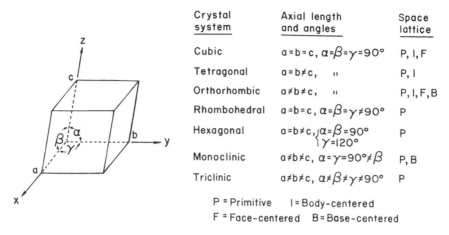

Crystal system	Axial length and angles	Space lattice
Cubic	$a=b=c, \alpha=\beta=\gamma=90°$	P, I, F
Tetragonal	$a=b\neq c,$ "	P, I
Orthorhombic	$a\neq b\neq c,$ "	P, I, F, B
Rhombohedral	$a=b=c, \alpha=\beta=\gamma\neq90°$	P
Hexagonal	$a=b\neq c, \begin{cases}\alpha=\beta=90°\\ \gamma=120°\end{cases}$	P
Monoclinic	$a\neq b\neq c, \alpha=\gamma=90°\neq\beta$	P, B
Triclinic	$a\neq b\neq c, \alpha\neq\beta\neq\gamma\neq90°$	P

P = Primitive   I = Body-centered
F = Face-centered   B = Base-centered

Figure 3-8. Crystal structure systems and space lattices.

it can be represented by a single cube. This is called a simple cubic *crystal system*. There are six crystal systems besides the simple cubic. The simple cubic crystal system is further divided into three space lattices: simple cubic (P), body-centered cubic (I), and face-centered cubic (F). Figure 3-8 shows the space lattices and seven crystal systems. Any material with a structure represented by one of the 14 space lattices is called crystalline. The cubic and hexagonal systems are common for metals and ceramics.

The face-centered cubic (fcc) structure is achieved by arranging atoms as shown in Figure 3-9. This structure is called *close-packed* in three dimensions. Because each atom touches 12 neighbors [hence coordination number (CN) = 12] rather than six as in simple cubic, it results in the most densely packed structure. The hexagonal close-packed (hcp) structure is characterized by repeating layers of every other plane, that is, the atoms in the third layer occupy sites directly over the atoms in the first layer. This can be represented as ABAB... packing while the fcc structure can be represented by three layers of planes ABCABC.... Both fcc and hcp have the highest packing efficiency; roughly three-fourths of the unit cell is occupied by the atomic volume.

Another common structure of metals is the body-centered cubic (bcc) in which an atom is situated in the center of the simple cube (Figure 3-10). Its packing is less efficient; only 68% of the cube is accounted for by the atomic volume. Some examples of crystal structures are given in Table 3-2.

Most materials used for implants are made of more than two elements except a few limited cases. When two or more different-size atoms are mixed together to form a solid, two factors should be considered: the types of site and the number of sites occupied.

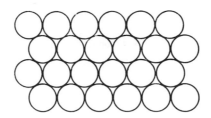

a. Two-dimensional

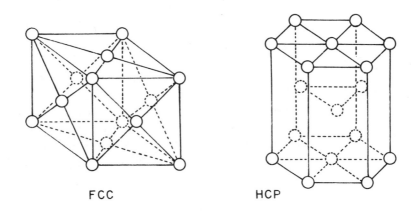

FCC   HCP

b. Three-dimensional

Figure 3-9. Close-packing arrangements: (a) two- and (b) three-dimensional.

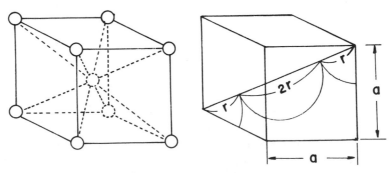

Figure 3-10. Body-centered cubic unit cell. Each unit cell has two atoms and $a = 4r/3^{1/2}$.

Table 3-2. Examples of Crystal Structures

Material	Structure
Cr	bcc
Co	hcp (below 460°C)
	fcc (above 460°C)
Fe	
Ferrite	bcc (below 916°C)
Austenite	fcc (916–1389°C)
Iron	bcc (above 1389°C)
Mo	bcc
Ni	fcc
Ta	bcc
Ti	hcp (below 900°C)
	bcc (above 900°C)
NaCl (salt)	fcc
Al_2O_3	hcp
Polyethylene	Orthorhombic
Polyisoprene	Orthorhombic

Consider the stability of the structure shown in Figure 3-11. In Figures 3-11a and b, because the interstitial atoms touch the larger atoms they are stable, but not in structure c. At some critical value the interstitial atom will fit the space between six atoms (only four atoms are shown in two dimensions), which will give the maximum interaction between atoms and consequently the most stable structure. Thus, at a certain (minimum) *radius ratio* of the host and interstitial atom the arrangement will be the most stable. These radius ratios are determined solely by geometric considerations, as shown in Figure 3-12.

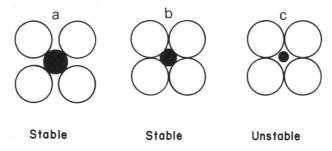

Figure 3-11. Possible arrangements of interstitial atoms.

Structural Geometry	Radius Ratio	Coordination Number
	0.155	3
	0.225	4
	0.414	6
	0.732	8
	1.0	12

Figure 3-12. Minimum radius ratios and coordination numbers for various structures.

3.3. CRYSTAL IMPERFECTIONS

Imperfections in crystalline solids, sometimes called defects, play a major role in determining the structure and physical properties of solids. Only three types of imperfections can be found—point, line, and three-dimensional grain boundary defects.

3.3.1. Point Defects

Point defects commonly appear as lattice *vacancies*, *interstitial* or *substitutional* atoms (Figure 3-13). The extra energy in the lattice created by the presence of an interstitial or substitutional atom is much greater than that associated with a vacant lattice site. For example, the energy of vacancy formation in copper is on the order of 1 eV compared to about 4 eV for an interstitial. Hence, vacancies are more common point defects and they are able to move relatively easily from site to site at low temperatures. The solid-state diffusion of atoms through a lattice would be impossible without the vacancies.

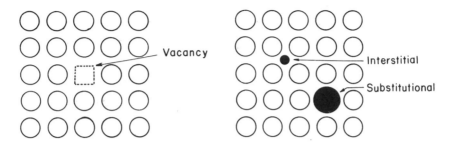

Figure 3-13. Types of point defects.

The vacancies can be created by rapidly cooling (quenching) a material from liquid, retaining a large number of vacancies during solidification. Also, they are produced by the mutual interaction of line defects (dislocations) during plastic deformation and by exposure to a rapidly fluctuating (cyclic) load. Bombardment with high-energy radiation will also generate large numbers of vacancies.

3.3.2. Line Defects (Dislocations)

Line defects differ from point defects in being extended in one dimension and are sometimes called *dislocations*. The main characteristic of the line defect is that the interatomic bonding is distorted and *elastic energy* is stored around the defect (Figure 3-14). Since the bonds are already locally distorted, it requires less additional energy to break them than it would if the lattice were perfect. This is one reason why most dislocations can move easily through a lattice under relatively small applied stress. This is also the reason why most materials are substantially weaker than the intrinsic interatomic bond strength would lead us to expect. The strength of the

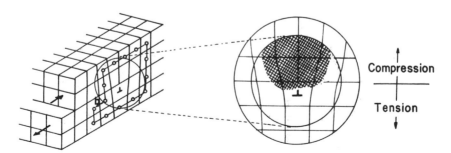

Figure 3-14. Edge dislocation. The elastic energy is stored around the dislocation in the form of tension and compression.

materials will increase if we can find ways to prevent the dislocations from moving about under applied load.

One of the line defects can be created by introducing an extra half plane of atoms as shown in Figure 3-14. The elastic energy in the form of tension and compression is stored around the end of the extra half plane which is symbolized by ⊥ and called an *edge dislocation*. Because of the extra half plane, *Burger's circuit* around the edge dislocation cannot be fully closed unlike a perfect lattice. The closure failure is known as Burger's vector (**b**) and for an edge dislocation its direction is perpendicular to the extra half plane of atoms.

Another type of line defect can be created by a shearing action which rearranges the plane of atoms in such a way that there is no extra half plane (Figure 3-15). Instead, the adjacent planes appear to be arranged in the form of a spiral ramp at the center. This is called a *screw dislocation* and its Burger's vector is parallel with the line of the actual defect and moves perpendicular to the line of action under applied shear stress. Unlike the edge dislocation, the screw dislocation is not constrained to move on a fixed lattice plane and therefore it can circumvent obstacles in its path more easily than the edge dislocation.

3.3.3. Grain Boundaries

Grain boundaries are created at the boundaries where crystals of different orientation meet resulting in a *polycrystalline solid*. Similar to the line defects, the grain boundaries are defects which have higher energy than

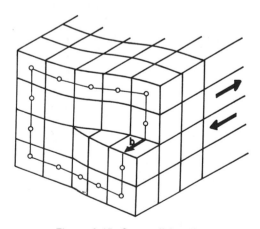

Figure 3-15. Screw dislocation.

the grains and are, therefore, more active with chemicals. This is the reason why the grain boundaries can be chemically etched more deeply than grains, thus allowing us to observe them under a microscope. Grain boundaries can be a source of considerable strengthening or weakening in a material depending on certain conditions. For example, most metallic materials deform more easily than ceramic materials, and thus the presence of the grain boundaries does not affect the mechanical properties of metals greatly. However, the grain boundaries can be a source of weakness for ceramic materials unless they are controlled rigorously.

In polymers the crystals (or grains) do not form in a manner similar to metals and ceramics. Due to their long chains, the crystallization is less complete and it is difficult to describe in terms of grains and grain boundaries. The long polymer chains are arranged in lamellar fashion with folds (Figure 3-16a), which results in spherulites and can be viewed between crossed-polarizing light during their formation (Figure 3-16b). We also describe polymer structure as a fringed micelle model in which the crystals are surrounded by a noncrystalline amorphous phase which is equivalent to the grain boundaries as shown in Figure 3-17.

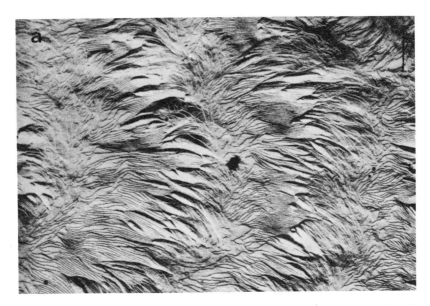

Figure 3-16. (a) Surface replica of a portion of a linear polyethylene spherulite. (From Ref. 7.) (b) Sequence of growth of spherulites in polypropylene. (From Ref. 8.)

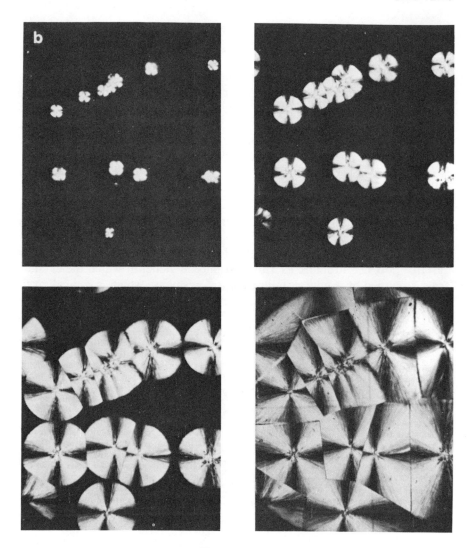

Figure 3-16b.

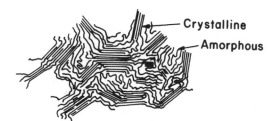

Figure 3-17. Fringed model of linear polymers.

3.4. NONCRYSTALLINE SOLIDS

A solid that does not possess a long-range order of crystallinity is called noncrystalline or amorphous. The structure is often described as being similar to a liquid, which has a short-range order but lacks a long-range order. The liquidlike structure can be retained by rapid cooling (quenching) so that the molecules are frozen before they can rearrange themselves. It is almost impossible to make metals noncrystalline due to the mobility and nondiscriminatory nature of the atoms to arrange themselves into a minimum energy state, that is, crystalline. On the other hand, some ceramics and polymers form noncrystalline solids due to their tendency to form three-dimensional networks and long-chain molecules which prevent long-range order.

3.4.1. Long-Chain Molecular Compounds

Polymers (see Chapter 10 for further study) have a very long-chain molecules which are also flexible and can be tangled easily. In addition, each chain can have side groups, branches, and copolymeric chains or blocks which can also interfere with the long-range ordering of chains. For example, the paraffin wax which has the same chemical formula of polyethylene (C_nH_{n+2}) will crystallize almost completely. However, when the chains become extremely long [from 40 to 50 *repeating units* ($-CH_2CH_2$) to several thousands as in linear polyethylene], they cannot be crystallized completely (up to 80–90% crystallization is possible). Also, branched polyethylene in which side chains are attached to the main backbone chain at positions which a hydrogen atom normally occupies, will not crystallize easily due to the steric hindrance of side chains resulting in a more noncrystalline structure. The partially crystallized structure (Figure 3-17) is called semicrystalline which is a most common occurrence for linear polymers.

Vinyl polymers other than polyethylene have a repeating unit $-CH_2-CHX-$ where X is some monovalent side group. There are three possible arrangements of side groups: atactic, isotactic, and syndiotactic (Figure 3-18). In atactic arrangements the side groups are randomly distributed; in isotactic and syndiotactic arrangements they are either alternating positions or in one side of the main chain. If side groups are small like polyethylene (X = H) and polyvinyl alcohol (X = OH) and the chains are linear, the polymer crystallizes easily. However, if the side groups are large as in polyvinyl chloride (X = Cl) and polystyrene (X = C_6H_6, benzene ring) and are randomly distributed along the chains (atactic), then a noncrystalline structure will be formed. The isotactic and syndiotactic polymers usually crystallize even when the side groups are large.

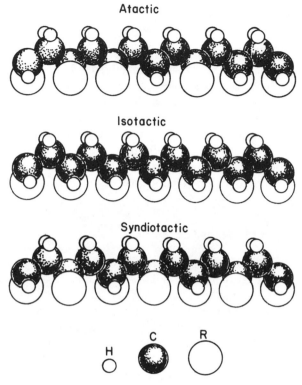

Figure 3-18. The possible arrangements of side groups in vinyl polymers: atactic (random), isotactic (same side), and syndiotactic (regularly alternating).

Copolymerization in which two or more homopolymers are combined always disrupts the regularity of polymer chains, thus promoting the formation of noncrystalline structures (Figure 3-19). The addition of *plasticizers* to prevent crystallization by keeping the chains separated from one another will result in a noncrystalline polymer which normally crystallizes. An example is celluloid, which is made of normally crystalline nitrocellulose plasticized with camphor. Plasticizers are also used to make rigid noncrystalline polymers like polyvinyl chloride (PVC) into a more flexible solid (a good example is the Tygon® tube).

Elastomers or rubbers are polymers which exhibit large stretchability at room temperature and can snap back to their original dimensions when the load is released. The elastomers are noncrystalline polymers which have an intermediate structure between long-chain molecules and three-dimensional networks (see next chapter for more details). The chains also have "kinks" or "bends" in them which straighten when a load is applied. For example,

Random

Alternating

Block

Graft

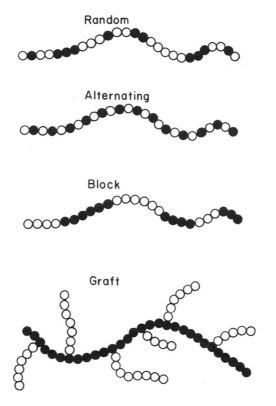

Figure 3-19. The possible arrangements of copolymers.

the chains of *cis*-polyisoprene (natural rubber) are bent at the double bond due to the methyl group interfering with the neighboring hydrogen in the repeating unit $[-CH_2-C(CH_3)=CH-CH_2-]$. If the methyl group is on the opposite side of the hydrogen, then it becomes a *trans*-polyisoprene which will crystallize due to the absence of the steric hindrance present in the *cis* form. The resulting polymer is a very rigid solid called *gutta-percha* which is not an elastomer.

Below glass transition temperature (T_g), natural rubber loses its elasticity and becomes a glasslike material. Therefore, all elastomers should have T_g well below room temperature for their flexibility. What makes the elastomers not behave like liquids above T_g is in fact due to the cross-links between chains which act as pinning points. Without cross-links the polymer would deform permanently as in the viscous liquid called latex before it is cross-linked with sulfur (vulcanization) by breaking double bonds (C=C) and forming CS—SC bond linkage between the chains. The more cross-links introduced, the more rigid the structure becomes like a network polymer.

3.4.2. Network Structures

Three-dimensional network solids are noncrystalline since the restrictions on the bonds and rigidity of subunits prevent them from crystallizing. The most common network polymer is phenolformaldehyde (Bakelite®) in

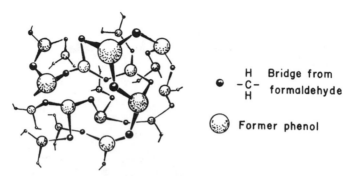

Figure 3-20. Network structure of phenolformaldehyde (Bakelite®).

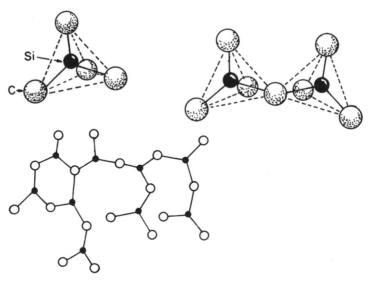

Figure 3-21. Schematic representation of silica glass. The subunit (SiO_4) is a tetrahedron with a silicon atom in the center.

which cross-links are formed by means of the phenol rings which are integral parts of each chain as shown in Figure 3-20. Inorganic glasses and oxides also form noncrystalline, three-dimensional network structures. For example, silica (SiO_2) forms glasses easily since the Si–O tetrahedron subunits are joined corner-to-corner but devoid of long-range order as in quartz (Figure 3-21). Commercial glasses contain a wide range of other oxides such as Na_2O, CaO, PbO, and Al_2O_3, which modify the basic Si–O network structure to bring about desired changes in melting temperature, formability, and physical properties.

PROBLEMS

3-1. Iron has a bcc structure with an atomic radius of 0.124 nm. Calculate its density. (Its atomic weight is 55.85 g/mol.)

3-2. Calculate the minimum radius ratio for coordination number 8.

3-3. How much sulfur is used to make one (mer) isoprene rubber completely cross-linked?

3-4. From $M_n = \Sigma N_i M_i / \Sigma N_i$ show that $M_n = 1/\Sigma(w_i/M_i)$ and from $M_w = \Sigma W_i M_i / \Sigma W_i$ show that $M_w = \Sigma w_i M_i$ where

N_i: moles of molecules with molecular weight of M_i
M_i: molecular weight of the ith species
w_i: weight fraction of molecules of the ith species
W_i: weight of material with molecular weight of M_i

Hint: $w_i = N_i M_i / W_t$ where W_t is the total weight of the sample. Also, $W_i = N_i M_i = w_i W_t$.

3-5. A polymer is made of the following fractional distribution. Calculate M_n, M_w, and M_w/M_n for the sample using the equations given in Problem 3-4.

w_i	0.05	0.10	0.30	0.30	0.15	0.10
M_i (g/mol)	1000	2000	4000	5000	6000	7000

3-6. Calculate the theoretical c/a ratio for hexagonal close-packing structure.

3-7. Titanium has an hcp structure below 882°C and a bcc structure above this temperature. Calculate the volume change during the allotropic transformation.

3-8. Complete the table:

	fcc	bcc	sc
Side of unit cell			
Body diagonal			
Face diagonal			

3-9. The density of Fe is 7.87 g/ml. Calculate its packing factor (efficiency) and number of atoms per milliliter.

REFERENCES

1. C. S. Smith, *A History of Metallurgy*, University of Chicago Press, Chicago, 1960.
2. W. G. Moffatt, G. W. Pearsall, and J. Wulff, *The Structure and Properties of Materials*, Volume 1, Chapters 1–3, Wiley, New York, 1964.
3. W. H. Bragg and W. L. Bragg, The reflection of x-rays by crystals, *Nature (London) 91*, 477, 1913.
4. R. D. Heidenreich, Electron microscope and diffraction study of metal crystal texture by means of thin section, *J. Appl. Phys. 20*, 993–1010, 1949.
5. P. B. Hirsch, R. W. Horne, and M. J. Whelan, Direct observations of the arrangement and motions of dislocations in aluminum, *Philos. Mag. 1*, 677–684, 1956.
6. B. Harris and A. R. Bunsell, *Structure and Properties of Engineering Materials*, Longmans, London, 1977.
7. F. W. Billmeyer, Jr., *Textbook of Polymer Science*, p. 151, Interscience, New York, 1962.
8. B. Maxwell, Modifying polymer properties mechanically, in: *Polymer Processing*, J. V. D. Fear (ed.), *Chem. Eng. Prog. Symp. Ser. 60*, 10–16, 1964.

BIBLIOGRAPHY

A. H. Cottrell, The nature of metals, in: *Materials*, D. Flanagan *et al.* (ed.), Freeman, San Francisco, 1967.
R. H. Krock and M. L. Ebner, *Ceramics, Plastics, and Metals*, Chapter 3, Heath, Boston, 1965.
L. Pauling, *The Nature of the Chemical Bonding*, Chapter 2, Cornell University Press, Ithaca, N.Y., 1960.
M. J. Starfield and M. A. Shrager, *Introductory Materials*, McGraw–Hill, New York, 1972.

THERMODYNAMICS OF STRUCTURAL CHANGES

In this chapter we will discuss some fundamentals of thermodynamics and then apply the knowledge to elucidate the rates of reactions, phase changes, diffusion, and surface properties.

4.1. THERMODYNAMIC RELATIONSHIPS

In this section the thermodynamics of materials and elementary engineering thermodynamics will be presented, leaving a rigorous development of the theory of thermodynamics to other classical texts.

4.1.1. Energy, Enthalpy, and the First Law of Thermodynamics

The first law of thermodynamics, that of energy conservation, states that if the system (object or region of interest) changes its state, then

$$\Delta E = q - w \tag{4-1}$$

where E is the internal energy, q is the total heat in the system, and w is the total work done by the system during changes of state. Note that E is a function only of the initial and final states of the system; hence, it is independent of the path between them while q and w are not. Temperature changes of a material by heating (input) are an intrinsic property of the material and depend on its capacity to absorb heat, which can be expressed as

$$C_P = (\partial q / \partial T)_P \qquad (4\text{-}2)$$

where C_P is the heat capacity at constant pressure. If the system works only against the atmosphere, the changes in internal energy can be given by

$$dE = dq - P\,dV \qquad (4\text{-}3)$$

and if the volume is held constant, then

$$C_V = (\partial E / \partial T)_V = (dq/dT)_V \qquad (4\text{-}4)$$

where C_V is the constant-volume heat capacity.

It is more convenient to define another property, enthalpy or heat content, which is defined as

$$H = E + PV \qquad (4\text{-}5)$$

Here again E is a state function and consequently H is also a single-valued function of state. The differential of H at constant pressure leads to the following relationship similar to equation (4-4):

$$C_P = (\partial H / \partial T)_P = (dq/dT)_P \qquad (4\text{-}6)$$

4.1.2. Entropy and the Second Law of Thermodynamics

The first law refers to the equivalence of heat and work in the form of energy. However, not all of the internal energy of a system is available to be converted into usable energy. The "unavailable energy" of the system is described as entropy. In a reversible closed system which cannot exchange energy in any form, the change can be expressed as

$$dS = q_{\text{rev.}} / T = (C_P/T)\,dT \qquad (4\text{-}7)$$

For an irreversible change which is a more common occurrence in nature,

$$dq_{\text{rev.}} \geq dq_{\text{irrev.}} \qquad (4\text{-}8)$$

Therefore,

$$dS \geq dq_{\text{irrev.}} / T \qquad (4\text{-}9)$$

This expression implies that the entropy tends to increase in all natural processes since they are irreversible. Note also that the entropy is maximal at equilibrium.

The statistical definition of entropy can be derived by considering reversible, isothermal expansion of an ideal gas and the reader should be able to prove

$$\Delta S = k \ln W \qquad (4\text{-}10)$$

where k is the Boltzmann constant and W is the relative probability of two states or the increase in randomness occurring during change. Table 4-1 gives entropies of some common substances. Note the increase in entropy with decreasing structural rigidity in general.

4.1.3. Free Energies: The Thermodynamic Potentials

Two types of free energies can be defined according to the convenience of describing a system. In both cases the "unavailable energy" which is defined in terms of entropy is considered maximal at equilibrium; thus the "available energy" or free energy can be expressed by

$$F = E - TS \qquad (4\text{-}11)$$

where F is the Helmholtz free energy. The Gibbs free energy (G) is defined in relation to the enthalpy:

$$G = H - TS \qquad (4\text{-}12)$$

Differentiating equation (4-11) and using the first law, we can write

$$dF = dq_{irrev.} - dq_{rev.} - dw - S\,dT \qquad (4\text{-}13)$$

Table 4-1. Entropies of Some Common Substances[a]

Substance	Entropy at 25°C 10^{-24} J/K per atom (or molecule)
Diamond	4.2
Iron	45
Platinum	69
Al_2O_3 (corundum)	85
Lead	108
Water	116
Mercury	128
Laughing gas (N_2O)	368

[a] From Ref. 1.

Further, for an isothermal process with no work done such as structural changes occurring at constant temperatures, we have

$$(dF)_{T,V} = dq_{irrev.} - dq_{rev.} \qquad (4\text{-}14)$$

But from equation (4-8), $q_{rev.} > q_{irrev.}$, hence

$$0 \geq (dF)_{T,V} \qquad (4\text{-}15)$$

Equation (4-15) indicates that for a reversible process there should be no free energy change ($dF = 0$). This occurs only when the system is in equilibrium. For a spontaneous change such as chemical reactions, the dF is negative and the free energy decreases. The same line of reasoning can be applied to the Gibbs free energy changes,

$$0 \geq (dG)_{T,P} \qquad (4\text{-}16)$$

which is the basis of thermodynamic stability.

4.2. RATES OF REACTIONS

The laws of thermodynamics do not provide information on the rates of reactions. Some materials such as woods and metals remain in metastable form for a long period of time unless they overcome an energy barrier which can be quantitated as an activation energy. The rate of a chemical reaction as a function of temperature was first empirically derived by Arrhenius:

$$\text{rate} = \text{constant} \times \exp(-E^*/RT) \qquad (4\text{-}17)$$

where E^* is the activation energy in units of joules per mole, R is the gas constant, T is the absolute temperature, and the constant is independent of temperature. If we plot $\ln(\text{rate})$ versus $1/RT$, the slope becomes $-E^*$.

Similar exponential dependence of temperature which specifies the energy distribution of gas molecules was described by the Boltzmann relation, in which the probability of finding a molecule at an energy level ΔE greater than the average energy at T is

$$\text{probability} = \text{constant} \times \exp(-\Delta E/kT) \qquad (4\text{-}18)$$

where k is the Boltzmann constant. The activation energy of a reaction can be depicted as shown in Figure 4-1 and the mechanical analog is shown in Figure 4-2. In both cases the changes in chemical and potential energies are

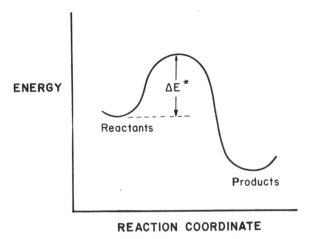

REACTION COORDINATE

Figure 4-1. Schematic representation of an activation energy barrier to a chemical reaction.

required for changing metastable to stable state, which can be thought of as "available energies" and are therefore related to the free energies. Therefore, equation (4-18) can be written in terms of the free energy of activation, ΔF^*, rather than internal energy changes:

$$probability = constant \times exp\left(-\Delta F^*/RT\right) \qquad (4\text{-}19)$$

and since $\Delta F^* = \Delta H^* - T\Delta S^*$,

$$probability = constant \times \left[exp\left(\Delta S^*/R\right)\right]\left[exp\left(-\Delta H^*/RT\right)\right] \quad (4\text{-}20)$$

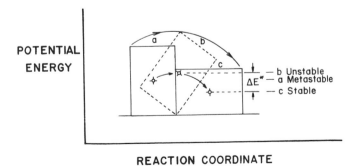

REACTION COORDINATE

Figure 4-2. A mechanical analog to the activation energy barrier using a brick.

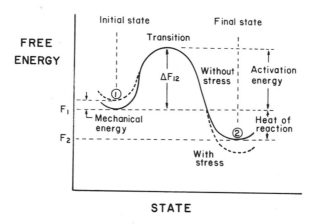

Figure 4-3. Changes in activation energy by the presence of a catalyst and an applied stress.

The last expression enables us to include the changes of entropy during a reaction which, in some instances such as nucleation, can be significant.

If the reaction is aided by the presence of a catalyst or by an applied stress, the activation energy can be lowered by the catalyst or the initial and final free energy states will be changed by the stress such that it becomes easier to overcome the barrier as shown in Figure 4-3. The major criticism of this rather simple but extremely useful activation energy theory is that it only applies to processes in which a single mechanism with a unique free energy of activation is controlling the reaction. It is conceivable that there is a spectrum of activation energies for some processes such as annealing to which the Arrhenius type of theory cannot be applied without modifications.

4.3. PHASE CHANGES

Properties of a material depend on its phase, which is defined as a physically distinct region of matter having characteristic atomic structure and properties which change continuously with thermodynamic variables such as temperature, pressure, and composition. The various phases of a system are, in principle, separable mechanically. Basic thermodynamic principles will be applied to understand some simple phase changes occurring in materials.

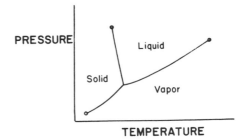

PRESSURE

Liquid

Solid

Vapor

Figure 4-4. P–T phase diagram of water.

TEMPERATURE

4.3.1. Single-Component Systems: Allotropy

Changes in phases are most familiar with water as shown in Figure 4-4. Each phase (vapor, liquid, and solid) is stable for a given pressure and temperature, that is, it is in thermodynamic equilibrium. Since two variables determine the phase, it is in a bivariant equilibrium. Three phases exist at one point for a particular temperature (0.0075°C) and pressure (4.58 mm Hg); hence, it is called an invariant point. If two phases are present (ice and water at 0°C, water and steam at 100°C, both at 1 atm) the equilibrium becomes univariant if either T or P is varied, the other becomes a specific value. The lines dividing the three phases represent the univariant equilibrium states.

Some elements or compounds can exhibit two or more phases in solid states. The process of temperature-induced phase change is called allotropic (or polymorphic) phase transformation. A good example is an iron which exhibits three crystal structures as shown in Figure 4-5. It seems, however, paradoxical to have different crystal structures in the solid state since thermodynamic considerations lead us to expect that the most probable structure is the closed-packed structure, fcc. This apparent paradox results

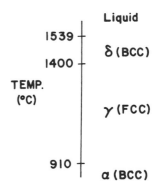

Liquid

1539

δ (BCC)

1400

TEMP.
(°C)

γ (FCC)

910

Figure 4-5. Phase diagram of pure iron at 1 atm.   α (BCC)

from the stored energy contribution by electrons themselves, especially in some of the transition elements.

The allotropic phase changes can be explained in terms of their temperature dependence of the free energies of various phases as shown in Figure 4-6. Differentiating equation (4-12) and substituting (4-6) and (4-7) at a constant pressure yields

$$dG = -S\,dT \qquad (4\text{-}21)$$

Hence, the free energy at any temperature T can be given as

$$G = G_0 - \int_0^T S\,dT \qquad (4\text{-}22)$$

where G_0 is the free energy at $0°$K. From equation (4-7), entropy at T ($°$K) is

$$S(T) = \int_0^T (C_P/T)\,dT \qquad (4\text{-}23)$$

Therefore, the temperature dependence of the free energy is given by

$$F = F_0 - \int_0^T \int_0^T (C_P/T)\,dT\,dT \qquad (4\text{-}24)$$

Equation (4-24) gives the forms of curves as shown in Figure 4-6 and the relative phases at any given temperature can be determined by the relative

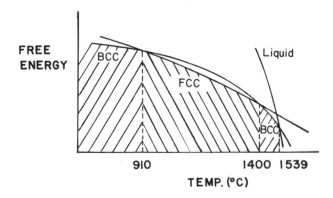

Figure 4-6. Schematic illustration of the free energy variations with temperature for iron phases.

specific heats of the different structures. The greater curvature of the bcc iron curve than that for fcc is caused by the fact that the heat capacity of bcc is greater than that of fcc.

4.3.2. Composition and Phase Stability

The same thermodynamic laws can be applied to mixing two or more components. Once they are mixed into a single homogeneous solution between two miscible components, it is said to be an irreversible process. Therefore, mixing always increases the entropy (see Section 4.1.2) and will be maximal near 50–50 composition. The increase in entropy of mixing, S_{mix}, is equivalent to a decrease of free energy, and consequently tends to mix. The entropy of mixing can be deduced from the Boltzmann equation (4-10). For a two-component system, we have to calculate the number of ways of arranging atoms A and B in separate pure crystals. If we are to arrange atoms on vacant lattice sites, the first A atoms we can insert on any of the N sites, the second atom can be put on any of the remaining $N-1$ sites, and so on. Since A atoms are not distinguishable from one another, the number of ways of arranging N atoms is given by

$$W_{AB} = N!/n!(N-n)! \tag{4-25}$$

where n is number of A atoms. Therefore, the entropy of mixing becomes

$$S_{mix} = k \ln \left[N!/n!(N-n)! \right] \tag{4-26}$$

Using Stirling's approximation ($\ln N! = N \ln N - N$), equation (4-26) can be rearranged to

$$S_{mix} = k \left[N \ln N - n \ln n - (N-n) \ln (N-n) \right] \tag{4-27}$$

Since $C_A = n/N$ and $C_B = 1 - C_A = (N-n)/N$,

$$S_{mix} = - Nk (C_A \ln C_A + C_B \ln C_B) \tag{4-28}$$

The S_{mix} is positive as we discussed previously since C_A and C_B are less than unity and are maximal at 50% as shown in Figure 4-7.

Another contribution to the free energy, enthalpy or internal energy, will determine the tendency of mixing. If there is an atomic size mismatch, then the enthalpy will increase by increasing the stored elastic strain energy. However, in some cases this increase is offset by having the energy of A–B bonds which is less than the energy of A–A or B–B bonds. This results in a

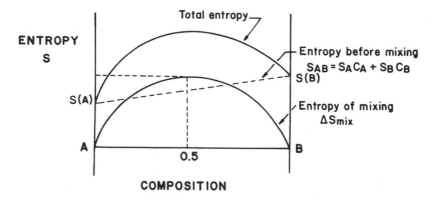

Figure 4-7. Schematic illustration of the entropy of mixing.

lower overall internal energy by mixing A and B. The reverse may be true in some cases, which have a large distortional strain energy and have a larger A–B bonding energy resulting in an increase in internal energy of mixing (Figure 4-8). If both entropic (Figure 4-7) and enthalpic or internal energies (Figure 4-8) are added according to the compositional variations, we will have a curve representing the free energy of mixing as shown in Figure 4-9. If E_{mix} is either zero (ideal solution) or a negative deviation from zero, the free energy of mixing is always lower than before mixing; thus, it tends to mix. On the other hand, if E_{mix} is positive, then the free energy of mixing curve will have two minima. This is because the initial rate of increase of S_{mix} is usually so great that $-TS_{mix}$ outweighs the E_{mix} contribution; thus, the resultant free energy will usually decrease with composition at first.

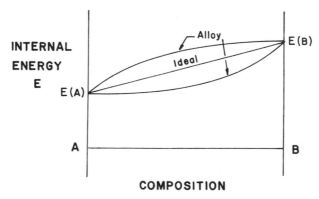

Figure 4-8. Variation of internal energy with composition in a two-component system.

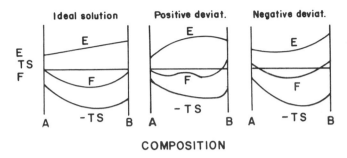

COMPOSITION

Figure 4-9. Variation of free energy with composition in a two-component system.

To find the stable arrangement for any given composition X, we need to consider the lowest possible total free energy as shown in Figure 4-10. Since the free energy of mixing of X_0 composition is greater than that of X_α and X_β composition, it will split up into a mixture of α and β phases. The relative amount of α and β phases for the given composition X_0 and temperature can be calculated by using a lever rule:

$$\frac{\alpha}{\alpha + \beta} = \frac{X_0 - X_\beta}{X_\beta - X_\alpha}; \qquad \frac{\beta}{\alpha + \beta} = 1 - \frac{\alpha}{\alpha + \beta} = \frac{X_0 - X_\alpha}{X_\beta - X_\alpha} \qquad (4\text{-}29)$$

The stable phase equilibrium diagram can be constructed in a similar method as for the allotropic phase transformation if the free energy versus composition curves for the various phases are known at various temperatures as shown in Figure 4-11. At each temperature the appropriate tangents are drawn to the free energy curves as was done in Figure 4-10. The points of tangency mark the constant composition of the two phases in the two-phase regions and the boundaries between two- and single-phase regions. It is interesting to note that at T_2 a single line touches all three free energy curves at their minimum in which all three phases coexist and a eutectic reaction results:

$$L \rightleftharpoons \alpha + \beta \qquad (4\text{-}30)$$

Figure 4-10. Free energy diagram for a two-component system showing phase separation into two terminal solutions α and β.

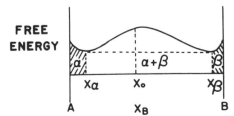

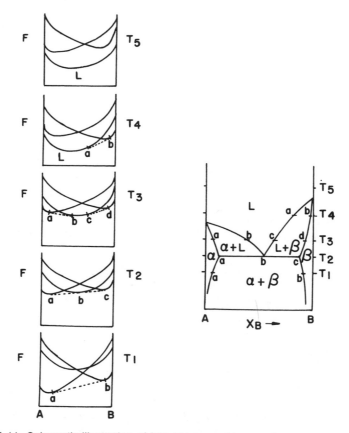

Figure 4-11. Schematic illustration of free energy and temperature versus composition.

The eutectic reaction only occurs at a fixed temperature T_2; therefore, it is an invariant point for the binary system. For a C-component P-phase system, it will have a total of $CP + 2$ variables, where 2 variables represent temperature and pressure. It can be shown that the number of independent variables at a constant pressure is given by

$$V = C - P + 1 \tag{4-31}$$

For a two-component system $V = 3 - P$; therefore, the three-phase equilibria are invariant, two-phase equilibria have one variable, and one-phase equilibrium has two variables, T and C.

An important binary phase diagram, the Fe–C system, is shown in Figure 4-12. It is important to remember that the phase diagram is de-

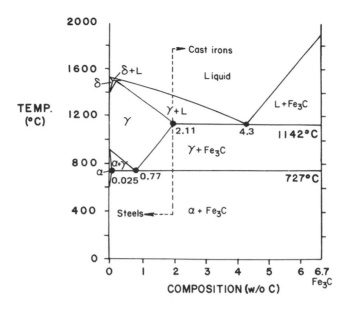

Figure 4-12. Fe–Fe$_3$C phase diagram.

termined under (closed) thermodynamic equilibrium states of phases for a
given temperature and composition. It does not give information on how
fast the system will approach the equilibrium nor on the structural distribution of the phases present. The reaction rate depends on the mechanism of
phase changes while the distribution of phases depends on the surface
energy and the strain energy of the phase transformation.

4.3.3. Mechanism of Phase Changes

There are two types of phase changes in solids; one is a diffusion-con-
trolled nucleation and growth, the other is a diffusionless transformation.
The best example of the latter is the martensitic phase transformation of
steel in which an fcc structure changes into a body-centered tetragonal by
shearing, distorting the lattice structure in the presence of excess carbon
atoms by the transformation (see Chapter 5). The majority of transformation, however, occurs by the nucleation and growth of new phase within old
phase.

The formation of the β phase starts by nucleating small regions of the
α phase as shown in Figure 4-13. The phase diagram tells us that the new
phase is thermodynamically stable in the old phase α, which is no longer in
a thermodynamically favorable state due to the decreased temperature. The

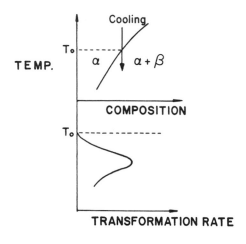

Figure 4-13. Schematic illustration of rate of phase transformation from α to $\alpha + \beta$.

transformation from α to β does not occur instantaneously because in order to precipitate the new crystals from the old it is necessary for atoms to diffuse over long distances. Therefore, when cooling through the equilibrium transformation temperature T_0, there will be a gradual change of phase, depending on the degree of undercooling. There will be competing forces which will control the degree of transformation for a given temperature. The total free energy change accompanying the formation of the new β phase, ΔF, is contributed by the surface and volume free energy changes:

$$\Delta F = \Delta F_s - \Delta F_v \qquad (4\text{-}32)$$

where $+ \Delta F_s$ opposes the new phase formation and $- \Delta F_v$ favors it. ΔF_s and ΔF_v are also proportional to the surface area and volume of the new phase nucleating out of the old. Therefore, if we assume the nucleus has a diameter r, then

$$\Delta F = Ar^2 - Br^3 \qquad (4\text{-}33)$$

When equation (4-33) is plotted as shown in Figure 4-14, there is a critical size of nucleus r^* associated with the nucleation energy ΔF_n. When the nucleus becomes larger than r^*, it will grow larger and for those smaller than r^* will disappear.

The probability of forming a stable nucleus $(r > r^*)$ is given by equation (4-19), that is, $\exp(- \Delta F_n/kT)$. Since the nucleation requires diffusion of atoms (see next section) to overcome the diffusion activation energy, the rate of (homogeneous) nucleation will be

$$dn/dt = \text{constant} \times \exp\left[-(\Delta F_d + \Delta F_n)/kT \right] \qquad (4\text{-}34)$$

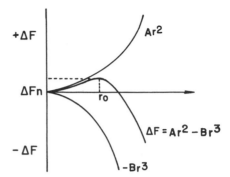

Figure 4-14. Free energy changes with the size of new phase formed for homogeneous nucleation.

where n is the number of nuclei formed. At high temperatures near the phase transition temperature, ΔF_n is very large and hence $\exp(-\Delta F_n/kT)$ and dn/dt are small. As the temperature decreases, the diffusion of atoms to the nuclei becomes slow and ΔF_n is still large so that dn/dt depends largely on $\exp(-\Delta F_d/kT)$. Therefore, the rate of nucleation decreases again due to the slow diffusion. Figure 4-15 shows that there is a maximum rate of nucleation where $\exp(-\Delta F_n/kT)$ and $\exp(-\Delta F_d/kT)$ curves are met.

The main problem of applying the homogeneous nucleation theory is that it does not apply to a real material system since the free energy of nucleation for the homogeneous process is much higher than the heterogeneous route since

$$\Delta F^* \text{ (homogeneous)} > \Delta F^* \text{ (heterogeneous)} \qquad (4\text{-}35)$$

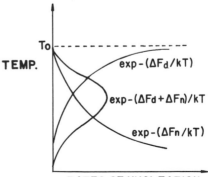

Figure 4-15. Temperature versus nucleation rate.

4.3.4. Time–Temperature Transformation of Steel

From the phase diagram of Fe–Fe$_3$C (Figure 4-12) one can see that the γ and α phases can contain a maximum 1.4 and 0.025 w/o of carbon at 1000 and 72 °C. The fcc phase *austenite* (γ) can only exist above the eutectoid temperature (727°C) in plain-carbon steels although by adding large amounts of alloying elements it can be made stable at room temperature such as 18-8 stainless steel which is an austenitic alloy. Table 4-2 gives the nomenclature for AISI and SAE steels and the strength of some important steels.

If we cool austenite below 727°C in equilibrium condition, it will transform into a soft, malleable α phase (bcc, ferrite) and a hard brittle Fe$_3$C compound (cementite). However, we can change the microstructure of steel by a controlled heat treatment as shown in Figure 4-16, in which log time versus temperature is plotted for various phases (hence called TTT curves). The TTT curves are S-shaped, which is the mirror image of the temperature versus rate of transformation curve of Figure 4-13. Above 550°C the austenite transforms into *pearlite*, which is a lamellar mixture of ferrite and cementite. At high temperatures the pearlite becomes coarse-

Table 4-2. Compositions and Strength of Some Steels and AISI and SAE Nomenclature

	Carbon (w/o)	Other alloying elements	Tensile strength (MPa)
Steel[a]			
Mild	0.1–0.2	0.5 Mn	300–500
Tool	0.6–1.20	0.5–1.0 Mn, < 0.3 Si	800–1300
18-8 S.S.	—	16–20 Cr, 8–10 Ni, 1.0 Mn	550
AISI or SAE number			
10 xx	.xx	—	
13 xx	.xx	1.5–2.0 Mn	
23 xx	.xx	3.25–3.75 Ni	
31 xx	.xx	1.10–1.40 Ni, 0.55–0.90 Cr	
40 xx	.xx	0.20–0.40 Mo	
41 xx	.xx	0.40–1.20 Cr, 0.08–0.25 Mo	
43 xx	.xx	1.65–2.00 Ni, 0.40–0.90 Cr, 0.20–0.30 Mo	
46 xx	.xx	1.40–2.00 Ni, 0.15–0.30 Mo	
51 xx	.xx	0.70–1.20 Cr	
61 xx	.xx	0.70–1.10 Cr, 0.10 V	
81 xx	.xx	0.20–0.40 Ni, 0.30–0.55 Cr, 0.08–0.15 Mo	
92 xx	.xx	1.80–2.20 Si	

[a] All steels contain a minimum of 0.5 w/o Mn.

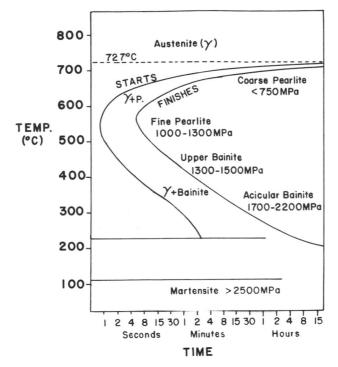

Figure 4-16. TTT diagram for a 1080 steel.

grained, which is considerably weaker than the fine pearlite formed at lower temperatures. In fact, the lamellar structure becomes so fine it cannot be resolved by an optical microscope.

Below 550°C the diffusion of carbon becomes more difficult and the rate-controlling step becomes the migration of iron atoms to form ferrite crystals. When the carbon content of carbide particles reaches below a critical level, they will then be precipitated out as discontinuous, needlelike crystallites, thus forming a so-called ε-carbide, which contains 8.4 w/o carbon instead of the 6.7 w/o for the Fe_3C phase. Again these carbides cannot be seen by an optical microscope and the bainite becomes stronger as the transformation temperature is decreased.

Below 250°C the diffusion of the dissolved carbon becomes too slow for the formation of carbide particles to occur and remove the excess carbon from solution. Consequently, when the bcc phase is formed (the thermodynamically stable state) it contains too much carbon; therefore, the bcc structure becomes a highly distorted body-centered tetragonal called *martensite*. This transformation occurs very fast as noted in the phase diagram (Figure 4-16) by complex shearing of the lattice; therefore, it contains a high strain energy, consequently making it very brittle and hard. To make the martensite more useful it must be *tempered* at sufficiently high temperatures to permit some of the excess carbon to precipitate into spheroidite which makes it softer yet very strong and tough.

4.4. DIFFUSION

Diffusion is the process by which matter is distributed by random thermal vibrations. Since diffusion occurs spontaneously, it increases the entropy, thereby decreasing the free energy. A simple diffusion model is shown in Figure 4-17 where the flux, J_x, of the diffusing species is from left to right due to the concentration difference, $C_s > C_x$. When a steady state is reached the concentration gradient dC/dx becomes constant. Therefore, the amount of flux is proportional to dC/dx:

$$J_x = - D_x(dC/dx) \qquad (4\text{-}36)$$

where the constant D is called diffusivity or diffusion coefficient in units of square centimeters per second. Equation (4-36) is known as Fick's first law.

In a more common diffusion, dC/dx will not be uniform and J_x will be time dependent. Under this transient condition, the dC/dx and J in one direction x can be expressed by

$$dC_x/dt = d/dx(D_x dC_x/dx) \qquad (4\text{-}37)$$

Equation (4-37) is called Fick's second law. If D is independent of concentration, then

$$dC_x/dt = D_x(d^2C_x/dx^2) \qquad (4\text{-}38)$$

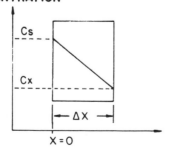

Figure 4-17. Schematic representation of a steady-state diffusion governed by Fick's first law.

If we consider an isotropic system $(D_x = D_y = D_z)$ in three dimensions, equation (4-38) can be written in a general form:

$$dC/dt = D\left(\partial^2 C/\partial x^2\right) \qquad (4\text{-}39)$$

This Laplacian partial differential equation is similar to the heat flow or conduction problem and can be solved if the boundary conditions are known.

Figure 4-18 illustrates two commonly encountered diffusion conditions. The first case of diffusion is the semi-infinite solid where the surface concentration C_s remains constant. The second case of diffusion is across a plane interface between semi-infinite blocks of solution and solvent. The solution to equation (4-39) for the first case is given by

$$(C_x - C_0)/(C_s - C_0) = 1 - \text{erf}\left[x/2(Dt)^{1/2}\right] \qquad (4\text{-}40)$$

and for the second case

$$C/C_0 = 1/2\left\{1 - \text{erf}\left[x/2(Dt)^{1/2}\right]\right\} \qquad (4\text{-}41)$$

where $\text{erf}[x/2(Dt)^{1/2}]$ is the Gaussian error function. Equation (4-41) is shown graphically in Figure 4-19 in which $1 - \text{erf}[x/2(Dt)^{1/2}]$ is plotted against $x/2(Dt)^{1/2}$. If C_0, C_s, and D are known in a material, C must be a function of $x/(Dt)^{1/2}$. Therefore, if it is desired to have a penetration depth doubled, then the diffusion time should be quadrupled.

The diffusion of atoms across lattices should overcome the energy barrier extended by the lattices as shown in Figure 4-20. Therefore, we can analyze the probability that the atoms will diffuse or jump over the energy barrier ΔF which will be equal to the atomic vibration frequency ν_0 multiplied by the probability that the atoms will have the necessary energy,

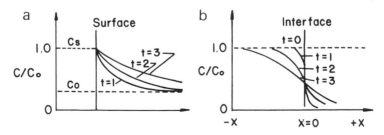

Figure 4-18. Two cases of diffusion: (a) semi-infinite solid; (b) coupled solid.

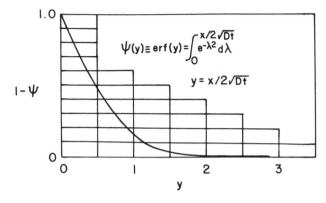

Figure 4-19. Gaussian error function for diffusion.

$\exp(-\Delta F/kT)$. Therefore, the jump frequency ν is given by

$$\nu = \nu_0 \exp(-\Delta F/kT) \qquad (4\text{-}42)$$

For a simple isotropic cubic lattice the diffusion coefficient is proportional to the jump frequency ν, hence

$$D = (1/6)a^2\nu \qquad (4\text{-}43)$$

where a is the lattice spacing and $1/6$ accounts for the three orthogonal directions for the cubic lattice. Substituting equation (4-42) into (4-43).

$$D = (1/6)a^2\nu_0 \exp(-\Delta F/kT) \qquad (4\text{-}44)$$

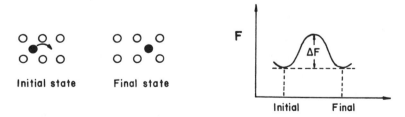

Figure 4-20. Energy barrier to the atomic jump.

and using equation (4-11)

$$D = (1/6)a^2\nu_0 \exp(\Delta S/k - \Delta E_d/kT) \tag{4-45}$$

which can be rewritten as

$$D = D_0 \exp(-\Delta E_d/kT) \tag{4-46}$$

where $D_0 = (1/6)a^2\nu_0 \exp(\Delta S/k)$ and ΔE_d is the diffusion activation energy.

There are four possible ways that atomic diffusions can take place as shown in Figure 4-21. The most common mechanism is vacancy diffusion. The vacancies are present in all crystals and the entropy of "mixing" full and vacant lattice sites can be calculated according to equation (4-26):

$$S_v = k \ln[N!/n!(N-n)!] \tag{4-47}$$

where N and n are available lattice sites and vacancies, respectively. Again using Stirling's approximation and the free energy equation,

$$F_v = nE_v - TS_v \tag{4-48}$$

And $\partial F/\partial n = 0$ for equilibrium; therefore,

$$n/(N-n) = \exp(-E_v/kT) \cong n/N \tag{4-49}$$

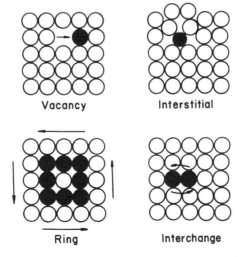

Vacancy   Interstitial

Ring   Interchange

Figure 4-21. Atomic mechanisms for solid diffusion.

4.5. SURFACE PROPERTIES

Understanding the surface properties of materials is very important, more so for implant materials which must be exposed to the body environment. The reader is advised to get familiar with this subject beyond the treatment given in this section.

4.5.1. Surface Tension

Surface tension is a direct result of the atomic or molecular force imbalance at two phases as shown in Figure 4-22. A molecule at the surface is attracted toward the interior which is surrounded by a uniform average field of neighboring molecules. Therefore, the surface has a higher free energy than the bulk and tends to contract to minimize its area. The extra surface free energy can be expressed by

$$dG = dw - \gamma\, dA \tag{4-50}$$

where A is surface area, w is work, and γ is surface tension.

The surface tensions of some substances are given in Table 4-3. The conventional units for surface tension are dynes per centimeter or ergs per square centimeter for surface energy but both units are exactly the same since 1 erg is equal to 1 dyne-cm. The SI unit N/m is equal to 1000 dynes/cm.

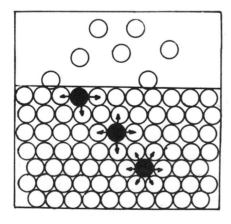

Figure 4-22. A two-dimensional representation of liquid–vapor interface.

Table 4-3. Surface Tension of Some Substances[a]

	Temp (°C)	Surface tension	
		Dynes/cm	N/m[b]
Isopentane	20	13.72	0.014
n-Hexane	20	18.43	0.018
Ethyl mercaptan	20	21.82	0.022
Benzene	20	28.86	0.029
Carbon tetrachloride	20	26.66	0.027
Water	20	72.75	0.073
Silver	970	800.00	0.800
Gold	1070	1000.00	1.000
Copper	1130	1100.00	1.100
NaF	1010	200.00	0.200
NaCl	1000	98.00	0.098
NaBr	1000	88.00	0.088

[a] From Ref. 3.
[b] 1 N/m = 1000 dynes/cm or ergs per cm^2.

4.5.2. Surface Tension Measurements of Solids

One method of measuring surface tension of solids is to suspend a fine wire in a furnace loaded with a known weight (w = mg) just sufficient to prevent any shrinkage in length. The surface energy can be approximated by

$$\gamma = mg/\pi d \qquad (4\text{-}51)$$

For a copper wire of 50-μm diameter the surface tension is about 1.4 N/m at 1250°C.

Another method is to determine the heat of solution of very fine particles of known size. Due to the increased surface energy of the particles, the heat of the solution will differ from that of the bulk material. This difference along with the known surface area can be used to calculate the surface energy exactly.

If a liquid is dropped on a flat solid surface, then the liquid drop will spread or make a spherical bubble as shown in Figure 4-23. At equilibrium the surface tension among the three phases in the solid plane should be zero because the liquid is free to move until force equilibrium is established; therefore,

$$\gamma_{GS} - \gamma_{LS} - \gamma_{GL} \cos \theta = 0 \qquad (4\text{-}52)$$

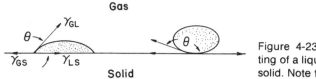

Figure 4-23. Wetting and nonwetting of a liquid on a flat surface of a solid. Note the contact angle θ.

hence

$$\gamma_{GS} = \gamma_{LS} + \gamma_{GL} \cos \theta \qquad (4\text{-}53)$$

The angle θ is called a contact angle and the wetting characteristics for a given liquid and solid can be generalized as

$$\theta = 0 \qquad \text{complete wetting}$$

$$0 < \theta < 90 \qquad \text{partial wetting} \qquad (4\text{-}54)$$

$$\theta > 90 \qquad \text{nonwetting}$$

Note that equation (4-53) only gives ratios rather than absolute values of the surface tension. However, the contact angle method can be used to measure the critical surface tension of a given liquid and a given solid which is close to the intrinsic surface tension. A series of homologous liquids is used to

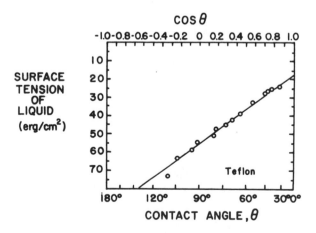

Figure 4-24. A Zisman plot of Teflon®. The contact angles of Teflon® with each of a series of pure liquid are presented. (From Ref. 2.)

Table 4-4. Contact Angle Value of Some Materials

Liquid	Substrate	Contact angle
Methylene iodine (CH_2I_2)	Soda lime glass	29
	Fused quartz	33
Water	Paraffin	107
Mercury	Soda lime glass	140

measure contact angles with the solid and these angles are plotted in a Zisman plot as shown in Figure 4-24. The surface tension values obtained by this method are in good agreement with results obtained by other methods. Table 4-4 gives some values of contact angles.

PROBLEMS

4-1. How many vacancy sites are there in a copper if the vacancy activation energy level is 1 eV at 1000°K?

4-2. Determine the surface area of 1 g of crystalline silica (SiO_2) particles in relation to their average diameter (density = 2.30 g/ml).

4-3. Determine the amount of pearlite in 100 g of 0.5% carbon steel which is cooled slowly from 1000°C. Also determine the amount of ferrite (α) and carbide (Fe_3C) present at 722°C and room temperature.

4-4. For 1080 steel:

a. What phases are present at 1600, 1400, 1000, 724, and 722°C?
b. Calculate the amount of each phase by using the lever rule and give their compositions.

4-5. Draw the free energy versus composition diagram for the 1080 steel of Problem 4-4 at the various given temperatures.

4-6. The surface tension of the liquids and the contact angle between each liquid and the solid polytetrafluoroethylene (Teflon®) are given:

Liquid	Surface tension (dynes/cm)	Contact angle (°)
A	63	103
B	50	80
C	28	45

a. Calculate the interfacial surface tension for each liquid and solid Teflon®.
b. What conclusions can be drawn from the data?

4-7. From the following data which were obtained for the diffusion of aluminum in silicone crystal (R. C. Miller and A. Savage, *J. Appl. Phys.* 27, 1430, 1956),

T (°C)	Diffusion coefficient (D, $\times 10^{-11}$ cm^2/s)
1380	31.1
1300	7.1
1250	4.1
1200	1.74

a. Plot the logarithm of D versus $1/T$ (°K^{-1}).
b. Determine D_0 and ΔE_d in equation (4-46).
c. Calculate the rate of diffusion at 800°C.

4-8. Derive equation (4-28).

4-9. C and Fe have atomic radii of 0.077 and 0.124 nm, respectively. Explain

a. Why C does not dissolve substitutionally in Fe.
b. Whether ferrite (α) or austenite (γ) Fe is more likely to dissolve C.
c. Which voids the C atoms are more likely to occupy.

REFERENCES

1. B. Harris and A. R. Bunsell, *Structure and Properties of Engineering Materials*, Longman, London, 1977.
2. R. E. Baier, *Guidelines for Physicochemical Characterization of Biomaterials*, Devices and Technology Branch, National Heart, Lung, and Blood Institute, NIH Publication 80-2186, p. 97, 1980.
3. W. J. Moore, *Physical Chemistry*, 3rd ed., Prentice–Hall, Englewood Cliffs, N.J., 1962.

BIBLIOGRAPHY

A. W. Adamson, *Physical Chemistry of Surfaces*, 2nd ed., Wiley, New York, 1969.
J. H. Brophy, R. M. Rose, and J. Wulff, *The Structure and Properties of Materials*, Volume 2, Wiley, New York, 1964.
A. H. Cottrell, *Theoretical Structural Metallurgy*, Chapter 2, St. Martin's Press, New York, 1955.
L. S. Darken and R. W. Gurry, *Physical Chemistry of Metals*, Chapter 18, McGraw–Hill, New York, 1953.
B. Harris and A. R. Bunsell, *Structure and Properties of Engineering Materials*, Chapter 3, Longmans, London, 1977.
D. H. Kaelble, *Physical Chemistry of Adhesion*, Wiley, New York, 1971.
W. J. Moore, *Physical Chemistry*, 3rd ed., Prentice–Hall, Englewood Cliffs, N.J., 1962.
P. G. Shewmon, *Diffusion in Solids*, Chapter 1, McGraw–Hill, New York, 1963.

5

STRENGTH AND STRENGTHENING MECHANISMS

The strength of a material depends on various intrinsic and extrinsic factors. Although the most important is the chemical composition, the properties can be manipulated by altering the structure. Various mechanical, thermal, and surface treatments are used to create the desired balance of strength, hardness, and ductility. This chapter explains structure–property relationships and strengthening mechanisms for metals, ceramics, polymers, and composites.

5.1. STRENGTHS OF PERFECT AND REAL MATERIALS

The nature of a material's atomic and molecular bonding will determine the type of failure mode exhibited. Figure 5-1a shows plastic shear deformation of crystalline solids, such as metals, where the interatomic bonds are broken and rearranged but the lattice structure is not destroyed. Figure 5-1b shows cleavage (fracture) across a well-defined crystallographic plane. Ceramics and some high-strength metals fail in this manner because they cannot deform plastically. Thermoplastic polymers show a localized drawing of chains which become highly oriented in the direction of applied stress as shown in Figure 5-1c. New secondary bonds are formed but the final fracture will be made by breaking the bonds of the primary backbone chain. Figure 5-1d shows the cleavage of a glassy material which occurs by breaking primary bonds across an arbitrary plane.

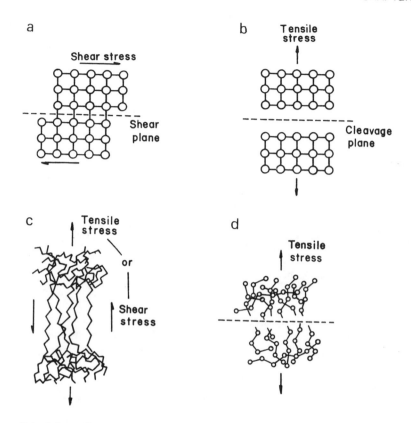

Figure 5-1. Schematic representation of various failure modes at atomic and molecular level. (a) Plastic shear deformation in ductile crystals; (b) cleavage across well-defined crystallographic planes; (c) alignment of chains before fracture; (d) cleavage of glassy solids by breaking primary bonds.

This simplified description of the deformation and fracture of various materials serves to illustrate what we can expect of the strengths of solids. The exact calculation of the theoretical fracture strength of a perfect material is very complicated but a simple model can be used to approximate this value (Figure 5-2). Assuming the cohesive energy force curve can be approximated by a sine curve, then

$$\sigma = \sigma_{max} \sin 2\pi x / \lambda \tag{5-1}$$

where σ_{max} is the theoretical cohesive strength. The work done for fracture

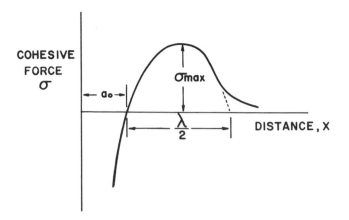

Figure 5-2. Cohesive force versus distance between atoms.

per unit volume is the area under the curve; therefore,

$$E = \int_0^{\lambda/2} \sigma_{max} \sin \frac{2\pi x}{\lambda}\, dx = \frac{\sigma_{max} \lambda}{\pi} \tag{5-2}$$

If all the work is used to create two new surfaces, then

$$\frac{\sigma_{max} \lambda}{\pi} = 2\gamma \tag{5-3}$$

hence,

$$\sigma_{max} = \frac{2\pi\gamma}{\lambda} \tag{5-4}$$

where γ is the energy of creating a new surface. Assuming Hooke's law (Section 2.1.1) can be applied for the initial part of the curve, the stress can be written as

$$\sigma = \frac{Ex}{a_0} \tag{5-5}$$

where a_0 is the equilibrium distance (lattice spacing). Taking derivative of equation (5-1) and noting $\cos(2\pi x/\lambda)$ is close to unity since x is small, then

$$\frac{d\sigma}{dx} = \sigma_{max} \frac{2\pi}{\lambda} \tag{5-6}$$

Equating equations (5-5) and (5-6) and substituting into equation (5-4) will give the theoretical strength of the material:

$$\sigma_{max} = \left(\frac{E\gamma}{a_0} \right)^{1/2} \tag{5-7}$$

Generally the σ_{max} is about 1/10th of the modulus of elasticity,

$$\frac{\sigma_{max}}{E} = 0.1 = \frac{\tau_{max}}{G} \tag{5-8}$$

For all materials σ_{max} is greater than τ_{max} since the shear is easier than the cleavage of a plane. Also the σ_{max}/τ_{max} ratio can be used as an indicator of whether a material will behave as ductile or brittle:

$$\sigma_{max}/\tau_{max} > 10 \qquad \text{ductile}$$
$$\sigma_{max}/\tau_{max} \leq 1 \qquad \text{brittle}$$

Therefore, silver, gold, and copper will always behave as ductile while silicon, diamond, Al_2O_3, and NaCl will always be brittle (cf. Table 5-1). The

Table 5-1. Theoretical Strengths of Various Materials[a]

Material	Young's modulus, E (GPa)	σ_{max} (GPa)	σ_{max}/E	Shear modulus (GPa)	τ_{max} (GPa)	τ_{max}/G	σ_{max}/τ_{max}	Behavior
Graphite	10	1.4	0.14	2.3	0.12	0.05	12	Brittle
Silicon	190	32	0.17	57	13.7	0.24	2.3	
Diamond	1200	205	0.17	505	121	0.24	1.7	
Al_2O_3	460	46	0.10	147	17	0.12	2.7	
NaCl	44	4.3	0.10	24	2.8	0.12	1.5	
Silver	120	24	0.20	20	0.8	0.04	30	Ductile
Gold	110	27	0.25	19	0.8	0.04	36	
Copper	190	40	0.21	31	1.2	0.04	33	
Tungsten	400	86	0.22	150	16.5	0.11	5.5	Transitional
Iron	210	46	0.22	60	6.6	0.11	7.0	
Silica (glass)	75	16	0.21	—	—	—	—	Brittle
Polyethylene	240 (theor. max.)	33	0.14	—	—	—	—	Ductile

[a] From Ref. 1.

transition metals which have bcc crystal structures such as tungsten and iron often fail in brittle manner at low temperatures or if they contain impurities. However, they fail in ductile fashion if they are in pure crystalline form.

Some materials can be made to approach their theoretical strengths; for example, extremely fine glass fibers freshly drawn from the melt. However, if the fibers are exposed to the atmosphere even for short periods of time, their strength decreases greatly. This indicates that fiber strength is highly dependent on surface condition. Metal whiskers (fine crystal filaments) show an increase in strength with a decrease in diameter, indicating that the decreased number of dislocations in smaller-diameter whiskers results in greater strength. These experimental results indicate that the difference between theoretical and observed strength is due to structural irregularities.

5.2. STRENGTH AND STRENGTHENING OF METALS AND SIMPLE IONIC SOLIDS

Some ionic crystals such as NaCl, LiF, and MgO and most metals fail by shearing interatomic bonds due primarily to the movement of dislocations (see Section 3.3.2). Dislocations exist in crystals due to growth faults but they are usually produced by applied stress at dislocation sources. Additional increases in stress will produce more plastic deformation requiring increased interactions between dislocations resulting in a stronger material with less ductility. Some basic deformation characteristics and dislocation energetics will be used to explain the strengthening mechanisms of the plastically deformable crystalline solids.

5.2.1. Deformation of Crystalline Solids

When a stress is applied to a crystalline solid, it will deform elastically first, then plastically after yielding. The plastic deformation is achieved by the gliding planes of atoms over one another, a phenomenon called *slip* which occurs by the dislocation motion over some specific crystallographic planes and directions.

Slip lines result from the relative displacement of crystal planes which ranges from 0.01 to 1 μm. The combination of a slip direction and the plane containing it is called a *slip system*. Table 5-2 shows some observed slip systems in various crystals. Slip will take place more easily along directions where dislocation motion requires the least energy. Therefore, slip will take place along dislocations of the shortest Burger's vector, that is, in the most densely packed planes. Higher energy is needed to produce slip on planes of lower packing density since the atomic distances are longer. As the material is further stressed, more slip occurs and the dislocations interact and tangle.

Table 5-2. Slip Systems of Some Crystals[a]

Structure	Slip plane	Slip direction	Number of slip systems	
fcc Cu, Al, Ni, Pb, Au, Ag, γFe,...	$\{111\}$	$\langle 1\bar{1}0 \rangle$	$4 \times 3 = 12$	
bcc αFe, W, Mo, β brass	$\{110\}$	$\langle \bar{1}11 \rangle$	$6 \times 2 = 12$	
αFe, Mo, W, Na	$\{211\}$	$\langle \bar{1}11 \rangle$	$12 \times 1 = 12$	
αFe, K	$\{321\}$	$\langle \bar{1}11 \rangle$	$24 \times 1 = 24$	
hcp Cd, Zn, Mg, Ti, Be,...	(0001)	$\langle 11\bar{2}0 \rangle$	$1 \times 3 = 3$	
Ti	$\{10\bar{1}0\}$	$\langle 11\bar{2}0 \rangle$	$3 \times 1 = 3$	
Ti, Mg	$\{10\bar{1}1\}$	$\langle 11\bar{2}0 \rangle$	$6 \times 1 = 6$	
NaCl, AgCl	$\{110\}$	$\langle 1\bar{1}0 \rangle$	$6 \times 1 = 6$	

[a] From Ref. 2.

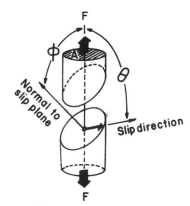

Figure 5-3. Schematic representation of a slip
plane in a crystalline solid when a tensile force is
applied (compare with Figure 5-1).

This makes further slip difficult, hence the strength increases while the
ductility decreases since most of the "available" plastic deformation has
been used up.

Since the slip occurs by the action of shear stress on the slip plane, the
component of stress normal to the slip plane does not influence slip.
Consider a single crystal of cross-sectional area A under a tensile force F
(Figure 5-3). If ϕ is the angle between the normal axis of the slip plane and
θ is the angle between the slip direction and the force, then the shear stress
resolved in the slip direction can be written:

$$\tau = \frac{F\cos\theta}{A/\cos\phi} = \sigma\cos\phi\cos\theta \qquad (5\text{-}9)$$

The stress required to initiate slip is called *critical resolved shear stress*
(CRSS) which is constant for a material at a given temperature. The CRSS
varies widely with materials from about $G/10^5$ to $G/10^4$ at room temperature which is considerably lower than the theoretical shear strength as given
in equation (5-8). This extreme weakness of single crystals led investigators
to propose the dislocation theory in the 1930s.

5.2.2. Energetics of Dislocations

Dislocations increase the free energy of the crystal since they create
higher strain energy around them than the corresponding perfect crystal
lattices. The energy of a screw dislocation with Burger's vector **b** and length
l (Figure 5-4) can be calculated since the shear strain, ϵ_s, is

$$\epsilon_s = \frac{b}{2\pi r} \qquad (5\text{-}10)$$

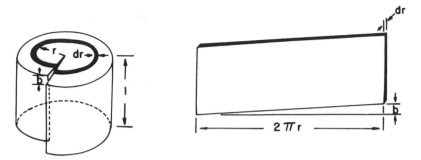

Figure 5-4. Schematic illustration of a screw dislocation.

Therefore, the elastic strain energy per unit volume of the annular region is

$$\frac{dE}{dV} = \frac{1}{2}(\tau\epsilon_s) = \frac{1}{2}(G\epsilon_s) = \frac{G}{2}\left(\frac{b}{2\pi r}\right)^2 \qquad (5\text{-}11)$$

By knowing $dV = 2\pi lr\,dr$ one can integrate equation (5-11):

$$E = \int_{r_0}^{R} \frac{Glb^2}{4\pi} \frac{dr}{r} = \frac{Glb^2}{4\pi} \ln\left(\frac{R}{r_0}\right) \qquad (5\text{-}12)$$

where $b = |\mathbf{b}|$ and E_0 is the core energy of the dislocation. If we choose a reasonable ratio for R/r_0, say $\ln R/r_0 = 4\pi$, since the effective range for the screw dislocation is immediately around the dislocation, then the energy of a screw dislocation is given by

$$E = Glb^2 \qquad (5\text{-}13)$$

Similarly, the energy of an edge dislocation is given by

$$E = \frac{1}{1-\nu} \frac{Glb^2}{4\pi} \ln\left(\frac{R}{r_0}\right) + E_0 = \frac{Glb}{1-\nu} \qquad (5\text{-}14)$$

where ν is Poisson's ratio. Equations (5-13) and (5-14) show that the energy of dislocations is proportional to b and length, l. A curved dislocation will have line tension, T, which can be approximated as

$$T = \frac{\partial E}{\partial l} \cong Gb^2 \qquad (5\text{-}15)$$

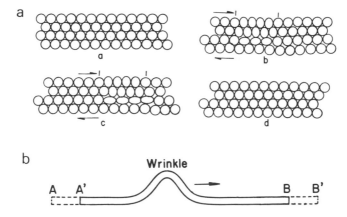

Figure 5-5. Schematic illustrations for the slip of a crystal due to dislocation movement (a) and wrinkle analogy of a carpet to slip (b).

5.2.3. Dislocation Movement

Moving a carpet across a floor offers a useful analogy for slip by dislocation movements (Figure 5-5). By making wrinkle (or ruck) much less force is required to displace the carpet than if the whole carpet had to be moved at once. The creation of a slip step by edge and screw dislocations by shear stress is illustrated in Figure 5-6.

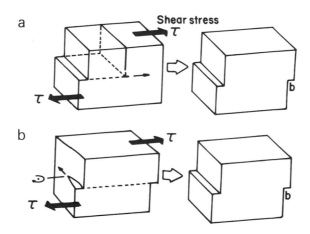

Figure 5-6. Schematic illustrations of creating a slip step by edge (a) and screw (b) dislocation movement.

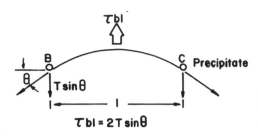

Figure 5-7. Force balance along a dislocation line when it meets obstacles (B and C) and starts to bulge through between them.

If precipitate particles with a higher shear strength than the matrix material form along the slip planes, then the dislocation line will be pinned at the particles (Figure 5-7). The applied shear stress of τ will result in the force τbl normal to the dislocation line segment l. The force is balanced by the parallel component of the dislocation line tension T, hence,

$$\tau bl = 2T \sin \theta \qquad (5\text{-}16)$$

By substituting equation (5-15) into (5-16),

$$\tau = \frac{2Gb}{l} \sin \theta \qquad (5\text{-}17)$$

This equation shows that a dislocation which does not meet any obstacles should have very little resistance to its movement while a higher stress is required for the smaller value of l. The maximum shear stress will be encountered by the dislocation line when $\theta = 90°$:

$$\tau_{max} = \frac{2Gb}{l} \qquad (5\text{-}18)$$

Dislocations can avoid the obstacles by changing to another slip plane as shown in Figure 5-8. The screw dislocation can change its slip plane by *cross-slip*, while the edge dislocation can *climb* up or down by the diffusion of vacancies.

It is also possible that two dislocations can interact by changing their total free energy. For example, two parallel edge dislocations with length l and distance x apart can exert a force,

$$F = \frac{Gb}{2\pi(1-\nu)} \frac{l}{x} \qquad (5\text{-}19)$$

Figure 5-9 shows several simple interactions between parallel edge disloca-

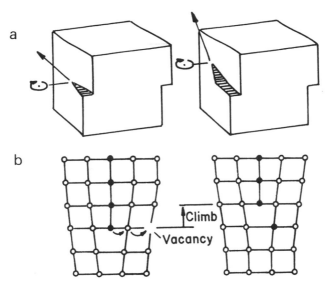

Figure 5-8. Schematic illustrations of a cross-slip of a screw dislocation (a) and climb of an edge dislocation (b).

tions. Like dislocations will repel or attract each other depending on the angle between slip plane and line joining the dislocations. Unlike dislocations on the same or nearby plane attract and either they will annihilate each other or become a perfect lattice if the dislocations are on an adjacent plane.

Screw dislocations of opposite signs with the same Burger's vector will attract with a force per unit length,

$$F = \frac{Gbl}{2\pi r} \qquad (5\text{-}20)$$

and will annihilate each other, leaving a perfect lattice. If the dislocations have the same sign and Burger's vector, then they will repel with the same magnitude of force per unit length.

Dislocations also can interact with any other structural features, such as point defects and grain boundaries in addition to other dislocations. In general, whenever the dislocation motions are impeded, the material becomes more resistant to slip, making it stronger.

Dislocations can be produced during a deformation process through the regenerative operation of certain sources. One of these, the Frank–Read

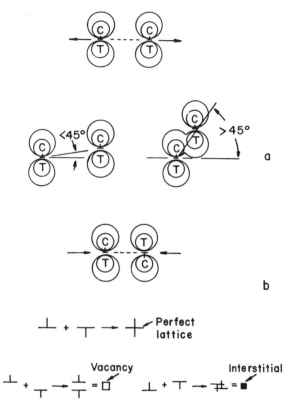

Figure 5-9. Schematic illustrations of interactions of edge dislocations: (a) like and (b) unlike dislocations.

source, is formed by "bowing out" of dislocation lines between obstacles as shown in Figure 5-10. The shear stress necessary to operate the Frank–Read source is given in equation (5-18).

5.2.4. Strengthening of Metals and Simple Ionic Solids

The main principle behind the following strengthening methods is that increased interaction of dislocations will increase the strength of the material. Cold-work is defined as

$$CW = \frac{A_o - A_f}{A_o} \tag{5-21}$$

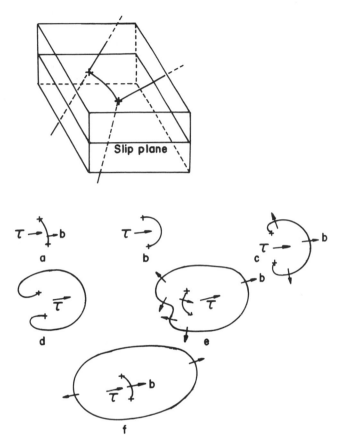

Figure 5-10. Schematic illustrations of Frank–Read source of dislocations.

where the subscripts refer to the original and final cross-sectional areas before and after cold-work. Work-hardening or strain-hardening can be achieved by rolling, wire drawing, and other cold-working processes. The major part of the stored energy is in the form of increased dislocation density. The increased shear strength of the crystals is proportional to the square root of dislocation density, ρ_d; hence from equation (5-18),

$$\tau = AGb\sqrt{\rho_d} \qquad\qquad (5\text{-}22)$$

where the constant A is about 0.2.

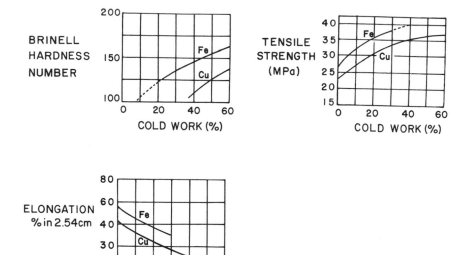

Figure 5-11. Cold-work versus mechanical properties of iron and copper.

Cold-work generally increases hardness and yield strength (Figure 5-11) as well as chemical reactivity, which will lead to a higher rate of corrosion. The harmful effect of cold-working may be removed by heat treatment process of recovery, recrystallization, and grain growth.

Solute-hardening can be achieved by adding solute or impurity atoms which will interact with dislocations. If the solute atom is larger than the solvent atom, compressive fields are developed and if it is smaller, tensile fields are created as shown in Figure 5-12. Impurities which distort the crystal lattice in tetragonal fashion cause rapid hardening. For example, the strength of NaCl is increased almost fourfold by adding a few percent of $PbCl_2$. The rate of hardening by the divalent cations is up to 100 times that of monovalent cations. This strengthening is partly due to the solute effect and partly to the precipitation-hardening effect.

Precipitation-hardening or *age-hardening* can be achieved by forming a solid solution and then quenching and aging below the phase separation temperature. An alloy must demonstrate decreasing solid solubility with decreasing temperature in order to be strengthened by this treatment. The binary phase diagram of the Al–Cu system (Figure 5-13) can be used to demonstrate this process. First, the alloy is completely solutionized in κ

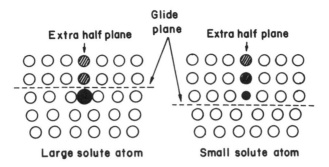

Figure 5-12. Preferred sites for large and small solute atoms around an edge dislocation.

phase to dissolve all precipitate particles. The mixture is quenched into $\kappa + \theta$ region to prevent the appearance of θ phase. This produces a supersaturated solid solution which is in a metastable equilibrium. Upon aging at or above room temperature, fine precipitates of stable θ phase form as the system reaches thermodynamic equilibrium. The rate and length of aging can be manipulated to create material with the desired combination of properties (Figure 5-14). The increase in hardness is due to increases in

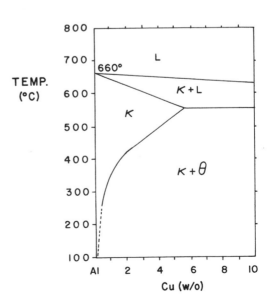

Figure 5-13. Phase diagram for aluminum-rich end of Al–Cu system.

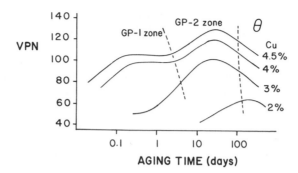

Figure 5-14. Hardness versus aging time for various Al–Cu alloys at 130°C.

lattice strain energy as precipitate particles form in the matrix. Interactions between dislocations and precipitates result in higher strength (Figure 5-15). As discussed previously, the resistance to shear stress will be increased by decreasing the effective length of dislocations between precipitates (cf. equation 5-18). This can be achieved by making finer precipitates. Overaging occurs when the precipitates cluster and coalesce into larger particles, thus reducing effectiveness of preventing dislocation movements and generations.

Selected gas diffusion through the surface of an alloy can increase the surface hardness due to the dispersion of fine precipitates. This is called *diffusion-hardening*. An example is nitriding martensitic steel with ammonia gas at 500–600°C. The diffusionless martensitic phase transformation is another method used to increase the strength of an alloy (see Section 4.3.4).

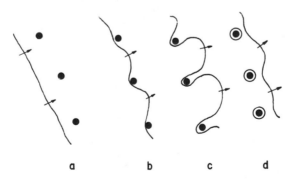

Figure 5-15. Interaction of dislocation movements and precipitate particles.

5.3. STRENGTH AND STRENGTHENING OF CERAMICS AND GLASSES

The lack of slip systems in ceramics and glasses prevents dislocation motion and generation, resulting in a material that is hard and brittle. This is partly because of the ionic nature of bonding (Figure 5-16). Slip can occur only in diagonal directions but not horizontally due to the repulsion of like-charged ions. In glasses the brittleness is caused by the *lack of plastic deformation* exhibited by the three-dimensional network structure (Figure 3-20).

Some ionic crystals will behave differently, however, if tested under certain conditions. For example, NaCl crystals in water are ductile, but this is lost if tested in air for a prolonged time. It is believed the variations in moisture content cause precipitates to form on the crystal surface by a local dissolution and subsequent reprecipitation which originates fracture. The strength of brittle materials is lower in tension than in compression. These experimental observations led Griffith to investigate the relationship between strength and microflaw in brittle materials.[3]

5.3.1. Griffith Theory of Brittle Fracture

Griffith assumed that in a brittle material the stress concentration for an infinitesimal extension of the elliptical crack (Figure 5-17) can be expressed as

$$\frac{\sigma_{max}}{\sigma_{\infty}} = 1 + \frac{2c}{b} \qquad (5\text{-}23)$$

For the case of a narrow crack of length c with an elliptical tip whose radius

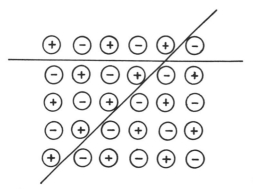

Figure 5-16. Two-dimensional representation of ionic crystals and the slip directions.

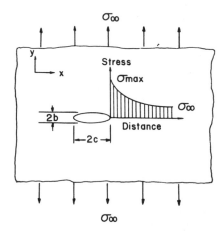

Figure 5-17. Stress concentration near an elliptical hole. (From Ref. 6.)

of curvature is r, equation (5-23) can be rewritten as

$$\frac{\sigma_{max}}{\sigma_\infty} = 2\left(\frac{c}{r}\right)^{1/2} \tag{5-24}$$

since $r = b^2/c$ and $c/b \gg 1$. For a glass, when r becomes atomic scale ($r = 0.1$ nm), the theoretical stress can be approximated by $E/10$, but if the crack is about 3 μm, the strength becomes $E/1000$.

When a crack begins propagating, it gains energy by creating new crack surfaces while releasing elastic energy. The elastic strain energy per unit thickness released by the propagating crack of $2c$ of a thin plate can be approximated by

$$U_e = \frac{\sigma^2}{2E} \pi c^2 \tag{5-25}$$

and the surface energy gained by the crack is

$$U_s = 4c\gamma \tag{5-26}$$

The crack will become self-propagating under an applied load when the change in U_e equals the change in U_s (Figure 5-18). Hence,

$$\frac{d}{dc}\left(\frac{\pi\sigma^2 c^2}{2E}\right) = \frac{d}{dc}(4c\gamma) \tag{5-27}$$

or

$$\sigma = \sigma_f = \left(\frac{2\gamma E}{\pi c}\right)^{1/2} \tag{5-28}$$

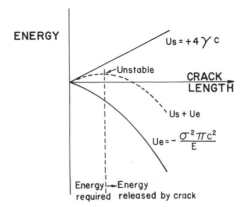

ENERGY

$U_s = +4\gamma c$

Unstable

CRACK LENGTH

$U_s + U_e$

$U_e = -\dfrac{\sigma^2 \pi c^2}{E}$

Energy required — Energy released by crack

Figure 5-18. Crack length versus energy (surface and elastic stored energy) diagram.

This is the Griffith fracture equation. In glasses and ceramics, the strength is governed by the flaw size, c, and its distribution as well as γ and E. The distribution of flaws determines the strength for a given material. It is also noted that the Griffith equation cannot be applied if c becomes less than the interatomic distance of atoms and in compressive load since the crack tends to be closed by the load. The size and amount of flaws can be reduced by drawing an extremely fine fiber. This will increase strength as given in Table 5-3.

5.3.2. Strengthening of Ceramics and Glasses

Elimination of the Griffith flaws is one method used to increase the strength of brittle materials. *Chemical etching* can round the crack tips, thus reducing the stress concentration. Hydrofluoric acid is frequently used because it dissolves silicates rapidly. *Fire-polishing* removes surface flaws by heating the material just above its T_g so that it can flow to close cracks or smooth crack tips.

A material's ability to withstand tensile stress can be improved by making surface layers compressive relative to the interior since applied force should overcome the compressive force before tensile force can take over. Surface compression can be introduced by *ion exchange*, *quenching*, and *surface crystallization*. The ion exchange can be accomplished by diffusion or electrical-migration techniques. Principally, larger ions are exchanged with smaller ions (as K^+ for Na^+) making the surface compressive due to the lattice straining. Use of electric fields can be advantageous since the foreign ions can be introduced at the surface at a lower temperature than possible with the diffusion process alone.

Table 5-3. Properties of Various Glasses and Alumina[a]

Materials and conditions	Particle or fiber diameter (μm)	Density (g/ml)	Young's modulus, E (GPa)	Tensile strength, σ_f (MPa)	σ_f/σ_{max}
Bulk soda glass, off the shelf	$>10^4$	2.5	70	7–140	1/1000
Ordinary-quality glass fiber (called E-glass)	10	2.5	70	1,500–2,000	1/10
E-rods, etched in HF to remove all surface defects	$\sim 10^3$	2.5	70	2,800	1/6
E-glass fiber prepared under exacting conditions	5–50	2.5	70	3,700	1/4
S-glass fiber	10	2.6	84	4,550	1/3
Silica fiber, tested in air	50	2.2	75	5,600	1/3
SiO_2 fiber, tested in vacuum	50	2.2	75	7,000	1/2
Sintered Al_2O_3 (bulk polycrystalline body)	$>10^4$	3.85–3.92	350	280	1/200
Fully dense Al_2O_3 (bulk polycrystalline body)	$>10^4$	3.98	370	700	1/60
Polycrystalline fiber	100	3.15	400	2,000	1/25
Single crystal rod, surface ground	10	3.98	370	500	1/100
Single crystal rod, flame-polished surface free of defects	10	3.98	400	5,000	1/9
Whisker	10	3.98	490	7,000	1/7
Whisker	1	3.98	490	21,000	1/2

[a] From Ref. 1.

Table 5-4. Various Strengthening Methods of Brittle Materials[a]

Treatment	Maximum strengthening	Examples
Chemical etching	30×	Soda lime, silicate glass
Fire-polishing	200×	Fused silica
Ion exchange	20×	Sodium aluminosilicate glass with potassium nitrate
Quenching	6×	Alkali silicate glass
Ion exchange and surface crystallization	22×	Lithium-sodium alumino-silicate glass ($Li^+ \rightleftharpoons Na^+$)
Surface crystallization	17×	Lithium-aluminosilicate glass
Second-phase particles	2×	Borosilicate glass with alumina

[a] From Ref. 4.

Glass-ceramics can be surface crystallized by changing the molar or specific volume of surface layers. Similar results can be achieved by rapid cooling (quenching) as the interior of the glass tries to shrink but is restrained by the rigid surface layer, making the surface in compression and the inside in tension.

Addition of a second crystalline phase can increase the strength by forming second-phase particles which can pin the propagating cracks. Table 5-4 gives a summary of the strengthening of brittle materials.

5.4. STRENGTH AND STRENGTHENING OF POLYMERS

Polymers may be classified according to nature of their polymerization. The *addition* or *free radical* polymers are produced when monomers with unsaturated bonds (double or triple bonds) are "free radicalized" for adding other monomers as in vinyl polymers,

$$
\begin{array}{ccc}
\text{H} \ \text{H} & \text{H} \ \text{X} & \left(\text{H} \ \text{H}\right. \\
| \ | & | \ | & | \ | \\
\text{C}=\text{C} \longrightarrow \text{C}-\text{C} & \left[\text{C}-\text{C}\right] \\
| \ | & | \ | & | \ | \\
\text{H} \ \text{X} & \text{H} \ \text{X} & \left.\text{H} \ \text{X}\right)_n
\end{array}
\qquad (5\text{-}29)
$$

where X can be H (polyethylene), Cl (polyvinyl chloride), benzene ring (polyvinyl benzene or polystyrene), etc.

Condensation or *step-reaction* polymers are produced by stepwise, intermolecular condensation of reactive groups such as

$$
\underset{\text{(amine)}}{\text{R-NH}_2} + \underset{\substack{\text{(carboxylic} \\ \text{acid)}}}{\text{R}'\text{COOH}} \rightarrow \underset{\text{(amide)}}{\text{R}'\text{CONHR}} + \text{H}_2\text{O}
\qquad (5\text{-}30)
$$

This particular process is used to make polyamide (nylon).

Another way of classifying polymers is according to their thermal behavior. The *thermoplastics* are "linear" polymers in which the chains are held together by the secondary van der Waals and hydrogen bonds. When enough thermal energy is applied, the chains break free from one another, and the material will flow and melt. Table 5-5 lists the cohesive energy densities of some polymers, revealing that in the absence of cross-links of main chains, it is the intermolecular forces that provide the restraints on molecular motion. The *thermosetting* polymers have a three-dimensional network structure (Figure 3-20) which prevents chains from being freed at high temperatures. They will burn instead of melt.

Table 5-5. Cohesive Energy Densities of Linear Polymers[a]

Polymer	Repeat unit	Cohesive energy density (cal/cm)
Polyethylene	$-CH_2CH_2-$	62
Polyisobutylene	$-CH_2C(CH_3)_2-$	65
Polyisoprene	$-CH_2C(CH_3)=CHCH_2-$	67
Polystyrene	$-CH_2CH(C_6H_5)-$	74
Polymethylmethacrylate	$-CH_2C(CH_3)(COOCH_3)-$	83
Polyvinyl acetate	$-CH_2CH(OCOCH_3)-$	88
Polyvinyl chloride	$-CH_2CHCl-$	91
Polyethylene terephthalate	$-CH_2CH_2OCOC_3COO-$	114
Polyhexamethylene adipamide	$-NH(CH_2)_6NHCO(CH_2)_4CO-$	185
Polyacrylonitrile	$-CH_2CHCN-$	237

[a] From Ref. 5.

The temperature dependence of mechanical properties is illustrated in Figure 5-19 which shows that the polymer exhibits brittle behavior at low temperature but ductile at high temperature. Similarly, at high strain rates the polymer is brittle while at low rates it is ductile. This is due to the viscoelastic nature of polymer properties (Section 2.2). The stress–relaxation modulus can be written in terms of temperature and time:

$$E_r(t, T) = \frac{\sigma(t, T)}{\epsilon(0)} \tag{5-31}$$

where $\sigma(t, T)$ is the stress which is a function of time and temperature and $\epsilon(0)$ is the constant tensile strain applied at time zero. Figure 5-20 shows the temperature dependence of E_r for polystyrene. Note that introducing cross-links and crystallinity results in a more stable polymer at higher temperatures.

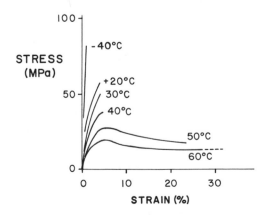

Figure 5-19. Stress–strain curves of polymethyl methacrylate at various temperatures. (From Ref. 7.)

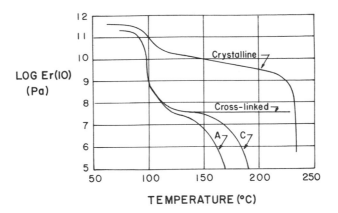

Figure 5-20. $E_r(10)$ versus temperature for crystalline isotactic polystyrene, cross-linked, and lightly cross-linked (A and C) polystyrene. (From Ref. 8.)

5.4.1. Strength of Polymers

The Griffith theory of fracture cannot be applied directly to ductile polymers, but it can be extended to cases where plastic flow takes place in a thin layer on the surface of a growing crack. The fracture still appears to be brittle in nature; therefore, the Griffith theory can be written in modified form as in equation (5-28):

$$\sigma_f = \left(\frac{EP}{c} \right)^{1/2} \tag{5-32}$$

where P is the plastic work of fracture energy and can be many times greater than the surface energy γ. For example, P values for polymethyl-methacrylate and polystyrene are about 490 and 2500 N/m, respectively.[9] Crystalline polymer tensile strength was predicted by using a thermodynamic theory which assumed that the material melts in the region where fracture takes place.[10] According to this theory the tensile strength is given by

$$\sigma = \Delta H_f \rho \left(\frac{1-2\nu}{3-5\nu} \right) \tag{5-33}$$

where ΔH_f is the heat of fusion, ρ is the density, and ν is Poisson's ratio. This theory cannot be applied to the amorphous polymers since they have no heat of fusion.

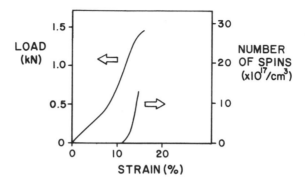

Figure 5-21. Load versus strain diagram of nylon 6 fibers simultaneously monitoring breakage of chains by EPR spectrometer. (From Ref. 11.)

The tensile strength of rigid amorphous polymers is related to their tensile moduli by

$$\sigma = E/30 \tag{5-34}$$

which assumes that fracture does not involve simultaneous fracture of a large number of bonds but instead the bonds are broken consecutively. Therefore, the tensile strength is much less than the elastic modulus and the flaw theory of Griffith is not required.

Actual observation of bond rupture during deformation can be made by using an electron paramagnetic resonance (EPR) spectrometer which can detect free radicals produced by the rupturing bonds. Figure 5-21 shows the

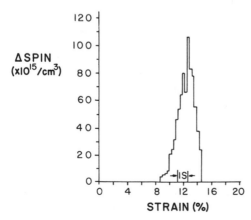

Figure 5-22. Number of nylon 6 fiber molecular chains broken at various strain levels, illustrating tautness of chain distribution. (From Ref. 12.)

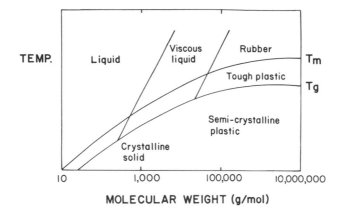

Figure 5-23. Approximate relations among molecular weight, T_g, T_m, and polymer properties.

simultaneous monitoring of the number of free radicals and the stress–strain curve of nylon 6 fibers. From this type of experiment, it can be predicted when the chains start to break and how many bonds are breaking for given strains. It was reported that there is a distribution of tautness in chains which are tied to the crystalline lamellae. When stress is applied, the most taut chains will break first, followed by the next taut ones, and so on, as shown in Figure 5-22. Therefore, strengthening can be achieved if one can devise a means to make the chains with equal tautness so that all the chains will bear the load at the same time.

5.4.2. Strengthening of Polymers

Several factors which affect the strength of polymers are molecular weight, chemical composition, side groups, cross-linking, crystallinity, copolymerization, and blending.

Thermal and mechanical stability increase with increased molecular weight as shown in Figure 5-23. The polymer chains increase in length with increasing molecular weight, resulting in increased entanglements which increases strength. When the molecular chains become more rigid due to the increased chemical complexity, the polymer increases in rigidity and becomes stronger as shown in Figure 5-24. This increase is due to "steric hindrance" by the side groups. Direct ties between chains such as ladder polymers result in increased strength. Cross-linking can also yield the same results in rubbers and soft thermoplastics.

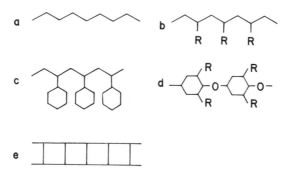

Figure 5-24. Modification of polymer chains. (a) Simple backbone chain as in polyethyl-ene. (b) Addition of simple side groups such as $-CH_3$ in polypropylene. (c) Addition of complex groups such as benzene ring polystyrene which will increase T_g further by steric hindrance. (d) Rigid chains with 180° bonds as in polyphenylene oxide cannot form randomly kinked structures and thus the thermal stability and mechanical strength are much higher than the two polymers with ethylene-type chain. (e) Direct linkages between two or more main chains form a ladder-type structure resulting in extremely high thermal stability and strength.

Increased crystallinity in semicrystalline polymers, like polyethylene and polyamide (nylon), also increases strength and thermal stability since the crystalline polymers have higher strengths and thermal stability than their noncrystalline counterparts. It is evident from their general relation-ship of melting temperature and glass transition temperature,

$$T_g = \frac{1}{2} \sim \frac{2}{3} T_m \qquad (5\text{-}35)$$

where T is the temperature (°K). The orientation of polymers can increase their strength in the oriented direction but not in the orthogonal direction as mentioned earlier. The strength of a fiber is related to its diameter (d) by

$$\sigma = A + \frac{B}{d} \qquad (5\text{-}36)$$

where A and B are constants.

Blending of polymers with fillers can increase strength if the fillers are stronger than the matrix polymer and have good adhesion between them. The relationship between the shear modulus of a filled material and the elastic constants of the polymer and filler was derived by Kerner[13]:

$$\frac{G}{G_1} = \frac{\dfrac{V_2 G_2}{(7-5\nu_1)G_1+(8-10\nu_1)G_2} + \dfrac{V_1}{15(1-\nu_1)}}{\dfrac{V_2 G_1}{(7-5\nu_1)G_1+(8-10\nu_1)G_2} + \dfrac{V_1}{15(1-\nu_1)}} \qquad (5\text{-}37)$$

where subscripts 1 and 2 refer to the matrix and filler phase, respectively, and V is volume fraction. Poisson's ratio of the material in the continuous phase is ν_1.

Equation (5-37) can be simplified for foams and polyblends of rubbers in a rigid polymer matrix since we can assume the shear modulus of the dispersed phase is zero:

$$\frac{1}{G} = \frac{1}{G_1}\left(1 + \frac{V_2 15(1-\nu_1)}{V_1(7-5\nu_1)}\right)$$

(5-38)

For rigid fillers at low volume fractions, equation (5-37) can be simplified to

$$G = G_1\left(1 + \frac{V_2 15(1-\nu_1)}{V_1(8-10\nu_1)}\right)$$

(5-39)

5.5. PROPERTIES OF COMPOSITES

Composites include a wide range of materials from the precipitation-hardened alloys and the polyblend discussed in the previous section to the "polyester or steel-belted" automobile tires. Most composite materials have been developed to enhance the mechanical and thermal properties. The strength of composites derives in part from the fact that fine fibers or whiskers of a material can be made nearly defect free for a greatly increased strength as given in Table 5-6.

The matrix in composites protects the surface of fibers, transfers the longitudinal force between fibers, and deflects cracks which may occur in the fibers.

If the fibers are oriented in the direction of the applied load and have perfect adhesion with the matrix, then for an elastic strain,

$$\sigma_c = \sigma_f V_f + \sigma_m V_m$$

(5-40)

where subscripts c, f, and m refer to composite, fiber, and matrix, respectively, and V is the volume fraction. When the elastic behavior of one or both components ceases, then one of three things occurs: (1) the fibers continue to deform elastically but the matrix deforms plastically, (2) both fibers and matrix behave plastically, and (3) the fibers break in brittle fashion which is followed by the failure of the composite since the matrix has insufficient strength to bear the increased stress.

If we assume that all the fibers have the same strength and that there are enough of them bearing a certain minimum portion of the applied load,

Table 5-6. Properties of Fibers Used for Composite Materials[a]

Class	Material	Tensile strength (GPa)	Young's modulus (GPa)	Density (g/cm³)	T_m (°C)
Whisker	Graphite	20.7	675.7	2.2	3000
	Al_2O_3	15.2	524.0	4.0	2050
	Iron	12.4	193.1	7.8	1540
	Si_3N_4	13.8	379.2	3.1	1900
	SiC	20.7	689.5	3.2	2600
	Boron	—	441.3	2.3	—
Glass,	Asbestos	5.9	186.2	2.5	500
ceramic,	Drawn silica	5.9	72.4	2.5	1700
polymer,	Boron glass	2.4	379.2	2.3	—
fibers	High-tenacity nylon 66	0.8	4.8	1.1	—
Metal	Carbon steel	3.9	206.9	7.8	—
wire	Molybdenum	2.1	365.4	10.3	2610
	Tungsten	2.9	344.8	19.3	3380

[a] From Ref. 14.

then the ultimate tensile strength of the composite will be reached when the fibers reach their maximum strain. Therefore, the maximum strain of the composite can be written as

$$\epsilon_{c\,max} = \epsilon_{f\,max} \tag{5-41}$$

Furthermore, the ultimate tensile strength of the composite can be written as

$$\sigma_{c\,max} = \sigma_f V_f + (\sigma_m)_{\epsilon_{f\,max}} V_m \tag{5-42}$$

where $(\sigma_m)_{\epsilon_{f\,max}}$ is the stress in the matrix when the strain is maximal as shown in Figure 5-25. If the elastic moduli of fibers and matrix are known, then $(\sigma_m)_{\epsilon_{f\,max}}$ can be obtained as

$$(\sigma_m)_{\epsilon_{f\,max}} = \epsilon E_m = \epsilon_{f\,max} E_m \tag{5-43}$$

Since $\sigma_f = E_f \epsilon_{f\,max}$,

$$(\sigma_m)_{\epsilon_{f\,max}} = (E_m/E_f)\sigma_f \tag{5-44}$$

Substituting equation (5-44) into (5-42),

$$\sigma_{c\,max} = \sigma_f V_f + \sigma_f (E_m/E_f) V_m \tag{5-45}$$

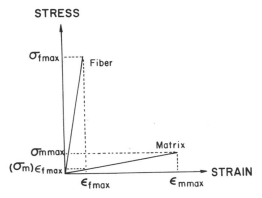

Figure 5-25. Stress–strain curves of fibers and matrix resin in a composite.

which can be reduced to

$$\sigma_{c\,max} = a + bV_f \tag{5-46}$$

where $a = \sigma_f(E_m/E_f)$ and $b = \sigma_f(1 - E_m/E_f)$.

In order to produce a composite that is stronger than the matrix material alone, it is necessary to have a certain critical volume of fibers; therefore,

$$\sigma_{c\,max} = \sigma_m \tag{5-47}$$

Then from equation (5-46),

$$\sigma_m = a + bV_{crit} \tag{5-48}$$

which can be rewritten as

$$V_{crit} = \frac{\sigma_m - \sigma_f(E_m/E_f)}{\sigma_f(1 - E_m/E_f)} \tag{5-49}$$

If V_f is less than V_{crit}, then equation (5-49) no longer applies since there are not enough fibers to control matrix elongation so that the fibers are under high strains and hence fracture will ensue. In this region the matrix properties are the limiting value of the composite as shown in Figure 5-26.

For many composites such as glass fibers in a polyester resin, V_{crit} is very small. In this case,

$$\sigma_m = (\sigma_m)_{\epsilon_{fmax}} = \sigma_f\left(\frac{E_m}{E_f}\right) \tag{5-50}$$

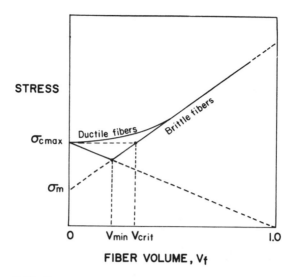

Figure 5-26. Stress–volume fraction of reinforcing fiber relationship.

which can be rewritten as

$$\frac{\sigma_m}{\sigma_f} = \frac{E_m}{E_f} \qquad (5\text{-}51)$$

If short fibers are embedded discontinuously in a matrix, the composite will take up the applied stress until the shear stress between fibers and matrix cannot sustain the load. The stress distribution of a fiber will look like that illustrated in Figure 5-27. The $l_c/2$ on both ends of the fiber are

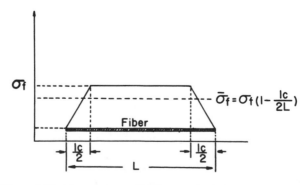

Figure 5-27. Linear stress distribution along a fiber with length L.

the ineffective portions of the fiber transferring the load less than σ_{fmax}. If we define B as the ratio of the area under the stress–distribution curve over the length $l_c/2$ to the area of the rectangle ($l_c/2 \cdot \sigma_{fmax}$), then the ineffective fiber length $l_c/2$ will support a reduced average stress by σ_{fmax}. In other words, the effective length $Bl_c/2$ is subjected to stress $B\sigma_{fmax}$.

The average stress in the discontinuous fiber can be written as

$$\overline{\sigma_f} = \frac{\sigma_f}{L}(L - l_c) + \frac{\sigma_f}{L}\frac{l_c}{2} \tag{5-52}$$

$$= \sigma_f\left(1 - \frac{l_c}{2L}\right) \tag{5-53}$$

When the matrix is a polymer the stress–strain curve of the matrix is nonlinear and there is no sharp division between elastic and plastic deformation. Therefore, the shear stress between the fiber and the matrix is not constant. The stress of the fiber builds up nonlinearly and the average fiber stress can be written as

$$\sigma_f = \sigma_f\left[1 - (1 - B)\frac{l_c}{2L}\right] \tag{5-54}$$

For an ideal plastic matrix which has a yield stress σ_{my}, B will be 0.5 and equation (5-54) will be the same as equation (5-53).

The strength of a composite will depend strongly on the fiber orientation with respect to the direction of load. For example, silicon fibers in aluminum matrix can have strength as high as 100 MPa along the fiber axis but only 75 MPa at 45 to 90° of its axis. In order to obviate the weakness of fiber composites in the transverse direction, it is common to laminate them in the orthogonal direction as seen in plywood.

PROBLEMS

5-1. Calculate the a_0 of equation (5-7) for Fe if $\gamma = 1200$ dynes/cm and $E = 210$ GPa.

5-2. The theoretical strength of a brittle material can be approximated by equation (5-7). Calculate the crack length of a glass which has $\sigma_{th} = 7$ GPa and $\sigma_f = 70$ MPa. Assume $a_0 = 0.2$ nm.

5-3. Calculate the fracture strength of polymethylmethacrylate (PMMA) if it contains microcracks of 10 μm. The elastic modulus is 3 GPa.

5-4. Calculate the shear modulus ratio of PMMA polyblend filled with 10 w/o of barium sulfate powders to the PMMA matrix. Assume $\nu = 0.4$ for PMMA and densities for PMMA and barium sulfate are 1.0 and 4.5 g/ml, respectively.

5-5. Derive an equation for the fiber composite modulus when the load is applied perpendicular to the fiber axis.

5-6. State the corollary to the fundamental principle of strengthening. Show that the following strengthening mechanisms obey the corollary:

a. Substitutional solute additions.
b. Cold-working.

5-7. A steel is tested in tension. In one case a crack propagates elastically, in another plastically. If $\gamma = 1.0$ J/m and $p = 1000$ J/m, what is the ratio of the crack size which can be tolerated for a given applied stress in the two cases?

5-8. Describe an experiment by which one could prove that fatigue cracks originate at the surface of a material. How do you improve the strength of a material if the surface cracks are the cause of weakening?

REFERENCES

1. B. Harris and A. R. Bunsell, *Structure and Properties of Engineering Materials*, Longmans, London, 1977.
2. H. W. Hayden, W. G. Moffatt, and J. Wulff, *The Structure and Properties of Materials*, Volume 3, Wiley, New York, 1965.
3. A. A. Griffith, The phenomena of rupture and flow in solids, *Philos. Trans. R. Soc. London Ser. A 221*, 163–198, 1920.
4. R. H. Doremus, *Glass Science*, Wiley, New York, 1973.
5. F. W. Billmeyer, Jr., *Textbook of Polymer Science*, 2nd ed., Wiley, New York, 1971.
6. C. E. Inglis, Stress in a plate due to the pressure of cracks and sharp corners, *Trans. Inst. Naval Arch. 55*, 219–230, 1913.
7. T. Alfrey, Jr., *Mechanical Behavior of High Polymers*, p. 516, Wiley, New York, 1948.
8. A. V. Tobolsky, *Properties and Structure of Polymers*, Wiley, New York, 1960.
9. J. J. Benbow and F. C. Roesler, Experiments on controlled fractures, *Proc. Phys. Soc. London Ser. B 70*, 201–211, 1956.
10. R. Fuerth, Relation between breaking and melting, *Nature (London) 145*, 741, 1940.
11. J. B. Park, K. L. DeVries, and W. O. Statton, Chain rupture during tensile deformation of nylon fibers, *J. Macromol. Sci. Phys. 15*, 205–227, 1978.
12. B. A. Lloyd, Fracture Behavior in Nylon 6 Fibers, Ph.D. thesis, University of Utah, 1972.
13. E. H. Kerner, The elastic and thermoelastic properties of composite media, *Proc. Phys. Soc. London Ser B 69*, 808–813, 1956.
14. R. Nicholls, *Composite Construction Materials Handbook*, Prentice–Hall, Englewood Cliffs, N.J., 1976.

BIBLIOGRAPHY

F. W. Billmeyer, Jr., *Textbook of Polymer Sciences*, 2nd ed., Part II, Wiley, New York, 1971.

L. J. Broutman and R. H. Krock, *Modern Composite Materials*, Chapter 1, Addison–Wesley, Reading, Mass., 1967.

L. W. Davis and S. W. Bradstreet, *Metal and Ceramic Matrix Composites*, Cahners, Boston, 1970.

R. H. Doremus, *Glass Science*, Chapter 15, Wiley, New York, 1973.

R. M. Gill, *Carbon Fibres in Composite Materials*, Part 2, Butterworths, London, 1972.

B. Harris and A. R. Bunsell, *Structure and Properties of Engineering Materials*, Chapter 8, Longmans, London, 1977.

H. W. Hayden, W. G. Moffatt, and J. Wulff, *The Structure and Properties of materials*, Volume 3, Wiley, New York, 1965.

L. Holliday (ed.), *Composite Materials*, Elsevier, Amsterdam, 1966.

W. D. Kingery, H. K. Bowen, and D. R. Uhlmann, *Introduction to Ceramics*, 2nd ed., Chapter 16, Wiley, New York, 1976.

F. A. McClintock and A. S. Argon, *Mechanical Behavior of Materials*, Chapters 4 and 8, Addison–Wesley, Reading, Mass., 1966.

R. Nicholls, *Composite Construction Materials Handbook*, Chapter 11, Prentice–Hall, Englewoods Cliffs, N.J., 1976.

L. E. Nielsen, *Mechanical Properties of Polymers*, Chapter 6, Reinhold, New York, 1962.

D. J. Williams, *Polymer Science and Engineering*, Chapter 11, Prentice–Hall, Englewood Cliffs, N.J., 1971.

A. V. Tobolsky, *Properties and Structure of Polymers*, Chapter 2, Wiley, New York, 1960.

STRUCTURE–PROPERTY RELATIONSHIPS OF BIOLOGICAL MATERIALS

The major difference between biological materials and biomaterials (implants) is *viability*. There are other important differences that distinguish living materials from artificial replacements. First, most biological materials are continuously bathed with body fluids, most of which are water as given in Table 6-1. Exceptions are the specialized surface layers of skin, hair, nails, hooves, and the enamel of teeth. Second, most biological materials can be considered as composites.

Structurally, biological tissues consist of a vast network of intertwining fibers with polysaccharide ground substances immersed in a pool of ionic fluid. Attached to the fibers are cells whose function is nutrition of the living tissues (fibers and ground substances). The ground substances have definite structural organization and are not completely analogous to solute suspended in a solution. Physically, ground substances behave as a glue, lubricant, and shock absorber in various tissues.

The structure and properties of a given biological material are dependent on the chemical and physical nature of the components present and their relative amounts. For example, neural tissues consist almost entirely of cells whereas bone is composed of collagenous fibers and calcium phosphate minerals with minute quantities of cells and ground substances as a glue.

An understanding of the exact role played by a tissue and its interrelationship with the function of the entire living organism is essential if biomaterials are to be used intelligently. Thus, to design an artificial blood vessel prosthesis, one has to understand not only the blood vessel wall

Table 6-1. Distribution of Various Tissues
and Physiological Condition of Western Man[a]

Muscle 43%, bone 30%, skin 7%, blood 7.2%
Organs: spleen (0.2%), heart (0.4%), kidneys (0.5%),
   lungs (1.0%), liver (2%), brain (2.3%), viscera (5.6%)
Water 60%, solids 40%
Average body weight: 70 kg (155 lbs)
Medium height: 1.8 m (5.91 ft)
Basic metabolic rate 68 kcal/h
pH: gastric contents (1.0), urine (4.5–6.0), intracellular
   fluid (6.8), blood (7.15–7.35)
pO_2 (mm Hg): interstitial (2–40), venous (40), arterial (100),
   atmospheric (160)
pCO_2 (mm Hg): alveolar (40)

---

[a] From Ref. 1.

structure–property relationship itself but also its systemic function. This is because the artery is not only a blood conduit, but a component of a larger system, including a pump (heart) and an oxygenator (lung).

6.1. STRUCTURE OF PROTEINS AND POLYSACCHARIDES

6.1.1. Proteins

As with polymeric materials, proteins are made of monomers called amino acids,

$$
\begin{array}{c}
\quad\;\; \text{H} \quad\;\; \text{O} \\
\quad\;\; | \quad\;\;\; \| \\
\text{H}_2\text{N}-\text{C}-\text{C}-\text{OH} \\
\quad\;\; | \\
\quad\;\; \text{R}
\end{array}
\qquad (6\text{-}1)
$$

Peptides are, in turn, polyamides formed by step-reaction polymerization between amino and carboxyl groups of amino acids whose basic chemical formula can be represented as

$$
\left(
\begin{array}{c}
\text{O} \quad \text{H} \quad \text{H} \\
\| \quad\; | \quad\; | \\
-\text{C}-\text{N}-\text{C}- \\
\quad\quad\quad | \\
\quad\quad\quad \text{R}
\end{array}
\right)_n
\qquad (6\text{-}2)
$$

where R is a side group. Depending on the side group the molecular structure changes drastically. Hydrogen is the simplest side group, forming

a   b

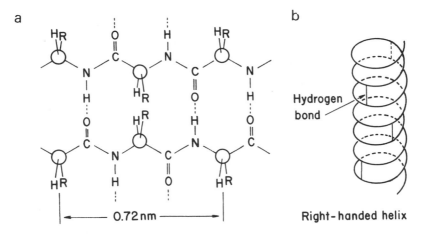

Right-handed helix

Figure 6-1. (a) Hypothetical flat sheet structure of a protein. (b) Helical arrangement of a protein chain.

glycine. The geometry of a peptide with the hypothetical flat sheet structure is shown in Figure 6-1a. The structure has a repeating distance of 0.72 nm and the side groups (R) are crowded. This crowding makes the flat structure impossible except for H side groups, that is, polyglycine. If the side groups are larger, then the resulting structure is a helix where the H bonds occur between different parts of the same chain and hold the helix together as shown in Figure 6-1b. Table 6-2 lists naturally occurring amino acids.

6.1.1a. Collagen

One of the basic constituents of protein is collagen, which has the general amino acid sequence -Gly-Pro-Hyp-Gly-X- (X = any amino acid) arranged in a triple α-helix. It has a high proportion of proline and hydroxyproline as given in Table 6-3. Since the presence of hydroxyproline is unique to collagen (elastin contains a minute amount), the determination of collagen content in a given tissue is readily done by assaying the hydroxyproline.

Three left-handed-helical peptide chains are coiled together to give a right-handed coiled helix with a periodicity of 2.86 nm. This triple super-helix is the molecular basis of tropocollagen, the precursor of collagen (see Figure 6-2). The three chains are held together strongly by H bonds between glycine residues and between hydroxyl (OH) groups of hydroxyproline. In addition, there are cross-links via lysine among the helices.

The primary factors stabilizing the collagen molecules are invariably related to the interactions among the α-helices. These factors are H bonding

Table 6-2. Amino Acids[a]

	R	Abbreviation	m.p. (°C)	Isoelectric point
Glycine	H	Gly	292	
Alanine	CH_3	Ala	297	
Valine*	$CH(CH_3)_2$	Val	315	
Leucine*	$CH_2CH(CH_3)_2$	Leu	337	
Serine	CH_2OH	Ser	228	
Threonine	CHOH $\vert$ CH_3	Thr	253	
Aspartic acid	$CH_3—COOH$	Asp	269	2.98
Glutamic acid	$CH—CH_2COOH$	Glu	247	3.22
Tyrosine	CH_2⟨◯⟩—OH	Tyr	342	5.67
Lysine*	$(CH_2)_4NH_2$	Lys	224	9.74
Arginine*	HNH $\vert$ $\parallel$ $(CH_2)_3—NC—NH_2$	Arg	230–244	10.76
Histidine*	$CH_2—C=CH$ HN N \ // C H			
Proline (imino acid)	H O $\vert$ $\parallel$ H—N—C—C—OH H_2C CH_2 \ / CH_2 = R	Pro	220	6.30
Hydroxyproline (imino acid)	H_2C CH_2 \ / HCOH = R	Hyp		
Cysteine	CH_2SH	Cys	—	5.02
Methionine*	$CH_2CH_2SCH_3$	Met	283	5.06
Cystine	COOH $\vert$ CH_2SSCH_2CH $\vert$ NH_2		258	5.06
Tryptophan*	H $CH_2—C=C$ \ NH	Trp	283	5.88
Phenylalanine*	CH_2—⟨◯⟩	Phe	283	5.48

[a] Basic formula:

$$HN—\overset{\displaystyle \overset{H}{\vert}}{C}—\overset{\displaystyle \overset{O}{\parallel}}{C}—OH$$
$$\vert$$
$$R$$

* Essential amino acid.

Table 6-3. Amino Acid Content of Collagen[a]

	Content (mol/100 mol amino acids)
Gly	31.4–33.8
Pro	11.7–13.8
Hyp	9.4–10.2
Acidic polar amino acids (Asp, Glu, Asn)	11.5–12.5
Basic polar amino acids (Lys, Aro, His)	8.5–8.9
Other amino acids	Residue

[a] From Ref. 2.

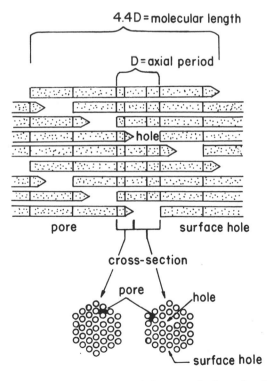

Figure 6-2. Diagrammatic representation of the organization of collagen macromolecules in a collagen fibril. Note the presence of both "holes" and "pores" in the hole-zone region and of "pores" only in the overlap region. (From Ref. 3.)

between C=O and NH groups, ionic bonding between the side groups of polar amino acids, and interchain cross-links. One of the secondary factors affecting the stability of collagen is steric rigidity which is related to the high pyrolidine content.

The collagen fibrils (diameter 20–40 nm) form fiber bundles of diameter 0.2–1.2 μm. Figure 6-3 shows scanning and transmission electron micrographs of collagen fibrils in bone, tendon, and skin. Note the straightness of tendon collagen fibrils compared to the skin fibrils.

The side groups of amino acids are highly nonpolar and hence hydrophobic; therefore, the chains avoid contact with water molecules and seek the greatest number of contacts with the nonpolar chains of amino acids. If

Figure 6-3. (A) Scanning electron micrograph of the surface of an adult rabbit bone matrix, showing how the collagen fibrils branch and interconnect in an intricate, woven pattern. 4800×. (From Ref. 3.) (B) Transmission electron micrographs of (a) parallel collagen fibrils in a tendon and (b) meshwork of fibrils in skin. 24,000×. (From Ref. 4.)

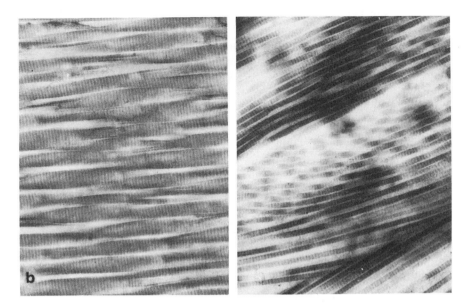

Figure 6.3. (*Continued*)

we destroy the hydrophobic contact by a solution (e.g., urea), the characteristic structure is lost resulting in microscopic changes such as shrinkage of collagen fibers. The same effect can be achieved by simply warming the collagen fibers. Another factor affecting the stability of collagen is the incorporation of water molecules into the intra- and interchain structure. If the water content is lowered, the structural stability decreases. If dehydrated completely (lyophilized), then the solubility also decreases (so called *in vitro* aging of collagen).

It is known that the acid mucopolysaccharides also affect the stability of collagen fibers by mutual interactions forming mucopolysaccharide–protein complexes. It is believed that the water molecules affect the polar region of the chains, making the dried collagen more disoriented than the wet state.

6.1.1b. Elastin

Elastin is another structural protein found in a relatively large amount in elastic tissues such as ligament, aortic wall, and skin. The chemical composition of elastin is somewhat different from that of collagen.

The high elasticity of elastin is due to the cross-linking of lysine residues via desmosine, isodesmosine, and lysinonorleucine as shown in

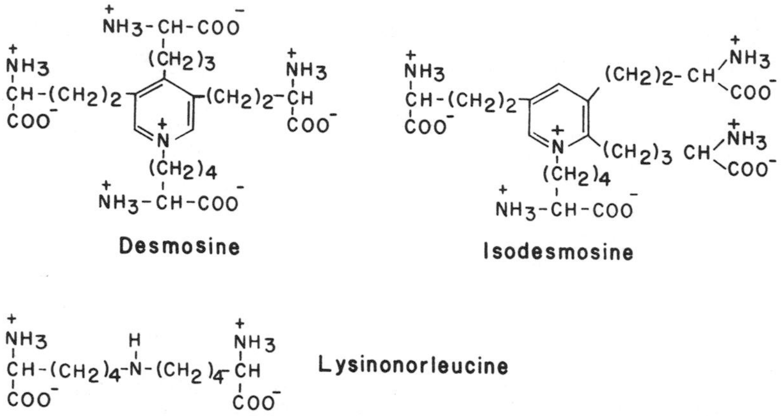

Figure 6-4. Structure of desmosine, isodesmosine, and lysinonorleucine.

Figure 6-4. The formation of desmosine and isodesmosine is dependent on the presence of copper and lysyl oxidase enzyme; hence, a deficiency of dietary copper may result in non-cross-linked elastin. This, in turn, will result in viscous nonelastic tissue similar to linear polymers and thus lacking its rubberlike elasticity, which can lead to rupture of the aortic walls.

Elastin is very stable at high temperatures in the presence of various chemicals due to the very low content of polar side groups (hydroxyl and ionizable groups). The specific staining of elastin in tissue by lipophilic stains such as Weigert's resorcin–fuchsin is due to the same reason. Elastin has a high percentage of amino acids with aliphatic side chains such as valine (6 times that of collagen). It also lacks all the basic and acidic amino

Table 6-4. Overall Composition of Elastin Protein (Residues/1000)

	Elastin	Microfibril	Tropoelastin
Glycine	324	110	334
Hydroxyproline	26	151	39
Cationic residues (Asp, Glu)	21	228	21
Anionic residues (His, Lys, Arg)	13	105	55
Nonpolar residues (Pro, Ala, Val, Met, Leu, Ile, Phe, Tyr)	595	356	541
Half-cystine	4	48	0

acids so that it has very few ionizable groups. The most abundant of these, glutamic acid, occurs only one-sixth as often as in collagen. Aspartic acid, lysine, and histidine are all below 2 residues per 1000 in mature elastin.

Up to now three major entities of elastin have been separated: tropoelastin, elastin, and microfibrils. The composition of these components is given in Table 6-4. The microfibrillar phase, which is a largely oriented crystalline structure, may constitute 5–10% of the total elastin, while the rest is amorphous.

6.1.2. Polysaccharides

Polysaccharides exist in tissues as a highly viscous material which interacts readily with proteins, including collagen, resulting in glycosaminoglycans or proteoglycans. These molecules readily bind both water and cations. They also exist at physiological concentrations not as viscous solids but as viscoelastic gels. All of these polysaccharides consist of disaccharide units polymerized into unbranched macromolecules as shown in Figure 6-5.

6.1.2a. Hyaluronic Acid and Chondroitin

Hyaluronic acid is made of residues of N-acetylglucosamine and D-glucuronic acid, but it lacks the sulfate residues. Hyaluronic acid of animals contains a protein component (0.33 w/o or more) and is believed to be chemically bound to at least one protein or peptide which cannot be removed. This, in turn, yields proteoglycan molecules which may behave differently from the pure polysaccharides. Hyaluronic acid is found in the vitreous humor of the eye, synovial fluid, skin, umbilical cord, and aortic walls. Chondroitin is similar to hyaluronic acid in its structure and properties and is found in the cornea of the eyes.

6.1.2b. Chondroitin Sulfate

This is the sulfated mucopolysaccharide which resists the hyaluronidase enzyme. It has three isomers as shown in Figure 6-5. Isomer A (chondroitin 4-sulfate) is found in cartilage, bone, and the cornea; isomer C (chondroitin 6-sulfate) can be isolated from cartilage, umbilical cord, and tendon; isomer B (dermatan sulfate) is found in skin and the lungs and is resistant to testicular hyaluronidase.

The chondroitin sulfate chains in connective tissues are bound covalently to a polypeptide backbone through their reducing ends. Figure 6-6 shows the proposed macromolecular organization of protein–polysaccharide

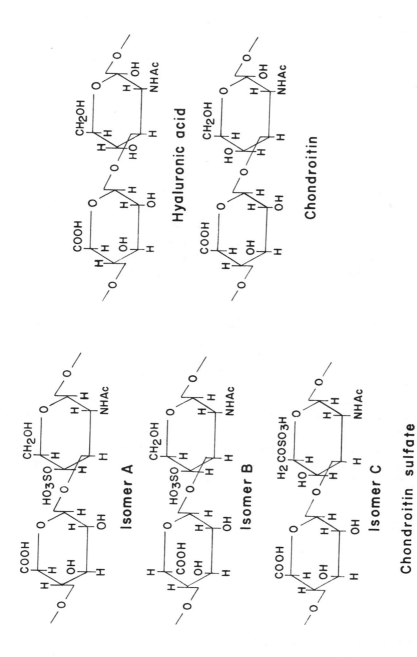

Figure 6-5. Structure of hyaluronic acid, chondroitin, and chondroitin sulfates.

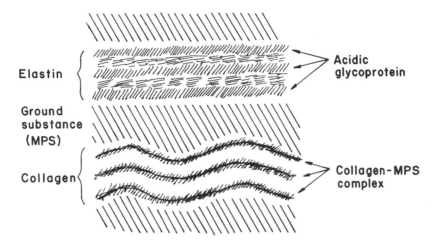

Figure 6-6. A schematic representation of mucopolysaccharide–protein molecules in connective tissues. Note the wavy nature of collagen fibers and the straighter form of elastin. Compare with Figure 6-36.

molecule from which one can imagine the nature of the viscoelastic properties of the ground substance. These complexes of protein and mucopolysaccharides (ground substance) play an important role in the physical behavior of connective tissues either as lubricating agents between tissues (e.g., joints) or between elastin and collagen microfibrils.

Similar structure-stabilizing function of the mucopolysaccharides is shown in Figure 6-6, where the collagen fibrils are stabilized by mucopolysaccharide–protein interactions.

6.2. STRUCTURE–PROPERTY RELATIONSHIP OF TISSUES

Understanding the structure–property relationship of various tissues is important since one has to know what is being replaced by the artificial materials (biomaterials). Also, one may want to use natural tissues as biomaterials (e.g., porcine heart valves). Measurement of the properties of any tissue is confronted with many of the following limitations and variations:

1. Limited sample size
2. Original structure can be changed during sample collection or preparation
3. Inhomogeneity

4. Cannot be frozen or homogenized without altering its structure or properties
5. Complex nature of the tissues makes it difficult to obtain fundamental physical parameters
6. Measurements of *in vitro* and *in vivo* properties are sometimes difficult to correlate if not impossible

The main objective of studying the structure–property relationships of tissues is to improve the performance of implants in the body. Therefore, one should always examine the type of physiological functions carried out by the tissues or organs under study *in vivo* and how one can best simulate the function or properties with as few parameters as possible. Keeping this in mind, let us study the structure–property relationships of tissues.

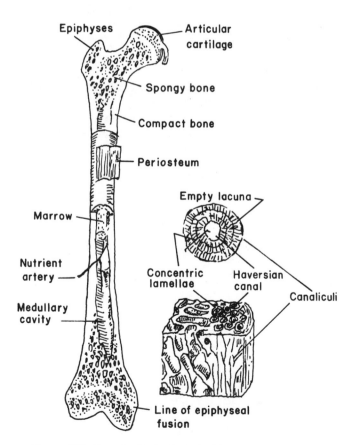

Figure 6-7. Organization of a typical bone. (From Ref. 5.)

6.2.1. Mineralized Tissue (Bone and Teeth)

6.2.1a. Composition and Structure

Bone and teeth are mineralized tissues whose primary function is "load-carrying." Teeth are in more extraordinary physiological circumstances since their function is carried out in direct contact with *ex vivo* substances, while functions of bone are carried out inside the body in conjunction with muscles and tendons. A schematic anatomical view of a long bone is shown in Figure 6-7.

Wet cortical bone is composed of 22 w/o organic matrix of which 90–96 w/o is collagen, and the rest is mineral (69 w/o) and water (9 w/o) as shown in Figure 6-8. The major subphase of the mineral consists of submicroscopic crystals of an apatite of calcium and phosphate, resembling hydroxyapatite crystal structure $[Ca_{10}(PO_4)_6(OH)_2]$. There are other mineral ions such as citrate $(C_6H_5O_7^{4-})$, carbonate (CO_3^{2-}), fluoride (F^-), and

WHOLE CORTICAL BONE

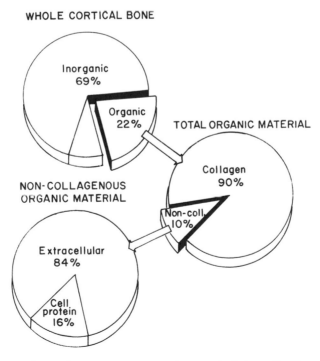

Figure 6-8. Distribution by weight of the constituents of whole cortical bone to illustrate the proportion of cell protein in the organic material. (From Ref. 3.)

hydroxyl ions (OH⁻) which may give some other subtle differences in microstructural features of the bone. The apatite crystals are formed as slender needles, 20–40 nm in length by 1.5–3 nm in thickness, in the collagen fiber matrix as shown in Figure 6-9. These mineral-containing fibrils are arranged into lamellar sheets (3–7 μm thick) which run helically with respect to the long axis of the cylindrical osteons (sometimes called Haversian systems). The osteon is made up of 4 to 20 lamellae which are arranged in concentric rings around the Haversian canal. Between these osteons the interstitial systems are sharply divided by the cementing line. The metabolic substances can be transported by the intercommunicating systems of canaliculi, lacunae, and Volkman's canals, which are connected with the marrow cavity. These various interconnecting systems are filled with body fluids, and their volume can be as high as $18.9 \pm 0.45\%$ according to one estimate for beef compact bone.[7] The external and internal surfaces of the bone are called the periosteum and endosteum, respectively, and both have osteogenic properties.

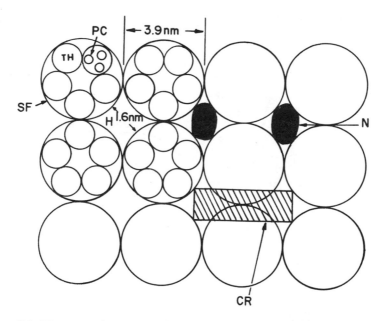

Figure 6-9. Diagrammatic cross-section through a model of the tetragonal arrangement of subfibrils in the collagen fiber, according to Miller and Perry. On the left are shown the elements of the collagen structure; on the right are shown calcium phosphate nuclei (N) and crystals (CR), which may develop in the fiber. SF, subfibrils; TH, triple helix; PC, protein chain; H, hole region between the subfibrils. (From Ref. 6.)

Figure 6-10. Scanning electron micrograph showing the mineral portion of osteon lamellae. The organic phase has been removed by ethylenediamine in a soxhlet apparatus. (From Ref. 7.)

It is interesting to note that the mineral phase is not a discrete aggregation of the calcium phosphate mineral crystals. Rather it is made of a continuous phase as evidenced in Figure 6-10 and by the fact that complete removal of the organic phase of the bone still gives very good strength.

The long bones like the femur are usually made of cancellous (or spongy) and compact bone. The spongy bone consists of three-dimensional branches or bony trabeculae interspersed by the bone marrow. More spongy bone is present in the epiphyses of long bone, whereas compact bone is the major form present in the diaphysis of the bone as shown in Figure 6-7.

There are two types of teeth, *deciduous* or primary and *permanent* of which the latter is more important for us from the biomaterials point of view. All teeth are made of two portions, the crown and the root, demarcated by the gingiva (gum). The root is placed in a socket called the alveolus in the maxillary (upper) or mandibular (lower) bones. A sagittal cross-section of a permanent tooth is shown in Figure 6-11 to illustrate various structural features.

The enamel is the hardest substance found in the body and consists almost entirely of calcium phosphate salts (97%) in the form of large apatite crystals.

The dentin is another mineralized tissue whose distribution of organic matrix and mineral is similar to that of regular compact bone. Consequently, its physical properties are also similar. The collagen matrix of the dentin might have a somewhat different molecular structure than that of normal bone, that is, being more cross-linked than that found in other tissues resulting in less swelling effect. Dentinal tubules (3–5 μm in diame-

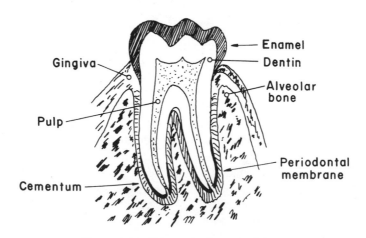

Figure 6-11. Schematic diagram of a tooth.

Table 6-5. Properties of Bone[a]

	Direction of test	Modulus of elasticity (GPa)	Tensile strength (MPa)	Compressive strength (MPa)
Leg bones	Longitudinal			
Femur		17.2	121	167
Tibia		18.1	140	159
Fibula		18.6	146	123
Arm bones	Longitudinal			
Humerus		17.2	130	132
Radius		18.6	149	114
Ulna		18.0	148	117
Vertebrae	Longitudinal			
Cervical		0.23	3.1	10
Lumbar		0.16	3.7	5
Spongy bone		0.09	1.2	1.9
Skull	Tangential	—	25	—
	Radial	—	—	97

[a] From Ref. 8.

ter) radiate from the pulp cavity toward the periphery and penetrate every part of the dentin. Collagen fibrils (2–4 μm in diameter) fill the dentinal tubules in the longitudinal direction and the interface is cemented by a protein–polysaccharide complex substance.

Cementum covers most of the root of the tooth with coarsely fibrillated bone substance but is devoid of canaliculi, Haversian systems, and blood vessels. The pulp occupies the cavity and contains thin collagenous fibers running in all directions and not aggregated into bundles. The ground substance, nerve cells, blood vessels, etc. are also contained in the pulp. The periodontal membrane anchors the root firmly into the alveolar bone and is made of mostly collagenous fibers plus glycoproteins (protein–polysaccharide complex).

6.2.1b. Mechanical Properties

As with most other biological materials, the properties of bone and teeth depend largely on the humidity, the mode of applied load (compressive or tensile, rate of loading, etc.), and the direction of the applied load with respect to the sample. Therefore, one usually studies the effect of the above-mentioned factors and correlates the results with structural features. For brevity of our discussion we will follow the same practice. Table 6-5 gives mechanical properties of various bones.

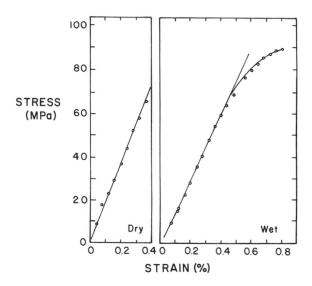

Figure 6-12. Effect of drying on the behavior of human compact bone. (From Ref. 9.)

The effect of *drying* the bone can be seen easily from Figure 6-12 where the dry sample shows a higher modulus of elasticity and compressive strength but lower toughness, fracture strength, and strain. Thus, wet bone, whose characteristics are similar to those *in vivo*, can absorb more energy and elongate more before fracture.

The effect of *anisotropy* is expected since the osteons are longitudinally arranged and the load is borne in that direction, as shown in Table 6-6. It is

Table 6-6. Ratios of Properties Measured, Parallel to
and across the Grain of Human Bone[a]

Young's modulus		Tensile strength		Compressive strength		Remark	Reference[b]
1/r	1/t	1/r	1/t	1/r	1/t		
1.92	1.92			1.54	1.61	DF	1
2.33	2.08			1.12	1.22	WF	2
		12.5	12.5			DT	
		9.8	9.8			WT	

[a] Abbreviations: l, longitudinal; r, radial; t, tangential direction; D, dry; W, wet; F, femur; T, tibia.
[b] 1, W. T. Dempster and R. T. Liddicoat, Compact bone as a nonisotropic material, *Am. J. Anat. 91*, 331, 1952. 2, W. T. Dempster and R. F. Coleman, Tensile strength of bone along and across the grain, *J. Appl. Physiol. 16*, 355, 1961.

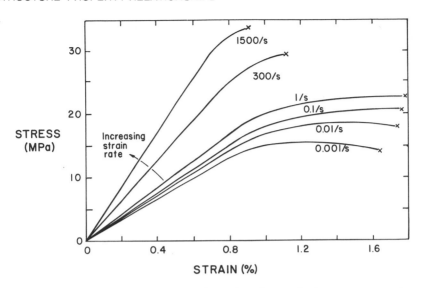

Figure 6-13. Stress as a function of strain and strain rate for human compact bone. (From Ref. 10.)

obvious that Young's modulus and the tensile and compressive strengths in the longitudinal direction are much higher than those in the radial or tangential directions. The tangential and radial directions differ little in mechanical properties.

The effect of the *rate of loading* on the bone is shown in Figure 6-13 and the data are summarized in Table 6-7. As can be seen Young's modulus, ultimate compressive and yield strength increase with increased rate of loading. However, the failure strain and the fracture toughness of the bone reach a maximum and then decrease. This implies that there is a critical rate of loading.

The effect of *mineral content* on the mechanical properties is given in Table 6-8. More mineralized bone has a higher modulus of elasticity and bending strength but lower toughness, illustrating once again that the organic phase of the bone exhibits the energy absorption capacities by straining or yielding to the applied load.

6.2.1c. Modeling of Mechanical Properties of Bone

As mentioned earlier, bone is a composite material, and many researchers have proposed a composite model based on two components, the mineral and the organic phase (review Section 5.5). If one assumes that the

Table 6-7. Summary of Data from Compression Tests
of Human Compact Bone at Various Rates of Strain[a]

Strain rate (1/s)	Ultimate compressive strength (psi)[b]	Energy absorption capacity (in.-lb/in.^{3})	Elastic modulus (psi)	Maximum strain to failure (%)
0.001	21,800	270	2.2×10^6	1.65
0.01	26,000	310	2.5×10^6	1.75
0.1	29,000	340	2.6×10^6	1.8
1	32,000	350	3.2×10^6	1.78
300	40,500	300	4.3×10^6	1.10
1500	46,000	260	5.9×10^6	0.95

[a] From Ref. 10.
[b] 1 psi = 6897 Pa.

load is independently borne by the two components (collagen and mineral, hydroxyapatite), then the total load (P_t) is borne by mineral (P_m) and collagen (P_c):

$$P_t = P_m + P_c \qquad (6\text{-}3)$$

Since $\sigma = P/A = E\epsilon$, thus,

$$P_c = A_c \times E_c \times \epsilon_c \qquad (6\text{-}4)$$

where A, E, and ϵ are area, modulus, and strain, respectively. The strain of collagen can be assumed to be equal to that of mineral, that is, $\epsilon_c = \epsilon_m$; thus,

$$P_c = P_m \frac{A_c E_c}{A_m E_m} \qquad (6\text{-}5)$$

therefore,

$$P_m = \frac{P_t A_m E_m}{A_m E_m + A_c E_c} \qquad (6\text{-}6)$$

Table 6-8. Properties of Three Different Bones with Varying
Mineral Contents[a]

Type of bone	Work of fracture (J/m^2)	Bending strength (MPa)	Young's modulus (GPa)	Mineral content (w/o)	Density (g/cm^3)
Deer antler	6190	179	7.4	59.3	1.86
Cow femur	1710	247	13.5	66.7	2.06
Whale tympanic bulla	200	33	31.3	86.4	2.47

[a] From Ref. 11.

As a simple example, one can assume a load of 1000 N acting on a cross-sectional area of 10 mm^2 bone; hence, $\sigma_t = 100$ MPa. Since the modulus of elasticity of collagen and bone are about 0.1 and 17 GPa, respectively, and the volume fraction of each component is about the same, the fraction of load borne by the collagen and mineral phase become 0.006 and 0.994, respectively. This indicates that most of the load is carried out by the mineral phase at a normal loading condition. Actually, the strength of the demineralized bone is about 5–10% of the whole bone.

If we express equation (6-3) in terms of Young's modulus, it becomes

$$E_t = E_m V_m + E_c V_c \qquad (6\text{-}7)$$

where V is the volume fraction. The above equation is valid if we assume the fibers are oriented parallel to the direction of loading. However, if the fibers are arranged in the perpendicular direction, then one can derive the following equation:

$$\frac{1}{E_t} = \frac{V_m}{E_m} + \frac{V_c}{E_c} \qquad (6\text{-}8)$$

Since not all collagenous fibers are exactly oriented in the same direction, one can propose another model:

$$\frac{1}{E_t} = \frac{x}{E_m V_m + E_c V_c} + (1 - x)\left(\frac{V_m}{E_m} + \frac{V_c}{E_c}\right) \qquad (6\text{-}9)$$

where x is that portion of bone which conforms to the parallel direction and $(1 - x)$ that which is perpendicular.

The rheological properties of bone naturally render themselves to viscoelastic representation, one of which is shown in Figure 6-14. The differential equation for the three-element model can be derived as follows and can be applied under various testing conditions. The total strain (ϵ) and stress (σ) can be written as

$$\epsilon = \epsilon_1 + \epsilon_2, \qquad \sigma = \sigma_1 = \sigma_2 \qquad (6\text{-}10)$$

where subscripts 1 and 2 indicate element 1 (Voigt model with η and E_1) and element 2 (a spring, E_2).

Element 1

From equation (2-27),

$$\sigma_1 = E_1 \epsilon_1 + \eta \frac{d\epsilon_1}{dt} \qquad (6\text{-}11)$$

Element 2

$$\sigma = E_2 \epsilon_2, \qquad \text{hence,} \quad \frac{d\epsilon_2}{dt} = \frac{1}{E_2}\frac{d\sigma}{dt} \qquad (6\text{-}12)$$

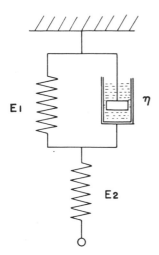

Figure 6-14. Three-element viscoelastic model of bone.

By solving equations (6-10), (6-11), and (6-12) simultaneously, we get

$$\frac{d\sigma}{dt} + \frac{(E_1 + E_2)}{\eta}\sigma = \frac{E_1 E_2}{\eta}\epsilon + E_2\frac{d\epsilon}{dt} \qquad (6\text{-}13)$$

By solving the differential equation (6-13) according to the testing (boundary) conditions, one can get a reasonable approximation of the viscoelastic behavior of the three-element model.

1. Static Testing: $d\epsilon/dt$, $d\sigma/dt$ are zero; therefore, equation (6-13) becomes

$$\sigma = \frac{E_1 E_2}{E_1 + E_2}\epsilon = E_s\epsilon \qquad (6\text{-}14)$$

Equation (6-14) represents purely elastic behavior and E_s is called the static modulus of elasticity.

2. Creep Testing: Instantaneously stretch the sample and hold the sample with a constant load.

The instantaneous elastic strain is solely due to element 2, thus $\epsilon_0 = \sigma/E_2$ and if we use this as our initial condition, the following solution to equation (6-13) can be obtained:

$$\epsilon = \frac{\sigma}{E_s} + \sigma\left(\frac{1}{E_2} - \frac{1}{E_s}\right)\exp\left(-\frac{E_1}{\eta}t\right) \qquad (6\text{-}15)$$

At $t = \infty$, $\epsilon_\infty = \sigma/E_s$, thus,

$$\frac{\epsilon_\infty}{\epsilon_0} = \frac{\sigma/E_s}{\sigma/E_2} = \frac{E_2}{E_s} = \frac{E_1 + E_2}{E_1} = 1 + \frac{E_2}{E_1} > 1 \qquad (6\text{-}16)$$

This indicates that the $\epsilon_\infty/\epsilon_0$ is dictated by the relative modulus of element 1 and 2 for a given constant stress and is always greater than one, that is, the length increases with time.

3. Stress–Relaxation Testing: The sample is stretched and held at a constant length while monitoring the stress changes with time.

The instantaneous stress on the body will be $\sigma_0 = E_2\epsilon$ and $d\epsilon/dt = 0$, hence equation (6-13) becomes

$$\frac{d\sigma}{dt} + \frac{E_1 + E_2}{\eta}\sigma = \frac{E_1 E_2}{\eta}\epsilon \tag{6-17}$$

The solution to the above equation is

$$\sigma = E_s\epsilon + (E_2 - E_s)\epsilon \exp\left(-\frac{E_1 + E_2}{\eta}t\right) \tag{6-18}$$

At $t = \infty$, $\sigma_\infty = E_s\epsilon$, therefore,

$$\frac{\sigma_\infty}{\sigma_0} = \frac{E_s\epsilon}{E_2\epsilon} = \frac{E_1}{E_1 + E_2} < 1 \tag{6-19}$$

The stress always decreases with time, that is, stress relaxation.

4. Constant Strain Rate Testing: $\epsilon = Kt$ or $d\epsilon/dt = K$ (constant). From equation (6-13),

$$\frac{d\sigma}{dt} + \frac{E_1 + E_2}{\eta}\sigma = E_2 K\left(1 + \frac{E_1 t}{\eta}\right) \tag{6-20}$$

By noting that $\sigma = 0$ at $t = 0$, the above equation can be solved:

$$\sigma = E_s\epsilon + \frac{K\eta E_2^2}{(E_1 + E_2)^2}\left[1 - \exp\left(-\frac{E_1 + E_2}{\eta}\frac{\epsilon}{K}\right)\right] \tag{6-21}$$

If $K = 0$, that is, static testing, we will have $\sigma = E_s\epsilon$ and if $K = \infty$, then $\sigma = E_2\epsilon$. Since $E_2 > E_s$, the model fits the observation that increasing the strain rate increases the modulus.

There are several shortcomings of the viscoelastic model: it describes only one-dimensional behavior, the assumptions are not realistic, that is, bone is highly anisotropic and inhomogeneous, and it cannot predict the fracture point.

Others have proposed an interesting phenomenological type of bone model after demineralization by using 0.05 to 0.5 N HCl solution.[12]

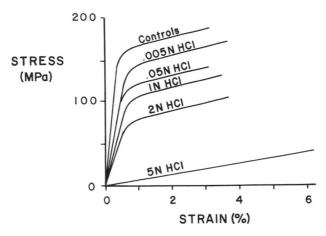

Figure 6-15. Average stress–strain curves for each group of specimens of bovine bone after exposure to different strengths of HCl solution. (From Ref. 12.)

Average stress–strain curves for each group of bovine bone specimens are shown in Figure 6-15. Since there is no variation of the slope of the plastic zone while the modulus of elasticity decreases with increased demineraliza- tion, the "elastic–perfectly plastic mineral" and "elastic collagen" com- posite structure was proposed as shown in Figure 6-16. This model also has the same kind of shortcomings as the spring–dashpot model.

Another interesting view of the properties of bone is the porosity model in which the strength varies according to the following equation[7]:

$$\sigma = \sigma_0 \exp\left(-nV\right) \tag{6-22}$$

where V is the volume fraction porosity and n is in the range of 4–7. The elastic modulus also can be expressed as

$$E = E_0\left(1 - 1.9V + 0.9V^2\right) \tag{6-23}$$

It is estimated that $18.9 \pm 0.45\%$ porosity can exist in beef compact bone due to the various canals in bone as mentioned previously and one can arrive at similar values of tensile strength and modulus as reported in the literature by using equations (6-22) and (6-23).

Before going into next section, some of the physical properties of teeth are given in Table 6-9. As can be expected, the strength is highest for enamel, and dentin is intermediate between bone and enamel. Thermal expansion and conductivity are higher for enamel than for dentin as given in Table 6-9.

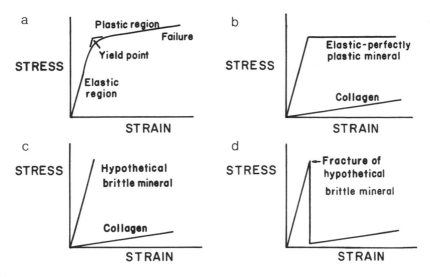

Figure 6-16. Burstein's proposed model of bone based on the results given in Figure 6-15. (a) Typical tension stress–strain curve for bovine femoral cortical bone. (b) The upper curve represents the hypothesized elastic–perfectly plastic behavior of the mineral phase. The lower curve represents the elastic behavior of collagen. (c) The upper curve represents a hypothetical elastic–brittle characteristic of the mineral phase. The lower curve represents the elastic collagen material. (d) Summation of the hypothetical behavior of bone mineral and collagen in tension, if the bone behaves in a brittle manner. (From Ref. 12.)

6.2.1d. Remodeling Mechanisms of Bone

Understanding the remodeling process of bone is of paramount importance in implant design and material selection. This can be illustrated easily by the fact that a bone plate that is too stiff results in thinning of the cortical bone underneath it after the fracture heals due to the "stress shielding" effect.[13] This functional adaptation of bone was first introduced

Table 6-9. Physical Properties of Teeth

	Density (g/cm^3)	Modulus of elasticity (GPa)	Compressive strength (MPa)	Coefficient of thermal expansion (cm/cm per °C)	Thermal conductivity (W/m-°K)
Enamel	2.2	48	241	11.4×10	0.82
Dentin	1.9	13.8	138	8.3×10	0.59

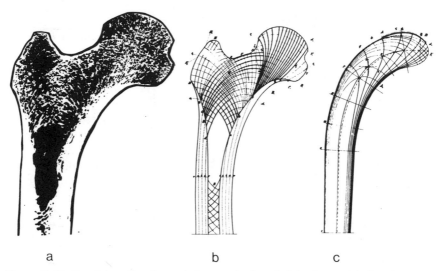

a   b   c

Figure 6-17. Frontal section through the proximal end of the femur (a), line drawing of the cancellous structure (b), and Culmann's crane (c). (From Ref. 14.)

by Wolff of Germany as the "law of bone transformation" in the 1870s in which he emphasized that the remodeling of cancellous bone structure followed mathematical rules corresponding to the principal stress trajectories as shown in Figure 6-17.

Although the exact mechanism of bone remodeling (micro as well as macro remodeling) is as yet incompletely understood, there are several proposed theories. One is the piezoelectricity theory in which bone modifies its structure by sensing mechanical stresses generated by dynamic loading and unloading cycles *in vivo* which in turn generate the electricity which triggers the remodeling activities. The piezoelectric properties of bone first discovered by Fukuda and Yasuda in the late 1950s and later by Bassett, Brighton, and their co-workers independently verified the theory. In fact, the original study of Fukuda and Yasuda led to experiments on "electric callus" formation by applying electricity (only the negative pole induced a callus which corresponds to the side of electronegativity when bone is bent). More recent studies[15] using a microelectrode show that the generation of electricity in bone is much more complicated than originally thought and have emphasized that there are strain-generated potentials (SGP) as shown in Figure 6-18. As can be seen, there are large, nonlinear electric fields near Haversian canals. This field is radial and is 30 to 1000 times larger than the average electric field determined from simultaneous macroscopic measurements.

"Electric callus" formation by direct or indirect usually (oscillating

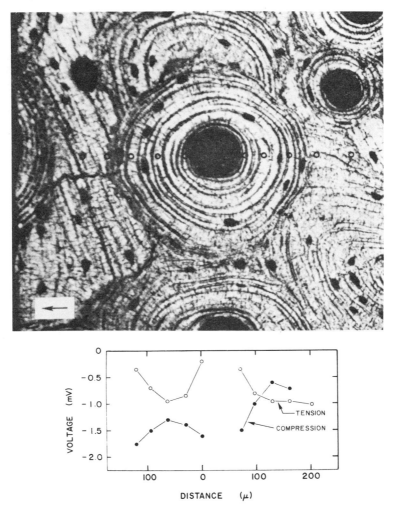

Figure 6-18. Tension and compression data for the same osteon. Osteon No. 6-12, specimen No. WH-90-1, human. (From Ref. 15.)

magnetic field) stimulation of bone has recently been used to successfully overcome nonunion of long bones.[16]

Streaming potentials of positive and negative ions in the tissue fluids are thought to be more important factors corresponding to the electrical potentials *in vivo* rather than the piezoelectric properties of bone.[17] It is conceivable that more rigorous activities on bone will lead to higher kinetic activities of ions which will induce higher streaming potentials which, in turn, will signal more bone mineral deposition. The exact mechanism, however, cannot be pinpointed by this theory.

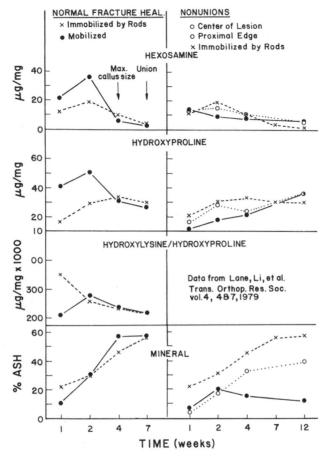

Figure 6-19. An overview of the metabolic interrelationships within the callus. (From Ref. 18.)

The remodeling process can be mediated by biomechanical activities of enzymes and other chemical species such as amino acids and minerals as shown in Figure 6-19. The study of these chemical activities is beyond the scope of this text and interested readers are referred to Urist.[19]

Bone remodeling can be separated into two categories—*surface* and *internal remodeling*.[20] Surface remodeling is the resorption and deposition of bone material on the external surfaces of bone (periosteal surfaces), while internal remodeling is the reinforcement and resorption in the endosteal surfaces, resulting in changes in the bulk density of the bone. In cancellous bone, this latter process can be accomplished by increasing the thickness of each trabecula. In cortical bone, internal remodeling occurs by changing the diameter of the lumina of the osteons and by totally replacing osteons.

6.2.2. Collagen-Rich Tissues

Collagen-rich tissues function mostly in a load-bearing capacity. These tissues include skin, tendon, and cartilage. Special functions such as transparency for the lens of the eye can also be borne by collagenous tissues.

6.2.2a. Composition and Structure

The collagen-rich tissues are composed predominately of collagen (75 w/o dry, see Table 6-10). Collagen is made up of tropocollagen, a three-chain coiled superhelical molecule (Figure 6-20a). The collagen fibrils aggregate to form fibers as shown in Figure 6-20b. The fibrils and fibers are stabilized through intra- and intermolecular hydrogen bonding ($C\!=\!O-HN$) as shown in Figure 6-21.

The physical properties of tissues vary according to the amount and structural variations of collagen fibers. Several different fiber arrangements are found in tissues: (1) parallel fibers, (2) crossed fibrillar arrays, and (3) feltworks in which the fibers are more randomly arranged with some fibers going through the thickness of the tissue.[24]

6.2.2b. Physical Properties

The collagen-rich tissues can be thought of as a polymeric material in which the highly oriented crystalline collagen fibers are embedded in the ground substance of mucopolysaccharides and amorphous elastin (rubber). When the tissue is heated, its specific volume increases (density decreases, exhibiting glass transition temperature, $T_g = 40°C$) as does shrinkage (T_s) or equilibrium melting temperature as shown in Figure 6-22. The shrinkage temperature is considered a denaturation point for collagen.

The stress–strain curves of collagen fibers are similar to those of synthetic fibers as shown in Figure 6-23. The initial toe region represents alignment of fibers in the direction of stress; the steep rise in slope indicates that the majority of fibers are in parallel directions. Individual fibers may be breaking well ahead of the final catastrophic failure (see Figures 5-21 and 5-22). The highest slope of the stress–strain curve is about 1.0 GPa, which is close to the modulus of collagen fibers but the tensile strength is much lower. Table 6-11 lists some mechanical properties of collagen and elastic fibers. The modulus of elasticity of collagen can increase a great deal by increasing the rate of loading due to its viscoelastic properties.

Stress–relaxation behavior of rat tail tendon collagen fibers showed trends similar to bulk polyethylene at an initial strain of 3.5%, but rather

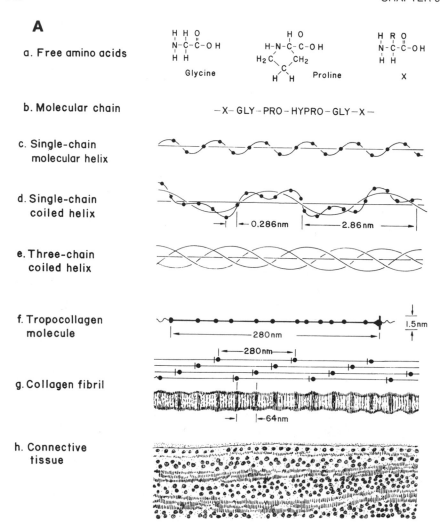

Figure 6-20. (A) Diagram depicting the formation of collagen, which can be visualized as taking place in seven steps. The starting materials (a) are amino acids, of which two are shown and the side chain of any others is indicated by R in amino acid X. (b) The amino acids are linked together to form a molecular chain. (c) This then coils into a left-handed helix (d, e). Three such chains then intertwine in a triple-stranded helix, which constitutes the tropocollagen molecule (f). Many tropocollagen molecules become aligned in staggered fashion, overlapping by a quarter of their length to form a cross-striated collagen fibril (g). (From Ref. 21.) (B) Diagram showing that the "reticular fibers" associated with the basal lamina of an epithelial cell (above) and the "collagen fibers" of the connective tissue in general (below) are both composed of unit fibrils of collagen. Those of the reticulum are somewhat smaller and interwoven in loose networks instead of in larger bundles. (From Ref. 22.)

B

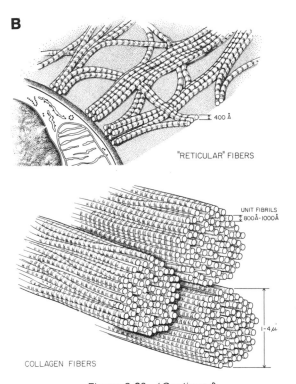

"RETICULAR" FIBERS

400 Å

UNIT FIBRILS
800Å-1000Å

1-4 μ

COLLAGEN FIBERS

Figure 6-20. (*Continued*)

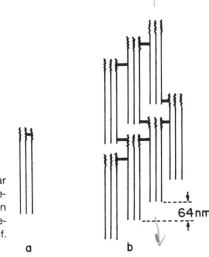

Figure 6-21. (a) Intra- and (b) intermolecular cross-links in collagen. Straight lines represent polypeptide chains in the helical region of tropocollagen molecules; wavy lines represent the N-terminal telopeptide. (From Ref. 23.)

a   b

64nm

Table 6-10. Composition of
Collagen-Rich Tissues

Component	% composition
Collagen	75 (dry), 30 (wet)
Mucopolysaccharides	20 (dry)
Elastin	< 5 (dry)
Water	60–70

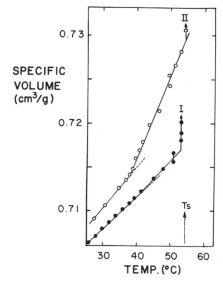

Figure 6-22. The glass and melting transitions in stressed kangaroo tail tendon in 0–0.9% saline. Curve I, first cycle of melting. Curve II, the melted fiber of curve I, cooled to room temperature overnight, then reheated. The break near 40°C represents T_g. (From Ref. 25.)

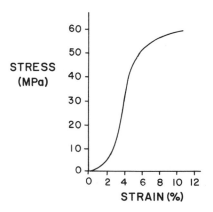

Figure 6-23. A typical stress–strain curve for tendon. (From Ref. 24.)

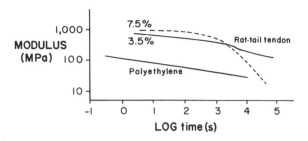

Figure 6-24. Stress–relaxation curves for collagen fibers. Solid and dashed curves are for rat tail tendon strained 3.5 and 7.5%, respectively.[26] The stress–relaxation curve of bulk crystallized polyethylene is included for comparison.[27]

faster relaxation occurred when strained 7.5%, which is well above the reversible extension range as shown in Figure 6-24.

Unlike tendon or ligament, *skin* has a feltwork of continuous fibers which are randomly arranged in layers or lamellae. The skin tissues also show mechanical anisotropy as shown in Figure 6-25. This is the reason behind the line of Langer, which follows the least resistant path of the dermis for easier surgery and better wound healing.

Another feature of the stress–strain curve of the skin is its extensibility with a small load despite its high collagen content. Upon stretching, the fibrous lamellae align with each other and resist further extension as shown in Figure 6-26. When the skin is highly stretched, the modulus of elasticity at these high-extension portions approaches that of tendon.

Cartilage is another collagen-rich tissue which has two main physiological functions. One is the maintenance of shape (ear, tip of nose, and rings around the trachea), and the other is to provide bearing surfaces at joints. It contains very large and diffuse polysaccharide–protein molecules which form a gel in which the collagen-rich molecules are entangled (see Figure 6-6). They can affect the mechanical properties of the cartilage by hindering movement through the interstices of the collagenous matrix network.

Table 6-11. Elastic Properties of Elastic
and Collagen Fibers

	Modulus of elasticity (MPa)	Tensile strength (MPa)	Ultimate elongation (%)
Elastic fibers	0.6	1	100
Collagen	1000	50–100	10

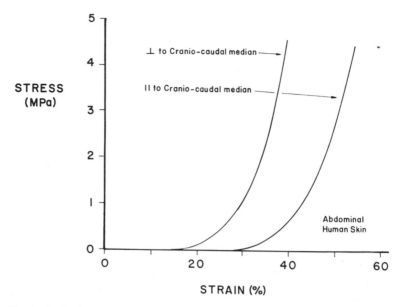

STRESS
(MPa)

⊥ to Cranio-caudal median

II to Cranio-caudal median

Abdominal
Human Skin

STRAIN (%)

Figure 6-25. Stress–strain curves of human abdominal skin. (From Ref. 28.)

Joint cartilage has a very low coefficient of friction (< 0.01). This is largely attributed to the squeeze-film effect between cartilage and synovial fluid.[30] The surface structure of the cartilage is particularly suitable for this type of lubrication, resulting in very low friction as shown in Figure 6-27. The lubricating function is carried out in conjunction with mucopolysaccharides, especially chondroitin sulfates as discussed in Section 6.1.2b. The modulus of elasticity (10.3–20.7 MPa) and tensile strength (3.4 MPa) are quite low. However, wherever high stress is required, the cartilage is replaced by purely collagenous tissue.

6.3. ELASTIC TISSUES

Elastic tissues exhibit a large deformation with a small load cyclically so that the least amount of energy is expended in the process. These tissues include blood vessels, ligamentum nuchae, and muscles (see Figure 6-28).

6.3.1. Composition and Structure

Elastic tissues contain a relatively large amount of elastin ("protein rubber") which gives the tissues their elastic properties. Usually, when

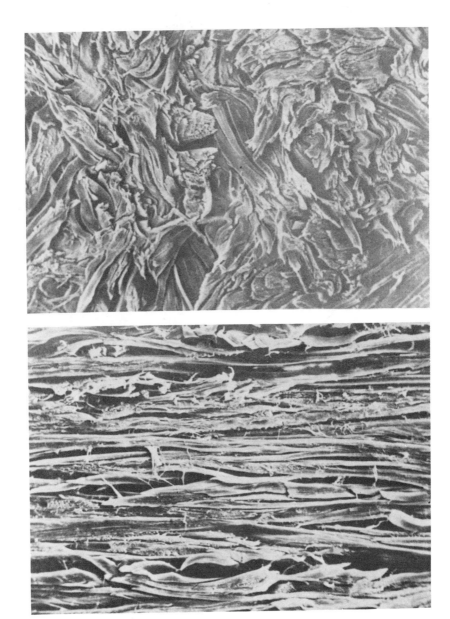

Figure 6-26. Scanning electron micrographs of dermal skin before and after stretching. Stretching direction is horizontal. 400×. (From Ref. 29.)

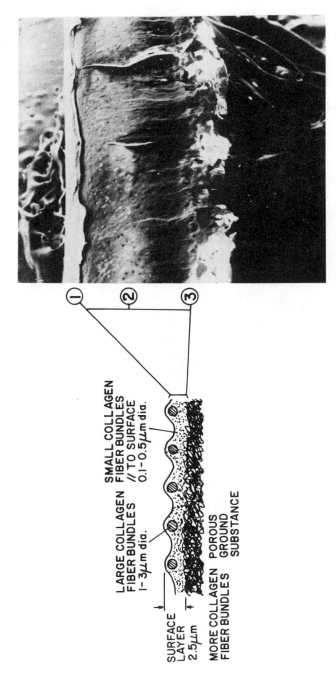

SURFACE LAYER 2.5 μm

LARGE COLLAGEN FIBER BUNDLES 1-3 μm dia.

SMALL COLLAGEN FIBER BUNDLES // TO SURFACE 0.1-0.5 μm dia.

MORE COLLAGEN FIBER BUNDLES

POROUS GROUND SUBSTANCE

Figure 6-27. Simple model for the surface of articular cartilage. Zone 1 provides, in addition to a relatively smooth bearing surface, a load diffusion effect. Zone 2 provides an area of deformability and energy storage. Zone 3 binds the cartilage tissue to the underlying bone and provides a degree of constraint. (From Refs. 30, 31.)

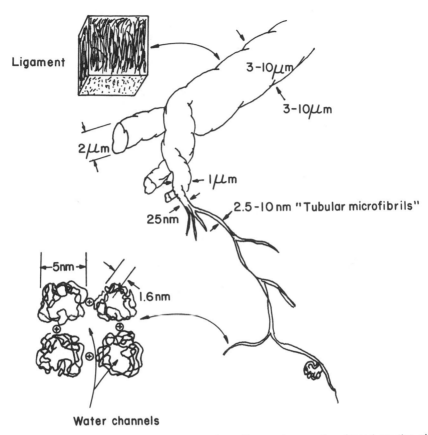

Figure 6-28. Physical structure of elastin from ligamentum nuchae based on the observations and conclusions of several investigators. (From Ref. 32.)

higher elasticity is required, more elastin is present in the tissue (up to 80 w/o dry, as in ligamentum nuchae).

One of the most important elastic tissues is the blood vessel, which has three distinct layers in cross-section (Figure 6-29): (1) the intima, in which the structural elements are oriented longitudinally; (2) the media, which is the thickest layer of the wall and whose components are arranged circumferentially; and (3) the adventitia, which connects the vessels to the matrix firmly via facia. The bonding between the intima and the media is formed by the internal elastic membrane (*elastica interna*), which is predominant in arteries of medium size. Between the media and the adventitia, a thinner external elastic membrane (*elastica externa*) can be found. The smooth muscle cells are found between adjacent elastic lamellae in helical array.

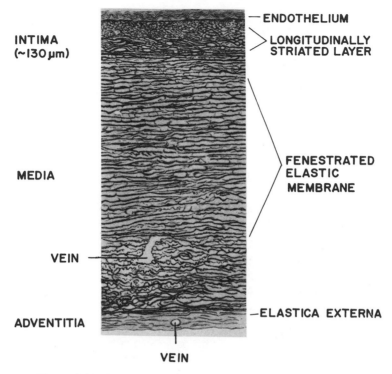

Figure 6-29. Structure of a blood vessel wall. (From Ref. 22.)

6.3.2. Properties of Elastic Tissues

There is disagreement in the literature as to whether elastin conforms to elasticity theory or not. Hoeve and Flory[33] found the volume of elastin to be independent of temperature in a 30% solution of ethylene glycol and water, and thermoelastic measurements in this solvent showed that it is in fact a rubber elastomer. However, Volpin and Ciferri[34] found that 90% of the elastic force can be contributed to the entropy. Calorimetric measurement during elastin stretching showed that there is an extremely large internal energy change associated with the stretch.[35] This is probably due to the unfolding of the hydrophobic region of the globular elastin molecule.

Figure 6-30 shows a stress–strain curve of bovine ligamentum nuchae at low extension. As can be seen, the modulus of elasticity and the amount of stored energy lost upon releasing the load are quite low. This is character-

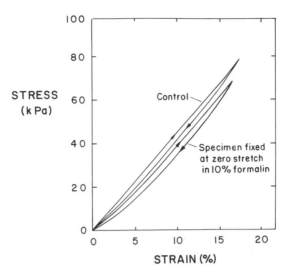

Figure 6-30. The stress–strain curve of elastin. The material is the ligamentum nuchae of cattle, which contains a small amount of collagen that was denatured by heating at 100°C for an hour. Such heating does not change the mechanical properties of elastin. The specimen is cylindrical with rectangular cross-section. Loading is uniaxial. The curve labeled "control" refers to native elastin. The curve labeled "10% formalin" refers to a specimen fixed in formalin solution for a week without initial strain. (From Ref. 4.)

istic of elastic tissues whose primary function is the restoration of a deformed state to the original shape with minimum energy loss. Therefore, in blood vessels the elastin is distributed along the walls, the greatest amount being in the arch of the aorta where the expulsed blood is temporarily stored as a "secondary pump." The amount of elastin decreases with decreasing size of the vessel as shown in Figure 6-31.

Due to the anisotropy of the arrangement of the blood vessel wall and its tubal structure, the intrinsic properties are not well defined. Early studies on the mechanical properties were done by pressurizing with saline solution a segment of the vessel after closing its branches and recording the pressure and diameter versus length changes as shown in Figure 6-32. However, the composition of the vessel walls changes along the length of the wall, and consequently, their physical properties also change (Figure 6-31). Another complicating factor is the existence of smooth muscle which is associated with arterial blood pressure regulation.

Table 6-12 shows that the mean pressure of the various blood vessels and the approximate tension developed at normal pressure are related as

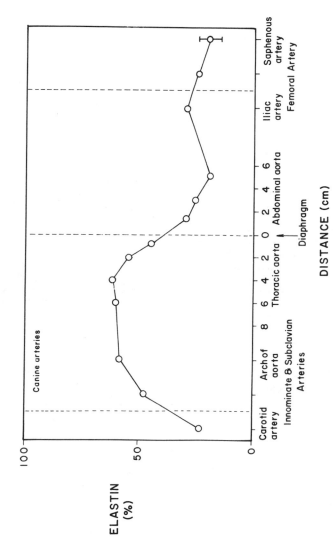

Figure 6-31. Variation of elastin percent per elastin and collagen along the major arterial tree. (From Ref. 36.)

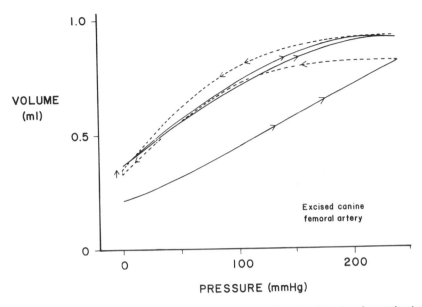

Figure 6-32. Repeated volume versus pressure tests of excised canine femoral artery. (From Ref. 37.)

given in the Laplace equation:

$$T = P \times r \qquad (6\text{-}24)$$

where T is the wall tension, P is the internal pressure, and r is the radius of the vessel. This theory is usually applied to a uniform, thin, isotropic tube in

Table 6-12. Wall Tension and Pressure Relationship of Various Sizes
of Blood Vessels[a]

	Mean pressure (mm Hg)	Internal pressure (dynes/cm^2)	Radius	Wall tension (dynes/cm^2)
Aorta, large artery	100	1.5×10	1.3 cm	170,000
Small artery	90	1.2×10	0.5 cm	60,000
Arteriole	60	8×10	62 μm–0.15 mm	500–1,200
Capillaries	30	4×10	4 μm	16
Venules	20	2.6×10	10 μm	26
Veins	15	2×10	200 μm	400
Vena cava	10	1.3×10	1.6 μm	21,000

[a] From Ref. 38.

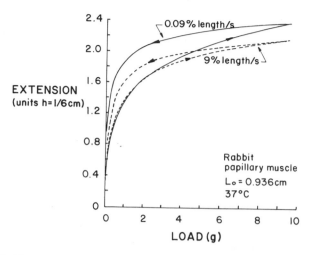

Figure 6-33. The length–tension curve of a resting papillary muscle from the right ventricle of the rabbit. Hysteresis curves at strain rates 0.09% length/s, 0.9% length/s, and 9% length/s. Length at 9 mg = 0.936 cm (37°C). (From Ref. 39.)

the absence of longitudinal tension. It is evident that none of the requirements can be met strictly by the blood vessel walls.

Muscle is another elastic tissue. This tissue is not discussed in detail in this book due to the "active" nature of the tissue whose importance is more obvious in conjunction with physiological function. As passive tissue the stress–strain curve shows an elastic behavior as shown in Figure 6-33. Table 6-13 gives mechanical properties of some of the nonmineralized tissues.

Table 6-13. Mechanical Properties of Some Nonmineralized Human Tissues[a]

	Tensile strength (MPa)	Ultimate elongation (%)
Skin	7.6	78
Tendon	53	9.4
Elastic cartilage	3	30
Heart values (aortic)		
Radial	0.45	15.3
Circumferential	2.6	10.0
Aorta		
Transverse	1.1	77
Longitudinal	0.07	81
Cardiac muscle	0.11	63.8

[a] From various sources.

6.4. CONSTITUTIVE EQUATIONS DESCRIBING MECHANICAL PROPERTIES OF SOFT TISSUES

6.4.1. Stress–Strain Relationship in Loading and Unloading

For an incompressible material the true stress (or sometimes called Eulerian stress) can be represented as follows:

$$\sigma_t = \frac{P}{A_t} = \frac{P\lambda}{A_r} = T\lambda \qquad (6\text{-}25)$$

where A_t and A_r are true and reference cross-sectional areas of the sample, λ is the stretch ratio (l/l_o), and T is "engineering stress" provided $A_r = A_o$ (original cross-sectional area). If one takes the slopes of the curves of Figure 6-33 and tries to fit them against a straight line, then

$$\frac{dT}{d\lambda} = A(T + B) \qquad (6\text{-}26)$$

By integrating,

$$T = C\exp(A\lambda) + B \qquad (6\text{-}27)$$

Note that A, B, and C are all constants and if the material behaves as a Hookean material, then

$$\frac{dT}{d\lambda} = \text{constant} \qquad (6\text{-}28)$$

Figure 6-34 is a more refined representation of the experimental data of

Figure 6-34. The variation of Young's modulus with load at a strain rate of 0.9% length/s, illustrating the method of determining the constants A_1, B_1. Near the origin, a different straight line segment is required to fit the experimental data. In this figure, Young's modulus, $dT/d\lambda$, is replaced by the difference $\Delta T/\Delta \lambda$, to which $dT/d\lambda$ is proportional because of equation (6-27); and T is written as P. For the ordinate, we used $\Delta \lambda = h = 1/6$ cm, and $\Delta T/\Delta \lambda$ is expressed as ΔP per unit of h. (From Ref. 39.)

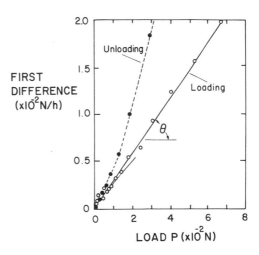

Figure 6-33 by breaking down the straight line into two regions:

$$\frac{dT}{d\lambda} = A_1(T + B_1) \qquad \text{for } 0 < T < T_1$$

$$\frac{dT}{d\lambda} = A_2(T + B_2) \qquad \text{for } T_1 < T < T_2$$

(6-29)

6.4.2. Strain Energy Function

For an isotropic material a strain energy function is taken to exist.[40] The energy function depends only on strains:

$$W = W(I_i) \tag{6-30}$$

where I_i is a strain invariant, $i = 1$, 2, and 3 for the three principal axes, and can be expressed (see Figure 6-35) as

$$I_1 = \Sigma(\lambda_i) = \lambda_1^2 + \lambda_2^2 + \lambda_3^2$$

$$I_2 = \Sigma(\lambda_i\lambda_j) = (\lambda_1\lambda_2)^2 + (\lambda_1\lambda_3)^2 + (\lambda_2\lambda_3)^2 \tag{6-31}$$

$$I_3 = (\lambda_1\lambda_2\lambda_3)^2$$

For an incompressible material,

$$\lambda_1\lambda_2\lambda_3 = 1 \tag{6-32}$$

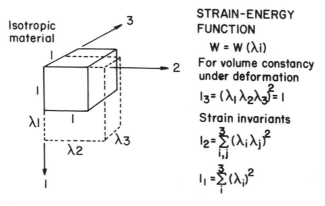

Figure 6-35. The definition of the principal extension ratios λ, for finite homogeneous deformation of an elastin cube. (From Ref. 41.)

Therefore, equation (6-31) can be rewritten:

$$I_1 = \lambda_1^2 + \lambda_2^2 + \frac{1}{\lambda_1^2 \lambda_2^2} = I(\lambda_1 \lambda_2)$$

$$I_2 = \frac{1}{\lambda_1^2} + \frac{1}{\lambda_2^2} + (\lambda_1 \lambda_2)^2 \qquad (6\text{-}33)$$

$$I_3 = 1$$

Veronda and Westmann proposed a strain energy function formula as follows:

$$W = A\left[\exp B(I_1 - 3) - 1\right] + C(I_2 - 3) + f(I_3) \qquad (6\text{-}34)$$

where A, B, and C are constants and f is a function which becomes zero for an incompressible material. For cat skin they give the following values:

$$A = 0.00394, \qquad B = 5.03, \qquad C = -0.01985 \qquad (6\text{-}35)$$

6.4.3. Contribution of Collagen, Elastin, and Mucopolysaccharide to the Mechanical Properties of Soft Tissues

As with other viscoelastic materials like polymers, soft tissues exhibit similar behavior which can be represented with traditional Voigt- and Maxwell-model series or in parallel fashion.[4] Although this type of analysis gives a very useful handle on the mechanical behavior of tissues, it does not explicitly explain the relative importance of constitutive components in tissues. One might, therefore, try to understand the overall behavior of tissues by remodeling the structure, as the example given in Figure 6-36. This model shows that the vein is made up of the three major force

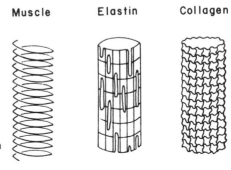

Muscle   Elastin   Collagen

Figure 6-36. Structure of the vein. (From Ref. 42.)

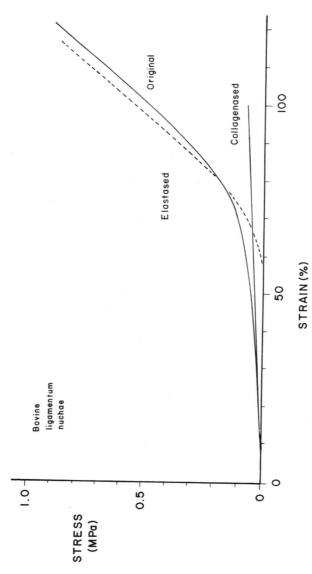

Figure 6-37. Stress–strain curves of bovine ligamentum nuchae after elastase and collagenase treatments. (From Ref. 43.)

components: smooth muscle fibers arranged in helical fashion with very short pitch; elastic fibers of elastin and collagen forming a crimped network structure which can be stretched out at high extension.

Others[43] have analyzed soft tissue behavior based on enzymolysis experiments in which a particular component is removed by an appropriate enzyme. Typical stress–strain curves obtained after removal of each major component are given in Figure 6-37. One can see that ligamentum nuchae shows rubberlike elasticity up to 50% elongation. However, if one removes the collagen component from the tissue by an enzyme (collagenase) or autoclaving, it behaves entirely like an elastomer up to 100% elongation. Conversely, if one removes the elastin component from the same tissue, then the remaining tissue behaves as collagenous tissue except the curve is shifted toward high extension which used to be taken up by the elastin. The removal of ground substance does not alter the basic stress–strain behavior. From these experiments one can deduce a simple model a shown in Figure 6-38.

In this model the elastic fibers (elastin) are stretched followed by the much stronger collagen fibers when two components are pulled together.

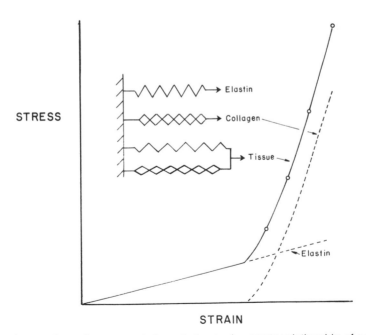

Figure 6-38. A schematic representation of structure–property relationship of connective tissues. The elastin is represented by a loosely knitted fabric. (From Ref. 44.)

With a little more imagination one can envision the distribution of fiber length and thus the response can be smoother than the two-fiber system as shown in Figure 6-38. From this kind of study relating the microstructure to the macrobehavior, one can understand fully the nature of interaction between components in a connective tissue.

PROBLEMS

6-1. Calculate x of equation (6-9) for bone with Young's modulus of 17 GPa by assuming $V_m = V_c$, $E_m = 100$ GPa, $E_c = 0.1$ GPa.

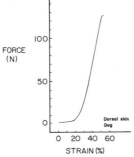

6-2. The force versus displacement curve in Figure P6-2 was obtained by tensile testing of canine skin. The skin specimen was cut by using a stamping machine which has a width of 4 mm. The thickness of the skin is about 3 mm and the length of the sample between the grips is 20 mm.

a. What is the tensile strength and fracture strain of the skin?
b. What are the moduli of elasticity in the initial and secondary regions?
c. What is the toughness of the skin?

Figure P6-2.

6-3. Ligamentum nuchae is made of elastin and collagen (others do not contribute to the mechanical properties). The relative amounts excluding water are 70, 25, and 5 w/o for elastin, collagen, and others, respectively, and assume they are homogeneously distributed.

a. Calculate the percent contribution of the elastin to the total strength assuming elastic behavior.
b. Compare the result with Figure 6-38.

6-4. If a bone contains 18.9% porosity, what is its modulus of elasticity if the modulus without pores is the same as that of hydroxyapatite (see Table 9-3)?

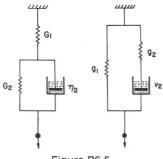

6-5. The viscoelastic properties of compact bone have been described by using the three-element models shown in Figure P6-5.

a. Derive differential equations for the models.
b. Solve part a for γ for stress–relaxation tests of each model.
c. By appropriate arrangement of the constants, express the parameters of one model from the other model.
d. What are the shortcomings of the models?

Figure P6-5.

REFERENCES

1. J. Black, *Biological Performance of Materials*, Dekker, New York, 1981.
2. M. Chvapil, *Physiology of Connective Tissue*, Chapter 2, Butterworths, London, 1967.
3. J. T. Triffit, The organic matrix of bone tissue, in: *Fundamental and Clinical Bone Physiology*, M. R. Urist (ed.), Chapter 3, Lippincott, Philadelphia, 1980.
4. Y. C. Fung, *Biomechanics: Mechanical Properties of Living Tissues*, Springer-Verlag, Berlin, 1981.
5. B. G. King and M. J. Showers, *Human Anatomy and Physiology*, Saunders, Philadelphia, 1963.
6. H. J. Hohling, B. A. Ashton, and H. D. Koster, Quantitative electron microscope investigation of mineral nucleation in collagen, *Cell Tissue Res. 148*, 11–26, 1974.
7. K. Pierkarski, Structure, properties and rheology of bone, in: *Orthopaedic Mechanics: Procedures and Devices*, D. N. Ghista and R. Roaf (ed.), Chapter 7, Academic Press, New York, 1978.
8. H. Yamada, *Strength of Biological Materials*, Williams & Wilkins, Baltimore, 1970.
9. F. G. Evans and M. Lebow, The strength of human compact bone as revealed by engineering technics, *Am. J. Surg. 83*, 326–331, 1952.
10. J. H. McElhaney, Dynamic response of bone and muscle tissue, *J. Appl. Physiol. 21*, 1231–1236, 1966.
11. J. D. Currey, What is bone for?—Property–function relationships in bone, in: *Mechanical Properties of Bone*, S. C. Cowin (ed.), pp. 13–26, ASME, New York, 1981.
12. A. H. Burstein, J. M. Zika, K. G. Heiple, and L. Klein, Contribution of collagen and mineral to the elastic–plastic properties of bone, *J. Bone Jt. Surg. 57A*, 956–961, 1975.
13. S. L.-Y. Woo, W. H. Akeson, R. D. Coutts, L. Rutherford, D. Doty, G. F. Jemmott, and D. Amiel, A comparison of cortical bone atrophy secondary to fixation with plates with large differences in bending stiffness, *J. Bone Jt. Surg. 58A*, 190–195, 1976.
14. J. Wolff, *Das Gesetz der Transformation der Krochen*, Hirchwild, Berlin, 1892.
15. W. Starkebaum, S. R. Pollack, and E. Korostoff, Microelectric studies of stress-generated potentials in four-point bending of bone, *J. Biomed. Mater. Res. 3*, 729–751, 1979.
16. C. T. Brighton, Z. B. Friedenberg, and J. Black, Evaluation of the use of constant direct current in the treatment of nonunion, in: *Electrical Properties of Bone and Cartilage*, C. T. Brighton, J. Black, and S. R. Pollack (ed.), pp. 519–545, Grune & Stratton, New York, 1979.
17. C. Eriksson, Streaming potentials and other water-dependent effects in mineralized tissue, *Ann. N.Y. Acad. Sci. 238*, 321–338, 1974.
18. A. Y. Ketenjian and C. Arsenis, Morphological and biochemical studies during differentiation and calcification of fracture callus cartilage, *Clin. Orthop. Relat. Res. 107*, 266–273, 1975.
19. M. R. Urist (ed.), *Fundamental and Clinical Bone Physiology*, Lippincott, Philadelphia, 1980.
20. S. C. Cowin, Continuum models of the adaptation of bone stress, in: *Mechanical Properties of Bone*, S. C. Cowin (ed.), pp. 193–210, ASME, New York, 1981.
21. J. Gross, Collagen, *Sci. Am. 204*, 121–130, 1961.
22. W. Bloom and D. W. Fawcett, *A Textbook of Histology*, 9th ed., Saunders, Philadelphia, 1968.
23. A. J. Bailey, C. M. Peach, and L. J. Fawler, Biosynthesis of intermolecular cross-links in collagen, in: *The Chemistry and Molecular Biology on the Intracellular Matrix*, Volume 1, E. A. Balazs (ed.), pp. 385–404, Academic Press, New York, 1970.
24. S. A. Wainwright, W. D. Biggs, J. D. Currey, and J. M. Gosline, *Mechanical Design in Organisms*, Arnold, London, 1976.

25. P. Mason and B. J. Rigby, Thermal transitions in collagen, *Biochim. Biophys. Acta 66*, 448–450, 1963.
26. B. J. Rigby, N. Hiraci, J. D. Spikes, and H. Eyring, The mechanical properties of rat tail tendon, *J. Gen. Physiol. 43*, 265–283, 1959.
27. G. W. Becker, Stress relaxation of polyethylene, *Kolloid Z. 175*, 99–110, 1961.
28. C. H. Daly, The Biomechanical Characteristics of Human Skin, Ph.D. thesis, University of Strathclyde, Scotland, 1966.
29. G. L. Wilkes, I. A. Brown, and R. H. Wildnauer, The biochemical properties of skin, *CRC Crit. Rev. Bioeng. 1*, 453–495, 1973.
30. P. S. Walker, J. Sikorski, D. Downson, M. Longfield, and V. Wright, Lubrication mechanism in human joints, in: *Lubrication and Wear in Joints*, V. Wright (ed.), pp. 49–61, Lippincott, Philadelphia, 1969.
31. J. McAll, Load deformation response of the microstructure of articular cartilage, in: *Lubrication and Wear in Joints*, V. Wright (ed.), pp. 39–48, Lippincott, Philadelphia, 1969.
32. A. S. Hoffman, A critical evaluation of the application of rubber elasticity principles to the study of structural proteins such as elastin, in: *Biomaterials*, A. L. Bement, Jr. (ed.), pp. 285–312, University of Washington Press, Seattle, 1971.
33. C. A. J. Hoeve and P. J. Flory, Elastic properties of elastin, *J. Am. Chem. Soc. 80*, 6523–6526, 1958.
34. D. Volpin and A. Ciferri, Thermoelasticity of elastin, *Nature (London) 225*, 382, 1970.
35. T. Weis-Fogh and S. O. Andersen, New molecular model for long-range elasticity of elastin, *Nature (London) 227*, 718–721, 1970.
36. R. D. Harkness, Mechanical properties of collageneous tissues, in: *Treatise on Collagen*, B. S. Gould (ed.), Volume 2, Part A, Chapter 6, Academic Press, New York, 1968.
37. D. H. Bergel, The static elastic properties of the arterial wall, *J. Physiol. (London) 156*, 445–475, 1961.
38. A. C. Burton, *Physiology and Biophysics of Circulation*, Chapter 7, Year Book Medical Publishers, Chicago, 1965.
39. Y. C. Fung, N. Perrone, and M. Anliker (ed.) *Biomechanics: Its Foundation and Objectives*, Prentice–Hall, Englewood Cliffs, N.J., 1972.
40. D. R. Veronda and R. A. Westmann, Mechanical characterization of skin—Finite deformities, *J. Biomech. 3*, 111–124, 1970.
41. J. D. C. Crisp, Properties of tendon and skin, in: *Biomechanics: Its Foundations and Objectives*, Y. C. Fung, N. Perrone, and M. Anliker (ed.), pp. 141–179, Prentice–Hall, Englewood Cliffs, N.J., 1972.
42. T. Azuma and M. Hasegawa, Distensibility of the vein: From the architectural view point, *Biorheology 10*, 469–479, 1973.
43. A. S. Hoffman, L. A. Grande, P. Gibson, J. B. Park, C. H. Daly, and R. Ross, Preliminary studies on mechanochemical–structure relationships in connective tissues using enzymolysis techniques, in: *Perspectives of Biomedical Engineering*, R. M. Kenedi (ed.), pp. 173–176, University Park Press, Baltimore, 1972.
44. J. B. Park, *Biomaterials: An Introduction*, Plenum Press, New York, 1979.

BIBLIOGRAPHY

R. Barker, *Organic Chemistry of Biological Compounds*, Chapters 4 and 5, Prentice–Hall, Englewood Cliffs, N.J., 1971.
J. Black, *Biological Performance of Materials*, Dekker, New York, 1981.
S. C. Cowin (ed.), *Mechanical Properties of Bone*, ASME, New York, 1981.

H. R. Elden (ed.), *Biophysical Properties of the Skin*, Wiley, New York, 1971.

H. Fleisch, H. J. J. Blackwood, and M. Owen (ed.), *Calcified Tissue*, Springer-Verlag, Berlin, 1966.

K. H. Gustavson, *The Chemistry of Reactivity of Collagen*, Academic Press, New York, 1956.

D. A. Hall, *The Chemistry of Connective Tissue*, Thomas, Springfield, Ill., 1961.

R. M. Kenedi (ed.), *Perspectives in Biomedical Engineering*, University Park Press, Baltimore, 1973.

H. Kraus, On the mechanical properties and behavior of human compact bone, *Adv. Biomed. Eng. Med. Phys.* 2, 169–204, 1968.

J. B. Park, C. H. Daly, and A. S. Hoffman, The contribution of collagen to the mechanical response of canine artery at low strains, in: *Frontiers of Matrix Biology*, Volume 3, A. M. Robert and L. Robert (ed.), pp. 218–233, Karger, Basel, 1976.

G. N. Ramachandran (ed.), *Treatise on Collagen*: Volume 2A, Chapter 6, "Mechanical Properties of Collagenous Tissues," by R. D. Harkness; Volume 2B, Chapter 3, "Organization and Structure of Bone," by M. J. Glimcher and S. M. Krane; Volume 1, Chapter 1, "Composition of Collagen and Allied Proteins," by J. E. Eastoe, Academic Press, New York, 1967, 1968.

J. W. Remington (ed.), *Tissue Elasticity*, American Physiological Society, Washington, D.C., 1957.

A. Viidik, Functional properties of collagenous tissues, *Int. Rev. Connect. Tissue Res.* 6, 127–215, 1973.

I. Zipkin (ed.), *Biological Mineralization*, Wiley, New York, 1973.

TISSUE RESPONSE TO IMPLANTS (BIOCOMPATIBILITY)

In order to implant a material, we first have to injure the tissue. The injured or diseased tissues will be removed to some extent in the process of implantation. The success of the entire operation depends on the kind and degree of tissue response (*biocompatibility*) toward the implants during the healing process. The tissue response toward the injury may vary widely according to the site, species, contamination, etc. However, the inflammatory reaction and cellular response toward the wound for both intentional and accidental injuries are the same regardless of the site.

7.1. WOUND-HEALING PROCESS

7.1.1. Inflammation

Whenever tissues are injured or destroyed, the adjacent cells respond to repair them. An immediate response to any injury is the inflammatory reaction. Soon after constriction of capillaries (which stops blood leakage), dilation occurs. At the same time there is a great increase of activities in the endothelial cells lining the capillaries. The capillaries will be covered by adjacent leukocytes, erythrocytes, and platelets. Concurrently with vasodilation, leakage of plasma from capillaries occurs. This fluid combined with the migrating leukocytes and dead tissues will constitute exudate. Once enough cells (see Table 7-1 for definitions of cells) are accumulated by lysis, the

Table 7-1. Definitions of Cells

	Definition
Chondroblast	An immature collagen (cartilage)-producing cell
Endothelial cell	A cell lining the cavities of the heart and the blood and lymph vessels
Erythrocyte	A formed element of blood containing hemoglobin (red blood cell)
Fibroblast	A common fixed cell of connective tissue that elaborates the precursors of the extracellular fibrous and amorphous components
Giant cell	
Foreign body giant cell	A large cell derived from a macrophage in the presence of a foreign body
Multinucleated giant cell	A large cell having many nuclei
Granulocyte	Any blood cell containing specific granules; included are neutrophils, basophils, and eosinophils
Leukocyte	A colorless blood corpuscle capable of ameboid movement; protects body from microorganisms and can be one of five types: lymphocytes, monocytes, neutrophils, eosinophils, and basophils
Macrophage	Large phagocytic mononuclear cell. Free macrophage is an ameboid phagocyte and present at the site of inflammation
Mesenchymal cell	Undifferentiated cell having similar role as fibroblasts but often smaller and can develop into new cell types by certain stimuli
Mononuclear cell	Any cell having one nucleus
Osteoblast	An immature bone-producing cell
Phagocyte	Any cell that destroys microorganisms or harmful cells
Platelet	A small circular or oval disk-shaped cell (3-μm diameter), precursor of blood clot

exudate becomes pus. It is important to know that pus can sometimes occur in nonbacterial inflammation.

At the time of damage to the capillaries, the local lymphatics are also damaged since the latter are more fragile than the capillaries. However, the leakage of fluids from the capillaries will provide fibrinogen and other *formed elements of the blood clotting system* which will quickly plug the damaged lymphatics, thus localizing the inflammatory reaction.

All of the reactions mentioned above—vasodilation of capillaries, leakage of fluid into the extravascular space, and plugging of lymphatics—will provide the classic inflammatory signs: redness, swelling, and heat which can lead to local pain.

When the tissue injury is extensive or the wound contains either irritants or bacteria, the inflammation may lead to extensive tissue destruction. Collagenase, a proteolytic enzyme capable of digesting collagen,[1] carries out the tissue destruction. The enzyme is released from granulocytes which in turn are lysed by the lower pH at the wound site. Local pH can drop to below 5.2 at the injured site from normal values of 7.4–7.6. If there is no drainage for the necrotic debris, lysed granulocytes, formed blood elements, etc., then the site becomes a severely destructive inflammation resulting in a necrotic abscess.

If the severe inflammation persists without the healing process occurring within 3–5 days, a *chronic* inflammatory process commences. This is marked by the presence of mononuclear cells called macrophages which can coalesce to form multinucleated giant cells (Figure 7-1). The macrophages are phagocytic and remove foreign material or bacteria. Sometimes the mononuclear cells evolve into histiocytes which regenerate collagen. This regenerated collagen is used to unite the wound or wall-off unremovable foreign materials by encapsulation.

In a chronic inflammatory reaction, lymphocytes occur as clumps or foci. These cells are the primary source of immunogenic agents which become active if foreign proteins are not removed by the body's primary defense. An autoimmune reaction is suggested as a foreign body reaction of nonproteinaceous materials like silica.

7.1.2. Cellular Response to Repair

Soon after injury, the mesenchymal cells evolve into migratory fibroblasts which move into the injured site while the necrotic debris, blood clots, etc. are removed by the granulocytes and macrophages. The inflammatory exudate contains fibrinogen which is converted into fibrin by enzymes released through blood and tissue cells (see Section 7.3). The fibrin scaffolds the injured site. The migrating fibroblasts use the fibrin scaffolds as a

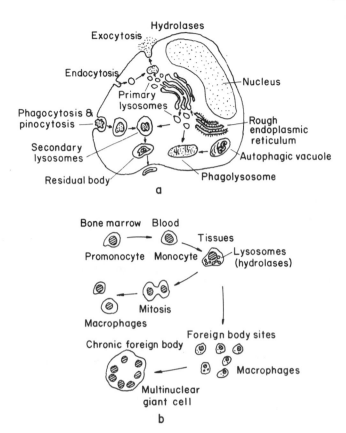

Figure 7-1. (a) The activated macrophage and (b) development of the multinuclear foreign body giant cell. (From Ref. 8.)

framework onto which collagen is deposited. New capillaries are formed following the migration of fibroblasts; then the fibrin scaffolds are removed by the fibrinolytic enzymes activated by the endothelial cells. The endothelial cells together with the fibroblasts liberate collagenase which limits the collagen content of the wound.

After 2–4 weeks of fibroblastic activities the wound undergoes remodeling by decreasing the glycoprotein and polysaccharide content of the scar tissue and lowering the number of synthesizing fibroblasts. A new balance of collagen synthesis and dissolution is reached and the maturation phase of the wound begins. The time required for the wound-healing process varies with various tissues although the basic steps described here can be applied to all connective tissue wound-healing processes.

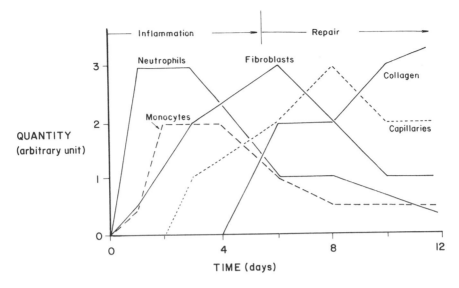

Figure 7-2. Soft tissue wound-healing sequence. (From Ref. 2.)

The healing of soft tissues, especially that of the skin wound, has been studied intensively since this is germane to all surgery. The degree of healing can be determined by histochemical or physical parameters. A combined method will give a better understanding of the overall healing process. Figure 7-2 is a schematic diagram of sequential events of the cellular response of soft tissues. The tensile strength of the wound is not proportional to the amount of collagen deposited in the injured site as shown in Figure 7-3. This indicates that there is a latent period for the collagen molecules (procollagen is deposited by fibroblasts) to polymerize. It may take additional time to align the fibers in the direction of stress and cross-link fibrils in order to increase the physical strength closer to a normal level. This collagen restructuring process requires more than 6 months to complete although the wound strength never reaches the original value. The wound strength can be affected by many variables, that is, nutrition, temperature, presence of other wounds, and oxygen tension. Other factors such as drugs, hormones, irradiation, and electrical stimulation all affect the normal wound-healing process.

The healing of bone fractures is regenerative rather than simple repair as seen in other tissues except liver. However, the extent of regeneration is limited in humans. The cellular events following fracture of bone are illustrated in Figure 7-4. When a bone is fractured, many blood vessels (including the adjacent soft tissues) hemorrhage and form a blood clot

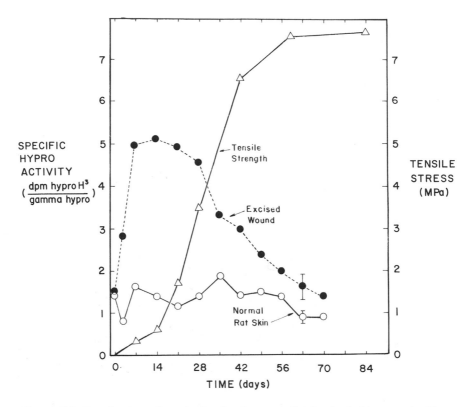

Figure 7-3. Tensile strength and rate of collagen synthesis of rat skin wounds. (From Ref. 3.)

around the fracture site. As in any wound repair, shortly after fracture the fibroblasts and osteogenic cells in the outer layer of the periosteum migrate and proliferate toward the injured site. These cells lay down a fibrous collagen matrix for callus formation. Osteoblasts evolve from the osteogenic cells near the bone surface to calcify the callus into trabeculae, forming a spongy bone. The osteogenic cells which migrate further away from an established blood supply become chondroblasts which lay down cartilage. Thus, after about 2–4 weeks the periosteal callus is made up of three parts as shown in Figure 7-5.

Simultaneously with the external callus formation a similar repair process occurs in the marrow cavity. Since there is an abundant supply of blood, the cavity turns into callus rather quickly and becomes fibrous or spongy bone.

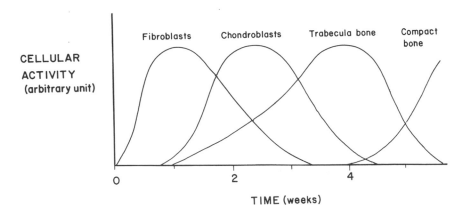

Figure 7-4. Sequence of events following bone fracture. (From Ref. 4.)

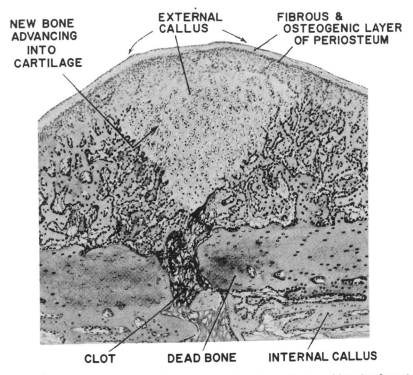

Figure 7-5. A drawing of a longitudinal section of fractured rib of a rabbit after 2 weeks, H & E stain. (From Ref. 5.)

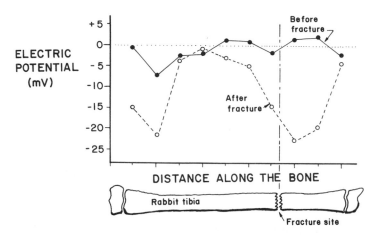

Figure 7-6. Skin surface of rabbit limb before and after fracture. Note that the fracture site has increased electronegative potential. (From Ref. 6.)

New trabeculae develop in the fracture site by appositional growth and the spongy bone turns into compact bone. This maturation process begins after about 4 weeks.

Some other interesting observations have been made on the healing of bone fractures in relation to the synthesis of polysaccharides on collagen. It is believed that the amount of collagen and polysaccharides is closely related to the cellular events following fracture. When the amount of collagen starts to increase, it marks the onset of the remodeling process. This occurs after about 1 week. Another interesting observation is the electrical potential (or biopotential) measured in the long bone before and after fracture as shown in Figure 7-6. The high electronegativity in the vicinity of the fracture marks an increased cellular activity in the tissues. Thus, there is a maximum negative potential in the epiphysis in the normal bone since this zone is the site of greatest activity (the growth plate is in the epiphysis).

7.2. BODY RESPONSE TO IMPLANTS

The response of the body implants varies widely according to the host site and species, the degree of trauma imposed during implantation, and all the variables associated with a normal wound-healing process. In addition, the implant's chemical composition and micro- and macrostructure also induce different bodily responses. The response has been studied in two different areas, that is, local (cellular) and systemic, although a single implant should be tested for both aspects. In practice, this has not been done simultaneously except in a few cases such as bone cement.

7.2.1. Cellular Response to Implants

Generally, the body's reaction to foreign materials is to expel them. The foreign material could either be extruded from the body if it can be removed or walled-off if it cannot be removed. If the material is a particulate or fluid, then it will be ingested by the giant cells (macrophages) and removed.

These responses are related to the healing process of the wound where the implant is present as an additional factor. A typical tissue response is that polymorphonuclear leukocytes appear near the implant followed by macrophages called foreign body giant cells. However, if the implant is chemically and physically inert to the tissue, then the foreign body giant cells may not be present near the implant. Instead, only a thin layer of

Table 7-2. Effect of Implantation on the Properties of Various Suture Materials[a]

Material	Wound tensile strength	Suture tensile strength	Tissue reaction
Absorbable plain catgut	Impaired	Zero by 3–6 days	Very severe
Chromic catgut	Impaired	Variable	Moderate (much less severe than plain catgut, but more than nonabsorbable materials)
Nonabsorbable silk	No effect	Well maintained	Slight
Nylon multifilament	No effect	Very low at 6 months	Moderately severe and prolonged
Monofilament	No effect	Well maintained	Slight
Polyethylene terephthalate	No effect	Well maintained	Very slight
PTFE (Teflon®)	No effect	Well maintained	Almost none

[a] From Ref. 7.

collagenous tissue encapsulates the implant. If the implant is chemically or physically irritating to the surrounding tissue, then the inflammation occurs in the implant site. The inflammation (both acute and chronic type) will delay the normal healing process, resulting in granular tissues. Some implants may cause necrosis of tissues by chemical, mechanical, and thermal trauma.

It is generally very difficult to assess tissue response to various implants due to the wide variations in experimental protocol. This is illustrated in Table 7-2, which describes tissue reactions for various suture materials. The variation of host tissue response with implant reactivity is shown schematically in Figure 7-7.

The degree of tissue response varies according to both the physical and the chemical nature of the implants.[8] Most pure metals evoke a severe tissue reaction. This may be related to the high-energy state or high free energy of pure metals, which tend to lower their free energy by oxidation or corrosion. Titanium and tantalum cause a minimal amount of tissue reaction due to the tenacious oxide layer which resists further diffusion of metal ions and gas (O_2) at the interface. In fact, this oxide layer makes them ceramiclike materials, which are very inert. Corrosion-resistant metal alloys such as cobalt–chromium, 316L stainless steel, and Ti-based alloys have similar effects on tissues.

Most ceramic materials investigated for their tissue compatibility have been oxides such as TiO_2, Al_2O_3, ZrO_2, $BaTiO_3$, and multiphase ceramics

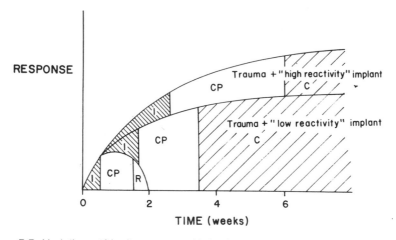

Figure 7-7. Variations of host response with implant "reactivity." I, insult; CP, cellular proliferation; R, reorganization; C, chronic host response. (From Ref. 8.)

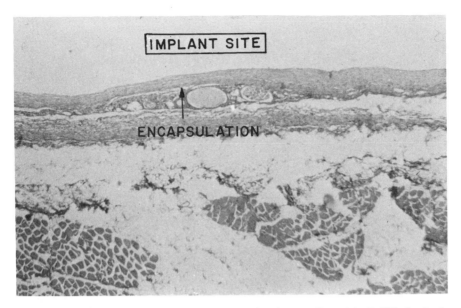

Figure 7-8. Optical micrograph of soft connective tissue adjacent to $BaTiO_3$ implants after 20 weeks (H & E stain). Arrow indicates encapsulation of dense collagenous tissue. 400×. (Courtesy of G. H. Kenner and J. B. Park.)

of $CaO-Al_2O_3$, $CaO-ZrO_2$, and $CaO-TiO_2$. These materials show minimal tissue reactions with only a thin layer of encapsulation as shown in Figure 7-8. Similar reactions are seen for carbon implants.

Some glasses (e.g., 45 w/o SiO_2, 24.5 w/o Na_2O, 24.5 w/o CaO, and 6.0 w/o P_2O_5) display direct bonding between implant and bone by dissolution of the silica-rich gel film at the interface as shown in Figure 7-9.

The polymers per se are inert toward tissues if there are no additives present (e.g., antioxidants, fillers, antidiscoloring agents, plasticizers). On the other hand, the monomers can evoke an adverse tissue reaction due to their reactivity. Thus, the degree of polymerization is somewhat related to the tissue reaction. Since 100% polymerization is almost impossible to achieve, there is present a range of different-size polymer molecules that can be leached out of the polymer. In this way, a particulate form of a very inert polymeric material can cause severe tissue reaction. This was amply demonstrated by polytetrafluoroethylene (Teflon®), which is quite inert in bulk form like rods or woven fabric but very reactive in tissues when made into

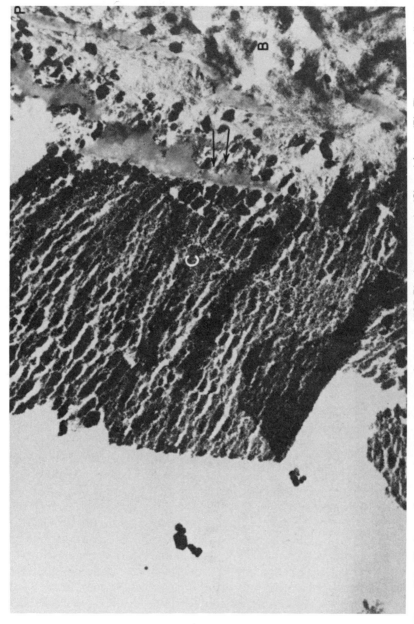

Figure 7-9. Electron micrograph of the interface between 45S5 Bioglass–ceramic (C) and bone (B). The arrows indicate region of gel formation, undecalcified. 10,000×. (From Ref. 9.)

IMPLANT : TISSUE

MINIMAL RESPONSE

Thin Layer of Fibrous Tissue

Silicone rubber, Polyolefins, PTFE (Teflon)
PMMA, most ceramics, Ti- & Co-based alloys

CHEMICALLY INDUCED RESPONSE

Acute, Mild Inflammatory Response

Absorbable sutures, Some thermosetting resins

Chronic, Severe Inflammatory Response

Degradable materials, Thermoplastics with
toxic additives, Corrosion metal particles

PHYSICALLY INDUCED RESPONSE

Inflammatory Response to Particulates

PTFE, PMMA, Nylon, Metals

Tissue Growth into Porous Materials

Polymers, Ceramics, Metals, Composites

NECROTIC RESPONSE

Layer of Necrotic Debris

Bone cement, Surgical adhesives

Figure 7-10. A brief summary of the tissue response to implants. (From Ref. 10.)

powder form. A schematic summary of tissue responses to implants is given in Figure 7-10.

There has been some concern about the possibility of tumor formation caused by the wide range of materials used in implantation. Although many implant materials are carcinogenic in rats, there is not a single documented case of a human tumor directly related to an implant. It may be too premature to pass final judgement since the latency for tumor formation in humans may be longer than 20 to 30 years. However, considering the number of surgical implants performed, the absence of any direct evidence of carcinogenesis tends to support the proposition that carcinogenesis is species-specific and that implant-related tumors may not form in man.

7.2.2. Systemic Effects of Implants

The systemic effect caused by implants has been well documented in hip joint replacement surgery. The polymethylmethacrylate bone cement (dough state) applied in the femoral shaft has been shown to lower the blood pressure significantly.[11] There is also concern about the systemic effect of biodegradable implants such as absorbable sutures and surgical adhesives, and the large number of wear and corrosion particles released by the metallic implants.

Table 7-3 indicates that the various organs have different affinities for different metallic elements. It is clear that the corrosion-resistant metal alloys are not completely stable structurally and some elements are released in the body. This has caused concern that the elevated ion levels in various organs may interfere with normal physiological functions.[8] The divalent metal ions may also inhibit activities of various biochemical substances such as enzymes and hormones.

Polymeric materials contain additives which cause cellular as well as systemic reactions to a greater degree than the pure polymer itself.[13] Even the well-accepted polymer dimethylsiloxane (Silastic® rubber) contains a filler, silica powder, to enhance the mechanical properties. Although the silica powder itself is reactive when implanted in a confined area, no problems have been apparent with its use. However, it is not certain whether a late complication may occur if a large amount of the silica is released into the tissue and retained in various organs.

7.3. BLOOD COMPATIBILITY

The single most important requirement for the blood interfacing implants is blood compatibility. Although blood coagulation is the most important factor for blood compatibility, the implants should not damage

Table 7-3. Concentrations of Metals in Tissue and Organs of the Rabbit after Implantation of Various Metals[a,b]

	Surrounding muscle		Liver		Kidney		Spleen		Lung		Control muscle	
	6 wk	16 wk	6 wk	16 wk	6 wk	16 wk	6 wk	16 wk	6 wk	16 wk	6 wk	16 wk
	Vitallium® (61.9% Co, 28–34% Cr, 4.73% Mo, 1.52% Ni, 0.61% Fe) (2 or 3 specimens per experiment)											
Chromium	30,25,0	0,0,15	0,5,5	5,5,10	0,5,5	5,5,10	0,0,5	5,5,10	0,0,10	5,5,10	0,5,5	0,10,0
Cobalt	25,45	0,30,0	5,10	5,5,10	0,0,105	5,10,70	5,5,20	5,5,300	0,0,300	10,5,10	0,0,50	0,10,0
Nickel	20,35	10	5,10	5,5,80	5,110	5,5,30	5,205	5,5,100	5,40	5,0,45	5,5	5
Titanium	5,5,10	0,10,0	0,0,5	0,0,10	0,0,10	0,0,10	0,0,5	0,0,15	0,0,5	0,0,15	0,0,10	0,0,5
Molybdenum	0,5,5	0,5,0	0,15,80	10,10,70	0,20,50	20,20,75	0,0,10	0,0,5	0,0,10	5,5,5	0,0,0	0,0,0
Iron	0,40,90	0,80,0	0,200,600	100,160,80	0,90,180	110,120,90	0,250,270	300,290,110	0,90,420	110,100,50	0,10,30	0,20,0
	316 stainless steel (17.8% Cr, 13.4% Ni, 2.3% Mo, 0.23% Cu, 66.27% Fe)											
Chromium	145,65	115,295	5,10	5,5	5,5	5,5	5,5	5,20	5,5	20,10	0,20	5,5
Cobalt	5,5	0,10	5,20	0,15	5,10	5,115	5,5	15,600	5,5	0,250	0,0	0,90
Nickel	15,50	200,70	20,20	0,95	0,10	10,10	5,10	1000,65	20,220	20,45	10,0	5,5
Titanium	20,10	25,10	0,10	5,5	0,5	5,65	0,5	10,50	0,10	5,5	5,10	10,10
Molybdenum	5,10	10,35	10,75	50,65	10,85	75,80	5,10	15,150	0,10	5,5	0,0	0,0
Iron	80,70	90,190	220,590	500,520	110,210	180,200	180,420	580,220	120,220	150,300	30,10	15,10

[a] From Ref. 12.
[b] Figures are in ppm dry ash, given to nearest 5 ppm.

BLOOD CLOTTING SYSTEM

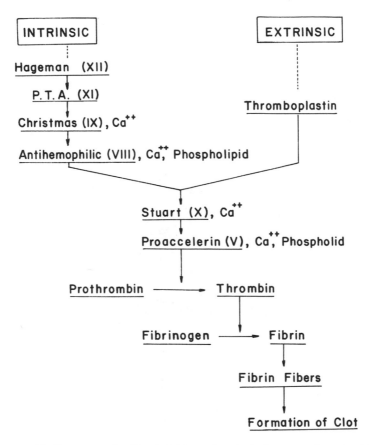

Figure 7-11. Two routes for blood clot formation (note the cascading sequence).

proteins, enzymes, or formed elements of blood (red blood cells, white blood cells, and platelets). The latter includes hemolysis (red blood cell rupture) and initiation of the platelet release.

Coagulated blood is called a clot. A clot that has formed inside blood vessels is referred to as a *thrombus* or an *embolus* depending on whether the clot is fixed or floating, respectively.

The mechanism and route of blood coagulation are not completely understood. A simplified version of blood clotting is proposed as a cascading sequence as shown in Figure 7-11. As discussed in Section 7.1.1, immediately after injury the blood vessels constrict to minimize the flow of blood. Platelets adhere to the vessel walls by coming into contact with the

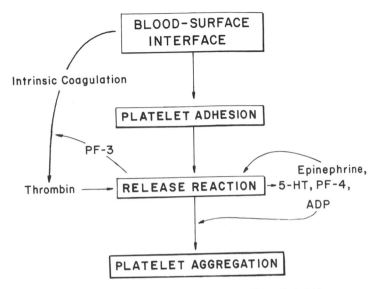

Figure 7-12. Platelet–surface interaction. (From Ref. 14.)

exposed collagen. The aggregation of platelets is achieved through release of adenosine diphosphate (ADP) from damaged red blood cells, vessel walls, and adherent platelets. A more detailed platelet–surface interaction is shown in Figure 7-12.

7.3.1. Factors Affecting Blood Compatibility

The surface roughness is an important factor affecting blood compatibility since the rougher the surface the more area is exposed to the blood. Therefore, rough surfaces promote faster blood coagulation than the highly polished surfaces of glass, polymethylmethacrylate, polyethylene, and stainless steel. Sometimes, thrombogenic materials with rough surfaces are used to promote clotting in porous interfaces to prevent initial leaking of blood and later tissue ingrowth through the pores of vascular implants.

The surface wettability, that is, its hydrophilic (wettable) or hydrohobic (nonwettable) character, was initially thought to be an important factor. However, the wettability parameter, contact angle with liquids, does not correlate consistently with blood clotting time.[15]

The surface of the intima (innermost layer contacting blood) of a blood vessel is negatively charged (1–5 mV) with respect to the adventitia (outermost layer). This phenomenon is partially attributed to the nonthrombogenic (or thromboresistant) character of the intima since the formed elements of blood are also negatively charged and hence are repelled from the surface of the intima. This was demonstrated using a solid copper tube (a thrombogenic material) implanted as an arterial replacement.[16] When the tube was negatively charged, clot formation was delayed compared with the control. In relation with this phenomenon, the streaming potential or zeta potential has been investigated since the formed elements of blood are flowing ionic particles *in vivo*.[17] However, it was not possible to establish a one-to-one direct relationship between clotting time and zeta potential.

The chemical nature of the material surface interfacing with blood is closely related to the electrical nature of the surface since the type of functional groups of the polymer (no intrinsic surface charge exists for metals and ceramics although some ceramics and polymers can be made piezoelectric) determines the type and magnitude of the surface charge. The surface of the intima is negatively charged largely due to the presence of mucopolysaccharides, especially chondroitin sulfate and heparin sulfate.

7.3.2. Nonthrombogenic Surfaces

There have been many attempts to identify nonthrombogenic materials. In these studies empirical approaches have often been used. These can be categorized as (1) heparinized or biologic surfaces, (2) surfaces with anionic

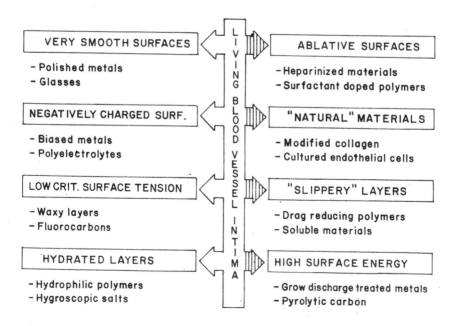

Figure 7-13. Approaches for producing thromboresistant surfaces. (From Ref. 18.)

radicals for negative electric charges, (3) inert surfaces, and (4) solution-perfused surfaces. Early attempts to obtain nonthrombogenic surfaces are shown in Figure 7-13.

Heparin is a polysaccharide with negative charges due to the sulfate groups as shown:

$$(7\text{-}1)$$

Initially, the heparin was attached to a graphite surface that had been treated with the quaternary salt benzalkonium chloride (GBH process). Later, a simpler heparinization was accomplished by exposing the polymer surface to a quaternary salt such as tridodecylmethylammonium chloride (TDMAC). This method was further simplified by making a TDMAC and heparin solution in which the implant can be immersed following drying.[12]

The heparinized materials showed a significant increase in thromboresistance compared with untreated controls. In an interesting application, a polyester fabric graft was heparinized. This reduced the tendency of initial bleeding through the fabric and a thin neointima was later formed. Many polymers were tested for heparinization including polyethylene and silicone rubber. Leaching of heparin into the medium is a drawback although some improvement was seen by cross-linking heparin with glutaraldehyde and covalently bonding it directly onto the surface.

Another approach involved coating the cardiovascular implant surface with other biological molecules such as albumin, gelatin (denatured collagen), and heparin. Some studies reported that albumin alone can be thromboresistant and decrease the platelet adhesion.

Negatively charged surfaces with anionic radicals (acrylic acid derivatives) have been produced by copolymerization or grafting.[19] Negatively charged electrets (polarized molecules) on the surface of a polymer (e.g., Teflon) were shown to enhance thromboresistance.

Hydrogels of both polyhydroxyethylmethacrylate (polyHEMA) and acrylamide are classified as inert materials since they do not contain highly negative anionic radical groups and are not negatively charged. These coatings tend to be washed away when exposed to the bloodstream as was also seen with the heparin coatings. Segmental or block polyurethanes were shown to be somewhat thromboresistant without surface modification.

Another method of making surfaces nonthrombogenic is perfusion of water (solution) through interstices of fabric which is interfaced with blood.[20] This approach has the advantage of minimizing damage to formed elements since no material will directly contact blood. The main disadvantage is the dilution of blood plasma although this is not a serious problem. Saline solution is often deliberately injected into the body, e.g., for fluid replacement during surgery.

PROBLEMS

7-1. What makes it so difficult to evaluate the tissue and blood compatibility of implants?

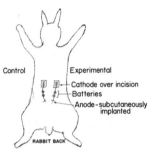

Figure P7-2.

7-2. In a study of the wound-healing process of the skin, electrical stimulation was used to accelerate the rabbit skin wound as shown in Figure P7-2. The mean current flow was 21 μA and the mean current density, 8.3 μA/cm^2. After 7 days, the load to fracture on the control samples was 797 g and on the stimulated experimental side, 1224 g on the average (J. J. Konikoff, *Ann. Biomed. Eng.* 4, 1, 1976).

a. Calculate the percent increase of strength by stimulation.
b. The width of the testing sample was 1.6 cm. Assuming a 1.8-mm thickness of the skin, calculate the tensile stress for both control and experimental samples.
c. Compared with the strength of normal skin (about 8 MPa) what percentages of the control and experimental skin wound strengths were recovered?
d. Compare the results of Part C with the results of Figure 7-2.
e. Calculate the corrosion rate in μm/year if a platinum electrode was used.

7-3. A biodegradable suture will have a strength 1/10th of the original after 3 weeks of implantation. If the strength of the implanted suture decreases according to $\sigma = \sigma_0 + b \ln t$, where $\sigma = 10$ MPa, $b = -2$ MPa/week, and $t =$ time in weeks, determine how long it was implanted.

7-4. In circulation of the blood, the formed elements are being destroyed by the blood pump and tube wall contacts. A bioengineer measured the rate of hemolysis (red blood cell lysis) as 0.1 g/100 liters pumped.

a. If the normal cardiac output for a dog is 0.1 liter/kg per min, what is the hemolysis rate?
b. If the animal weighs 20 kg and the critical amount of hemolysis is 0.1 g/kg body wt, how long can the bioengineer circulate the blood before reaching critical condition? Assume a negligible amount of new blood formation.

7-5. Some materials do not normally induce tissue reaction as a bulk. However, when implanted in powder form, they are no longer biocompatible. Explain why.

7-6. Sometimes the degree of tissue reaction toward an implant is represented by the thickness of the collagenous capsule (e.g., Figure 7-8).

a. State the fallacy these experimental results may have in deciding biocompatibility.
b. Is a thicker encapsulation better or worse for an implant?

7-7. Calculate the average concentration of metals when Vitallium® is implanted (Table 8-2) in surrounding muscle and kidney at 6 and 16 weeks. Normalize the averages according to their weight percentage in the alloy.

7-8. Explain why metals are generally less biocompatible than ceramics or polymers. What can be done to improve this disadvantage of metals as implant materials?

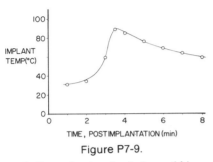

Figure P7-9.

7-9. The temperature changes due to the heat of polymerization of bone cement (polymethylmethacrylate–polystyrene copolymer powder plus methylmethacrylate monomer liquid) (placed in the canine femur as a 9-mm-diameter plug) were monitored at the interface between bone and cement as shown in Figure P7-9.

a. What will happen to the adjacent tissues by the heat generated by the polymerization?
b. Would the temperature increase or decrease by putting a metal cylinder in the middle similar to the situation in femoral hip replacement?
c. What problem will occur if the cement shrinks when it reaches ambient temperature?

7-10. Describe the major differences between normal wound healing and the tissue responses to "inert" and "irritant" materials. What factors besides the choice of material can affect the local tissue response to an implant?

REFERENCES

1. T. Gillman, On some aspects of collagen formation in localized repair and in diffuse fibrotic reactions to injury, in: *Treatise on Collagen*, B. S. Gould (ed.), Volume 2B, Chapter 4, Academic Press, New York, 1968.
2. R. Ross, Wound healing, *Sci. Am. 220*, 40–50, 1969.
3. E. E. Peacock, Jr., and W. Van Winkle, Jr., *Surgery and Biology of Wound Repair*, Chapters 1, 5, 6, 11, Saunders, Philadelphia, 1970.
4. L. L. Hench and E. C. Erthridge, Biomaterials—The interfacial problem, *Adv. Biomed. Eng. 5*, 35–150, 1975.
5. A. H. Ham and W. R. Harris, Repair and transplantation of bone, in: *The Biochemistry and Physiology of Bone*, G. Bourne (ed.), 2nd ed., Volume 3, pp. 337–399, Academic Press, New York, 1971.
6.. Z. B. Friedenberg and C. T. Brighton, Bioelectric potentials in bone, *J. Bone Jt. Surg. 48A*, 915–923, 1966.
7. J. K. Newcombe, Wound healing, in: *Scientific Basis of Surgery*, 2nd ed., W. T. Irvin (ed.), Churchill, London, 1972.
8. J. Black, *Biological Performance of Materials*, Dekker, New York, 1981.
9. L. L. Hench and H. A. Paschall, Histochemical responses at a biomaterials interface, *J. Biomed. Mater. Res. Symp. 5*, 49–64, 1974.
10. D. F. Williams and R. Roaf, *Implants in Surgery*, p. 233, Saunders, Philadelphia, 1973.

11. J. Charnley, *Acrylic Cement in Orthopaedic Surgery*, Churchill/Livingstone, Edinburgh, 1970.
12. A. B. Ferguson, Y. Akahoshi, P. G. Laing, and E. S. Hodge, Characteristics of trace ions released from embedded metal implants in the rabbit, *J. Bone Jt. Surg. 44A*, 323–336, 1962.
13. S. D. Bruck, *Blood Compatible Synthetic Polymers: An Introduction*, Thomas, Springfield, Ill., 1974.
14. E. W. Salzman, Surface effects in homeostasis and thrombosis, in: *The Chemistry of Biosurfaces*, M. L. Hair (ed.), Dekker, New York, 1971.
15. E. W. Salzman, Nonthrombogenic surfaces: Critical review, *Blood 38*, 509–523, 1971.
16. P. N. Sawyer and S. Srinivasan, The role of electrochemical surface properties in thrombosis at vascular interfaces: Cumulative experience of studies in animals and man, *Bull. N.Y. Acad. Med. 48*, 235–256, 1972.
17. B. C. Taylor, W. V. Sharp, J. I. Wright, K. L. Ewing, and C. L. Wilson, The importance of zeta potential, ultrastructure, and electrical conductivity to the *in vivo* performance of polyurethane-carbon black vascular prostheses, *Trans. Am. Soc. Artif. Intern. Organs 17*, 22–28, 1971.
18. R. E. Baier, The role of surface energy in thrombosis, *Bull. N.Y. Acad. Med. 48*, 257–272, 1972.
19. A. S. Hoffman, Applications of radiation processing in biomedical engineering—A review of the preparation and properties of novel biomaterials, *Radiat. Phys. Chem. 9*, 207–219, 1977.
20. D. R. Clarke and J. B. Park, Prevention of erythrocyte adhesion onto porous surfaces by fluid perfusion, *Biomaterials 2*, 9–13, 1981.

BIBLIOGRAPHY

C. O. Bechtol, A. B. Ferguson, and P. G. Laing, *Metals and Engineering in Bone and Joint Surgery*, Ballière, Tindall & Cox, London, 1959.
G. H. Bourne (ed.), *The Biochemistry and Physiology of Bone*, 2nd ed., Volume 3, Chapter 10, Academic Press, New York, 1971.
S. D. Bruck, *Blood Compatible Synthetic Polymers*, Thomas, Springfield, Ill., 1974.
J. Charnley, *Acrylic Cement in Orthopaedic Surgery*, Churchill/Livingstone, Edinburgh, 1970.
L. L. Hench and E. C. Erthridge, Biomaterials—The interfacial problem, *Adv. Biomed. Eng. 5*, 35–150, 1975.
S. F. Hulbert, S. N. Levine, and D. D. Moyle, *Prosthesis and Tissue: The Interfacial Problems*, Wiley, New York, 1974.
S. N. Levine (ed.), *Materials in Biomedical Engineering*, *Ann. N.Y. Acad. Sci. 146*, 1968.
H. I. Maibach and D. T. Rovee (ed.), *Epidermal Wound Healing*, Year Book Medical Publishers, Chicago, 1972.
M. R. Urist (ed.), *Fundamental and Clinical Bone Physiology*, Lippincott, Philadelphia, 1980.
L. Vroman, *Blood*, American Museum Science Books, B26, Doubleday, New York, 1971.

METALLIC IMPLANT MATERIALS

As mentioned in Chapter 1, metals have been used in various forms as implants. The first metal developed specifically for implant use was the "Sherman vanadium steel" from which fracture plates and screws were made.[1] Most metals used for manufacturing implants such as Fe, Cr, Co, Ni, Ti, Ta, Mo, and W can be tolerated by the body in minute amounts and sometimes are essential in red blood cell function (Fe) or synthesis of vitamin B_{12} (Co) but cannot be tolerated in large amounts.[2] The *biocompatibility* of the implant metals is of considerable concern because they can be corroded in the hostile environment of the body. As a consequence of corrosion, the material itself is wearing away and thus weakening the implant, and more importantly the corrosion products are released into the surrounding tissues, resulting in undesirable effects.

In this chapter we will study the metals presently used for implants: stainless steels, Co-based alloys, Ti and its alloys, and other metals.

8.1. STAINLESS STEELS

The first stainless steel used for implantation was the 18% Cr–8% Ni steel alloy (18-8 or type 302 in modern classification), which is stronger than the vanadium steel used to fabricate fracture plates and more resistant to corrosion.[3] Later, 18-8sMo stainless steel was introduced which contains molybdenum to improve the corrosion resistance in saltwater. This alloy became known as type 316 stainless steel. In the 1950s the maximum carbon content of 316 stainless steel was reduced from 0.08 w/o to 0.03 w/o for better corrosion resistance and became known as 316L.[4]

8.1.1. Types and Composition of Stainless Steels[5,6]

The chromium content of stainless steels should be at least 11 w/o to resist corrosion. Chromium is a reactive element but its alloys and itself can be *passivated* to give excellent corrosion resistance. There are many types of stainless steels which are grouped into four major classes according to their microstructures (Table 8-1). The AISI (American Institute of Steel and Iron) numbers in Table 8-1 are for wrought alloys.

The *martensitic chromium steels* (group I) can be hardened by heat treatment as in the martensitic carbon steels. These steels are not as corrosion resistant as other stainless steels but are often used to make surgical instruments (type 420) because of their hardenability.

Table 8-1. Types of Wrought Stainless Steels and Their Chemical Compositions[a]

AISI type	% C	% Cr	% Ni	% other elements	Remarks
			Group I: martensitic chromium steels		
410	0.15 max.	11.5–13.5	—	—	Turbine blades, valve trim
420	0.35–0.45	12–14	—	—	Cutlery, surgical instrument
431	0.2 max.	15–17	1.25–2.5	—	Improved ductility
440A	0.60–0.75	16–18	—	—	Very hard; cutters
			Group II: ferritic nonhardenable steels		
430	0.12 max.	14–18	0.5 max.	—	Auto trim, tableware
446	0.20 max.	23–27	0.5 max.	0.25 N max.	Resists O and S at high temperatures
			Group III: austenitic chromium–nickel steels		
301	0.15 max.	16–18	6–8	2 Mn max.	Strain-hardens
304	0.08 max.	18–20	8–12	1 Si max.	Continuous 18-8s
304L	0.03 max.	18–20	8–12	1 Si max.	Very low carbon
310	0.25 max.	24–26	19–22	1.5 Si max.	25-20, heat resistance
310X	0.08 max.	24–26	19–22	1.5 Si max.	Lower carbon
314	0.25 max.	23–26	19–22	1.5–3.0	Si for high-temp. oxidation
316	0.10 max.	16–18	10–14	2–3 Mo	18-8sMo, surgical implants
√316L	0.03 max.	16–18	10–14	2–3 Mo	Very low carbon, surgical implants
317	0.08 max.	18–20	11–14	3–4 Mo	Higher Mo
321	0.08 max.	17–19	8–11	Ti 4 × C (min)	Ti stabilized
			Group IV: age-hardenable steels		
322	0.07	17	7	0.07 Ti, 0.2 Al	

[a] From Ref. 5.

The group II *ferritic nonhardenable steels* have ferrite (αFe) structure since they fall outside the γ loop (austenitic phase) and cannot be hardened as shown in Figure 8-1. This group of stainless steels has good deep drawing characteristics and corrosion resistance, especially *stress corrosion*. The major use of this type of steel is in the chemical and food processing industries as containers.

The group III *austenitic stainless steels* are most widely used for implants especially type 316 and 316L. These are not hardenable by heat treatment but can be hardened by cold-working. This group of stainless steels is nonmagnetic and possesses better corrosion resistance than any of the others. The inclusion of molybdenum in types 316 and 317 enhances resistance to *pitting corrosion*. It was discovered that lowering the carbon content of type 316 stainless steels makes them more corrosion resistant to chloride solution such as physiological saline in the human body. Therefore, the ASTM (American Society of Testing and Materials) recommends type 316L rather than 316 for implant fabrication. The specifications of stainless steels for implant use are much stricter than those for industrial use as given in Table 8-2.

Nickel serves to stabilize the austenitic phase at room temperature. The austenitic phase stability at room temperature can be influenced by both Ni and Cr content as shown in Figure 8-2 for 0.10 w/o C stainless steels.[8]

The group IV stainless steels are age-hardenable at about 450 to 550°C. Their corrosion resistance is considered less than that of the 18-8 type.

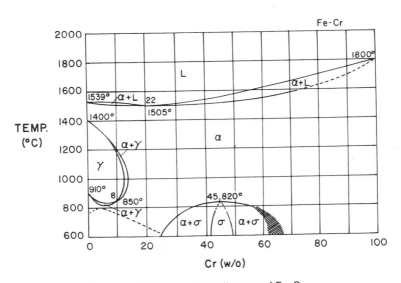

Figure 8-1. Binary phase diagram of Fe–Cr.

Table 8-2. Compositions of 316 (Grade 1)
and 316L (Grade 2) Stainless Steels[a]

| | Composition (w/o) | |
	Grade 1	Grade 2
Carbon	0.08 max.	0.030 max.
Manganese	2.00 max.	2.00 max.
Phosphorus[b]	0.030 max.	0.030 max.
Sulfur	0.030 max.	0.030 max.
Silicon	0.75 max.	0.75 max.
Chromium	17.00–20.00	17.00–20.00
Nickel	12.00–14.00	12.00–14.00
Molybdenum	2.00–4.00	2.00–4.00

[a] From Ref. 7.
[b] Slight variations are given (0.025 max.) for special
quality stainless steels (F138 and F139 of ASTM).

8.1.2. Properties of Stainless Steels

Table 8-3 gives the mechanical properties of stainless steels. As can be noted, a wide range of properties exists and one must be careful when selecting for its final use. The particular requirements for 316 and 316L stainless steels for implants specified by the ASTM are given in Table 8-4.

Even 316L stainless steels corrode inside the body.[9–11] Their use is, however, quite acceptable in temporary devices such as fracture plates, screws, and hip nails. Corrosion of stainless steels appears to occur via one or more of the following[12]:

1. Incorrect composition or metallurgical conditions. Although the addition of molybdenum increases the resistance of stainless steels to the

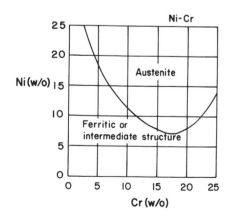

Figure 8-2. The effect of Ni and Cr contents on the austenitic phase of stainless steels containing 0.1 w/o C. (From Ref. 8.)

Table 8-3. Nominal Mechanical Properties of Stainless Steels[a,b]

AISI type	Condition	Tensile strength (lb/in.²)	Yield point (lb/in.², 0.2% offset)	Elongation (% in 2 in.)	Hardness Rockwell	Brinell
Type 410	Annealed	75,000	40,000	30	B82	155
	Hardened, tempered at 600°F	180,000	140,000	15	C39	375
	Hardened, tempered at 1000°F	145,000	115,000	20	C31	300
Type 420	Annealed	95,000	50,000	25	B92	195
	Hardened, tempered at 600°F	230,000	195,000	25	C50	500
Type 440A	Annealed	105,000	60,000	20	B95	215
	Hardened, tempered at 600°F	260,000	240,000	5	C51	510
Type 430	Annealed	75,000	45,000	30	B82	155
Type 446	Annealed	80,000	50,000	23	B86	170
Type 301	Annealed	110,000	40,000	60	B85	165
	Cold-worked, 1/2 hard	150,000	110,000	15	C32	320
Type 304	Annealed	85,000	35,000	55	B80	150
Type 304L	Annealed	80,000	30,000	55	B76	140
Type 310	Annealed	95,000	40,000	45	B87	170

[a] From Ref. 5.
[b] 1 psi = 6895 N/m² (Pa).

Table 8-4. Mechanical Properties of Stainless Steel Surgical Implants[a, b]

Condition	Ultimate tensile strength, min., psi (MPa)	Yield strength (0.2% offset), min., psi (MPa)	Elongation in 2 in. (50.8 mm), min., %	Rockwell hardness, max.
	Grade 1 (type 316)			
Annealed	75,000 (515)	30,000 (205)	40	95 HRB
Cold-finished	90,000 (620)	45,000 (310)	35	—
Cold-worked	125,000 (860)	100,000 (690)	12	—
	Grade 2 (type 316L)			
Annealed	73,000 (505)	28,000 (195)	40	95 HRB
Cold-finished	88,000 (605)	43,000 (295)	35	—
Cold-worked	125,000 (860)	100,000 (690)	12	—

[a] From Ref. 7.
[b] 1 psi = 6895 N/m^2 (Pa).

saline solution, too much of it can result in formation of the σ phase which may embrittle the alloy. The σ phase is a hard, brittle intermetallic compound formed at prolonged time above 480°C. The intergranular formation of chromium carbide (Cr_6C_{23}), which occurs in the 430–870°C range, can cause a deficiency of chromium next to the grain boundaries. This process is known as sensitization and may result in grain boundary corrosion. One solution for both of these problems is to use 316L stainless steel, which contains less carbon (0.03 w/o maximum). Another solution is treatment with heat: since carbide formation takes place at 600 to 950°C, elevating the temperature above 950°C and then quenching it will avoid carbide formation through the diffusion of carbons. This procedure should be the final step in any treatment of implant components.

2. Improper selection and handling of implant. This can arise by the intermixing of components from the variety of implants available. For example, a recent catalog of one company lists 11 types of bone plates, 6 types of intramedullary pins, and 25 types of (multicomponent) hip fracture devices. Other manufacturers also have a large number of products and the possibility of mixing components is large. The problem with intermixing is twofold: (1) the components may not fit together properly, resulting in (crevice) corrosion, and (2) the materials and the manufacturing process may not be identical, resulting in galvanic or grain boundary corrosion. A detailed inventory procedure and careful selection of devices should alleviate most of the problems.

3. Mismatch of components. This is fairly common in multicomponent devices such as screws and plates, nail and plate, etc., and can produce

Table 8-5. Effect of Cold-Working and Heat Treatment on the Mechanical Properties of Some Stainless Steels[a]

	18 chromium 8 nickel fully softened	Extra-low-carbon 18 chromium 10 nickel fully softened	Type 316 fully softened	Type 316 cold-worked
Yield strength (MPa)	200–230	200–250	240–300	700–800
Ultimate tensile strength (MPa)	540–700	540–620	600–700	1000
Ductility (elongation %)	50–65	55–60	35–55	7–10
Young's modulus (Gpa)	200	200	200	200
Hardness (V.P.N.)	175–200	170–200	170–200	300–350
Fatigue limit (Mpa)	230–250	—	260–280	300

[a] From Ref. 12.

fretting, crevice, or simple galvanic corrosion. Galvanic corrosion occurs due to a difference in composition of the two components, although both meet the specifications which are given in *ranges* rather than absolute values and also due to the slightly different metallurgical states because of differing manufacturing processes. For example, bone screws are usually in a somewhat more cold-worked condition than are fracture plates. Cold-working and heat treatment of stainless steels can change the mechanical properties as shown in Table 8-5. Generally, heat treatment decreases the strength while cold-working increases it by increasing the dislocation density (see Section 5.2).

Although crevice corrosion is not a materials problem, it is the most common type of corrosion in implants.[9] The solution is to implant the two components so that they are immovable with respect to each other; this may sometimes be difficult due to the nature of the bones. Another solution is to place a spacer made of plastics in the susceptible area. This may also solve the problem of fretting corrosion.

8.1.3. Manufacturing of Implants Using Stainless Steels

The austenitic stainless steels work-harden very rapidly (Figure 8-3) and therefore cannot be cold-worked without intermediate heat-treatments. These heat treatments must not induce, however, the formation of chromium carbide in the grain boundaries, which may cause corrosion as discussed previously. Great care should be taken if the stainless steel is to be welded since welding tends to cause sensitization.

Heat treatment may cause distortion of components, but this problem can be solved easily by controlling the uniformity of heating. Another

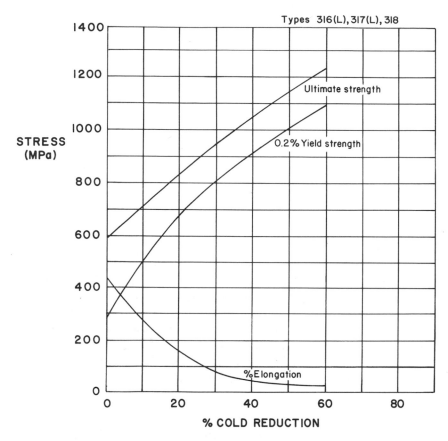

Figure 8-3. Effect of cold-work on the yield and ultimate tensile strength of 18-8 stainless steel. (From Ref. 13.)

undesirable effect of heat treatment is the formation of surface oxide scales which have to be removed either chemically (acid) or mechanically (sand-blasting). After the scales are removed, the surface of the component is polished to a mirror finish. The surface is then cleaned, degreased, and *passivated* in nitric acid (ASTM Standard F86). The component is washed and cleaned again before sterilizing and packing.

8.2. Co-BASED ALLOYS

These materials are usually referred to as cobalt–chromium alloys, one of which was first used in dentistry.[14] There are basically two types: one is the CoCrMo alloy, which is usually casted, and the other is the CoNiCrMo alloy, which is usually wrought by (hot) forging. The castable CoCrMo alloy has been in use for many decades in dentistry and recently in making artificial joints. The wrought CoNiCrMo alloy is a relative newcomer in the fabrication of the stems of heavily loaded joints such as the femoral hip stem.

8.2.1. Types and Composition of Co-Based Alloys

The ASTM has recommended four types of Co-based alloys for surgical implant applications: (1) cast CoCrMo alloy (F76), (2) wrought CoCrWNi alloy (F90), (3) wrought CoNiCrMo alloy (F562), and (4) wrought CoNiCrMoWFe alloy (F563). Their chemical compositions are summarized in Table 8-6. At the present time, only two of the four alloys are used extensively in implant fabrications, the castable CoCrMo and the wrought CoNiCrMo. From Table 8-6, their compositions are quite different. Most of the Co–Cr or Co-based alloys usually refer to the cast CoCrMo alloy, for which there are many names (Table 8-7).

8.2.2. Properties of Co-Based Alloys

The two basic elements of Co-based alloys form a solid solution of up to 65 w/o Co–35 w/o Cr as shown in Figure 8-4. Molybdenum is added to produce finer grains which results in higher strengths after casting or forging.[16,17]

One of the most promising wrought Co-based alloys is the CoNiCrMo alloy originally called MP35N (Standard Pressed Steel Corp.), which contains approximately 35 w/o Co and Ni each. The alloy was developed by Smith[18] to have a high degree of corrosion resistance in seawater (chlorine) under stress. Cold-working can increase the strength of the alloy[19] considerably as shown in Figure 8-5. However, there is considerable difficulty in cold-working, especially for making large devices such as hip joint stems. Only hot-forging can be used to mechanically work the alloy.[20]

Another way to increase strength involves hot isostatic pressing after the alloy is atomized into powder in a chamber as shown in Figure 8-6. After sieving to obtain uniform size and distribution of the powders, they are put into a hot isostatic pressing chamber (Figure 8-7). The chamber temperature is below the melting temperature of the alloy. The bonding of the metal particles occurs by solid-state diffusion resulting in fine grains as shown in Figure 8-8.

The homogeneous solid solution in the fully annealed condition shows an austenitic structure which has fcc lattice structure (Figure 8-9a). This

Table 8-6. Chemical Compositions of Co-Based Alloys[a]

	Cast CoCrMo (F76)		Wrought CoCrWNi (F90)		Wrought CoNiCrMo (F562)		Wrought CoNiCrMoWFe (F563)	
	Min.	Max.	Min.	Max.	Min.	Max.	Min.	Max.
Cr	27.0	30.0	19.0	21.0	19.0	21.0	18.00	22.00
Mo	5.0	7.0	—	—	9.0	10.5	3.00	4.00
Ni	—	2.5	9.0	11.0	33.0	37.0	15.00	25.00
Fe	—	0.75	—	3.0	—	1.0	4.00	6.00
C	—	0.35	0.05	0.15	—	0.025	—	0.05
Si	—	1.00	—	1.00	—	0.15	—	0.50
Mn	—	1.00	—	2.00	—	0.15	—	1.00
W	—	—	14.0	16.0	—	—	3.00	4.00
P	—	—	—	—	—	0.015	—	—
S	—	—	—	—	—	0.010	—	0.010
Ti	—	—	—	—	—	1.0	0.50	3.50
Co	Balance							

[a] From Ref. 7.

Table 8-7. Some Commercial Names of
Co-Based Alloys[a]

Zimaloy, Zimmer USA, Warsaw, Indiana
Vitallium, Howmedica, Rutherford, New Jersey
Vinertia, Deloro Surgical Ltd., Statton St. Margaret, England
Protasul, Sulzer Brothers, Switzland and Protek, Indiana
Orthochrome, Depuy, Warsaw, Indiana
Francobal, S. A. Benoist Girad and Cie, Heroville, France
CoCrMo, Orthopedic Equipment Co., Bourbon, Indiana
MP 35N (wrought CoNiCrMo), Standard Pressed Steel Corp.
HS-21 (cast CoCrMo) and HS-25 (wrought CoNiCrMo),
    Haynes-Stellite, Cabot Corp.

[a] From Ref. 22.

structure is retained after hot-forging at temperatures above 650°C but the grains are elongated (Figure 8-9b). However, cold-working below 650°C, the austenitic lattice structure is distorted and an ϵ phase with hcp structure appears according to Semlitch.[21]

The abrasive wear properties of the wrought CoNiCrMo alloy are similar to the cast CoCrMo alloy (0.14 mm/year); however, the former is not recommended for the bearing surfaces of a joint prosthesis because of its poor frictional properties with itself or other materials. The superior

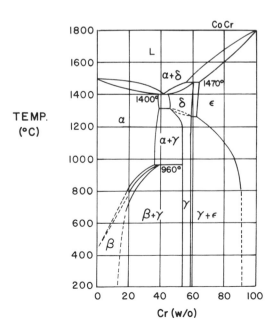

Figure 8-4. Phase diagram of
Co–Cr. (From Ref. 15.)

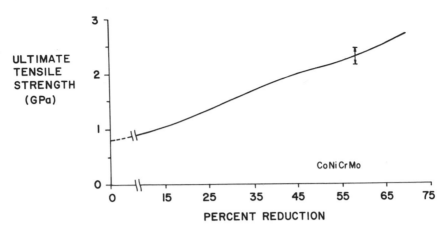

Figure 8-5. Relationship between ultimate tensile strength and the amount of cold-work for CoNiCrMo alloy. (From Ref. 19.)

fatigue and ultimate tensile strength of the wrought CoNiCrMo alloy make it suitable for applications that require long service life without fracture or stress fatigue. Such is the case for the stems of hip joint prostheses. This advantage is a significant one in light of the difficult and costly operation involving an implant that is deeply embedded in the femoral medullary canal.

Table 8-8 shows the mechanical properties required of Co-based alloys. As is the case with other alloys, the increased strengths are accompanied by decreased ductility.

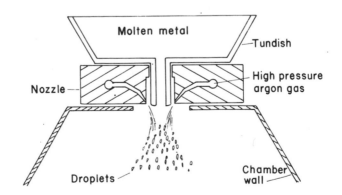

Figure 8-6. Powder atomization of CoCrMo alloy. (From Ref. 16.)

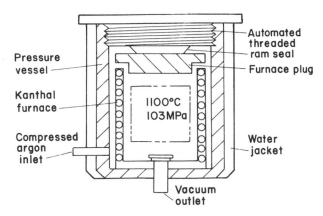

Figure 8-7. Hot isostatic pressing chamber. (From Ref. 16.)

Although both cast and wrought alloys have excellent corrosion resistance,[21,22] there has been some concern about the possibility of galvanic corrosion taking place at the tungsten inert gas (TIG) weld sites as shown in Figure 8-10.[23] Another manufacturer claims to have better weld results using an electron beam, welding on the solid CoCrMo cast femoral head which also has a larger weld area.[20]

Another interesting behavior is the rate of nickel release from the CoNiCrMo alloy and 316L stainless steel in 37°C Ringer's solution.[20] Although the cobalt alloy has a greater initial release of nickel ion into the solution, the rate of release after 5 days is about the same (3×10^{-10} g/cm^2 per day) for both alloys as shown in Figure 8-11. This is rather surprising since the nickel content of the CoNiCrMo alloy is about three times that of 316L stainless steel.

One final note is that the modulus of elasticity for the Co-based alloys does not change with changes in their ultimate tensile strength.[20] The values range from 220 to 234 GPa, which are higher than other materials such as stainless steels (200 GPa)[25]. The mode of load transfer from the implant to the bone may be affected by the modulus of elasticity of the implants, although it has not been clearly established what the relationship between the two is.[26-29]

8.2.3. Manufacturing of Implants Using Co-Based Alloys

The CoCrMo alloy is particularly susceptible to work-hardening so that the normal working procedure used with other metals cannot be employed.

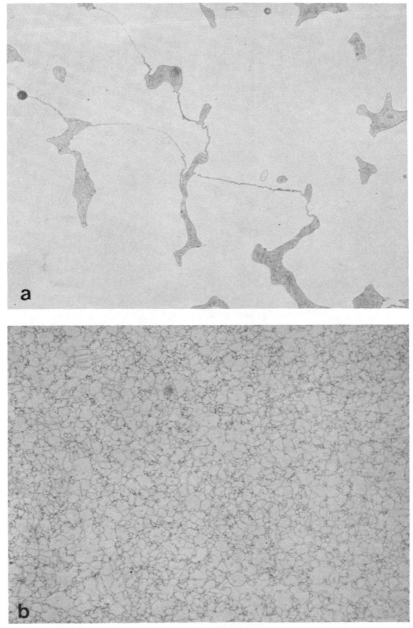

Figure 8-8. Microstructure of CoCrMo alloy. (a) Investment cast. Note that only portions of three grains are observed. 500×. (b) Hot isostatic pressed. The grain size is ASTM 12. 500×. (From Ref. 16.)

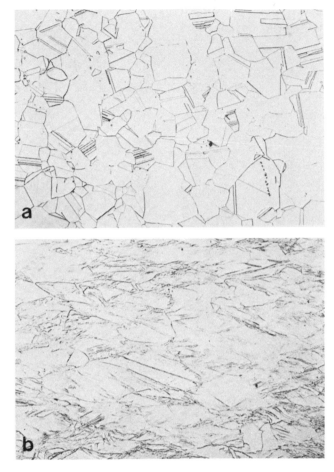

Figure 8-9. Microstructure of CoNiCrMo alloy. (a) Solution or heat-treated conditions; completely recrystallized austenitic grains. (b) Hot-forged condition; long austenitic grains with subgrains. (From Ref. 21.)

Instead the alloy is cast by a lost wax (or investment casting) method which involves the following steps (Figure 8-12):

1. A wax pattern of the desired component is made.
2. The pattern is coated with a refractory material, first by a thin coating with a slurry (suspension of silica in ethyl silicate solution) followed by investing completely after drying.
3. The wax is melted out in a furnace (100–150°C).

Table 8-8. Mechanical Property Requirements of Co-Based Alloys[a]

| | Cast CoCrMo (F76) | Wrought CoCrWNi (F90) | Wrought CoNiCrMo (F562) | | Wrought CoNiCrMoWFe (F563) | | | |
| | | | Solution annealed | Cold-worked and aged | Fully annealed | Cold-worked or cold-worked and aged | | |
						Medium hard	Hard	Cold-worked and aged extra hard
Tensile strength (MPa)	655	860	795–1000	1790[b]	600	1000	1310	1586
Yield strength (0.2% offset) (MPa)	450	310	240–655	1585[b]	276	827	1172	1310
Elongation (%)	8	10	50.0	8.0	50	18	12	—
Reduction of area (%)	8	—	65.0	35.0	65	50	45	—
Fatigue strength[c] (MPa)	310, 793[d]	—	—	—	340	400	500	400

[a] From Ref. 7.
[b] All are minimum requirements.
[c] From Ref. 20.
[d] From Ref. 17.

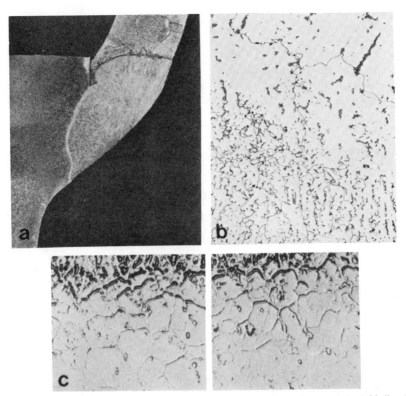

Figure 8-10. (a) Light micrograph of the welded junction of an original Muller hip prosthesis of compound design (hollow ball head is made of CoCrMo and the stem, wrought CoNiCrMo alloy). (b) Transition from coarse-grained CoCrMo cast structure (top) to the fine-grained dendritic weld zone. (c) Transition from the weld zone (top) to the fine grained CoNiCrMo alloy. (From Ref. 24.)

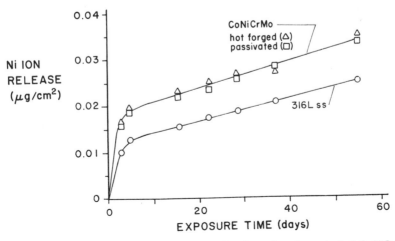

Figure 8-11. Nickel ion release versus time for hot-forged and passivated CoNiCrMo and 316L stainless steel in 37°C Ringer's solution. (From Ref. 20.)

4. The mold is heated to a high temperature, burning out any traces of wax- or gas-forming materials.
5. Molten alloy is poured with gravitational or centrifugal force. The mold temperature is about 800–1000°C and the alloy is 1350–1400°C.

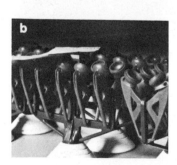

Figure 8-12. Lost wax casting of femoral joint prothesis. (a) Injection of wax into a brass mold. (b) Wax patterns assembled for a ceramic coating (note the hollow part of the femoral head). (c) Application of the ceramic coating. (d) A hot pressure chamber retrieves the wax, leaving behind a ceramic coating. (e) Pouring molten metals into the preheated ceramic mold. (Courtesy of Howmedica Inc., Rutherford, N.J.)

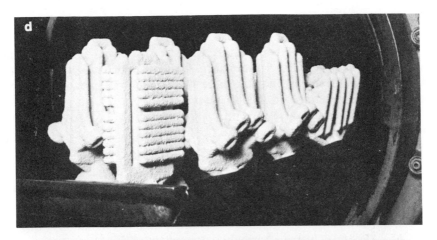

Figure 8-12. (*Continued*)

Controlling the mold temperature will affect the grain size of the final cast—coarse ones at higher temperatures, which will decrease the strength. However, it will result in larger carbide precipitates with greater distances between them, resulting in a less brittle material.

Hot-forging and hot isostatic pressing techniques of powdered CoCrMo alloys were discussed in Section 8.2.2.

8.3. Ti AND Ti-BASED ALLOYS

Attempts to use titanium began in the late 1930s and it was found to be tolerated well in cat femurs as well as stainless steel or Vitallium® (CoCrMo alloy).[30] Its low density (4.5 g/ml compared to 7.9 for 316 stainless steel, 8.3 for cast CoCrMo, and 9.2 for wrought CoNiCrMo alloys) and good mechanochemical properties are salient features for implant application. Detailed technical papers related to the science and technology of titanium can be found in a single volume.[31]

8.3.1. Composition of Ti and Ti-Based Alloys

There are four grades of unalloyed titanium for surgical implant applications as shown in Table 8-9. Impurity contents are the basis of the different grades; oxygen, iron, and nitrogen should be controlled carefully, especially so for oxygen, which has a great influence on ductility and strength.

One titanium alloy (Ti6Al4V) is widely used in the manufacture of implants[32] and its chemical requirements are given in Table 8-10. The main alloying elements are aluminum (5.5–6.5 w/o) and vanadium (3.5–4.5 w/o). Two other titanium alloys have been mentioned in the literature:

Table 8-9. Chemical Requirements of Titanium (ASTM F67)

	Composition (%)							
	Grade 1		Grade 2		Grade 3		Grade 4	
	Flat product	Bar and billet	Flat product	Bar and billet	Flat product	Bar and billet	Flat product	Bar and billet
Nitrogen, max.	0.03	0.03	0.03	0.03	0.05	0.05	0.05	0.05
Carbon, max.	0.10	0.10	0.10	0.10	0.10	0.10	0.10	0.10
Hydrogen, max.	0.015	0.0125[a]	0.015	0.0125[a]	0.015	0.0125[a]	0.015	0.0125[a]
Iron, max.	0.20	0.20	0.30	0.30	0.30	0.30	0.50	0.50
Oxygen, max.	0.18	0.18	0.25	0.25	0.35	0.35	0.40	0.40
Titanium	Balance							

[a] Bar only; max. hydrogen content for billet is 0.0100%.

Table 8-10. Chemical Requirements
of Ti6Al4V Alloy (ASTM F136)

	Composition (w/o)
Nitrogen, max.	0.05
Carbon, max.	0.08
Hydrogen, max.	0.0125 bars only
Hydrogen,[a] max.	0.0125 sheet, strip, or plate only
Iron, max.	0.25
Oxygen, max.	0.13
Aluminum	5.5–6.5, aim 6.0
Vanadium	3.5–4.5
Other elements	0.1 each max; or 0.40 total remainder
Titanium[b]	

[a] Specimen 0.032 in. (0.813 mm) and under may have hydrogen content as high as 0.0150 (150 ppm).
[b] The percentage of titanium is determined by difference; need not be determined or certified.

TiSnMoAl alloy for implants requiring extreme wear resistance[33] and Ti13V11Cr3Al alloy, which was not given any specific applications.[34]

8.3.2. Structure and Properties of Ti and Ti-Based Alloys

Titanium is an allotropic material which exists as an hcp structure (αTi) up to 882°C and as a bcc structure (βTi) above that temperature. Selective addition of alloying elements enables titanium to have a wide range of properties:[34]

1. Aluminum tends to stabilize the α phase, that is, it increases the transformation temperature to the β phase (Figure 8-13).
2. Chromium, columbium, copper, iron, manganese, molybdenum, tantalum, and vanadium stabilize the β phase by lowering the temperature of the transformation from α to β.

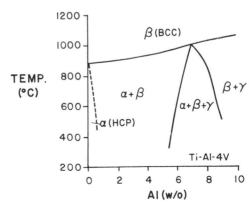

Figure 8-13. Part of phase diagram of Ti–Al–V at 4 w/o V. (From Ref. 35.)

3. Tin and zirconium do not affect the transformation temperature but do strengthen the room temperature phase.

The α alloys have single-phase microstructure (Figure 8-14a) which promotes good weldability. The stabilizing effect of the high aluminum content of this group of alloys gives excellent strength characteristics and oxidation resistance at high temperatures (300–600°C). These alloys cannot be heat-treated for strengthening since they are single-phased.

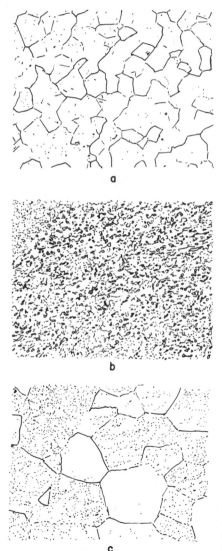

a

b

c

Figure 8-14. Microstructure of Ti alloys. (a) Annealed α alloy. (b) Ti6Al4V, α–β alloy, annealed. (c) β alloy, annealed. All 500×. (From Ref. 34.)

The addition of a controlled amount of β-stabilizers causes the higher-strength β phase to persist below the transformation temperature, which results in a two-phase system. As discussed in Section 5.2.4, the precipitates (α phase) will appear by heat treatment at the solid-solution temperature and subsequent quenching followed by aging at a somewhat lower temperature. The aging cycle causes the precipitation of some fine α particles from the metastable β, imparting a structure that is stronger than the annealed $\alpha-\beta$ structure (Figure 8-14b).

The higher percentage of β-stabilizing elements in the Ti13V11Cr3Al alloy results in a microstructure that is substantially β which can be strengthened by the heat treatment (Figure 8-14c).

The mechanical properties of the commercially pure titanium and Ti6Al4V alloy are given in Table 8-11. The modulus of elasticity of these materials is about 110 GPa, which is half the value of Co-based alloys. From Table 8-11 one can see that higher impurity content leads to higher strength and reduced ductility. Their strength varies from a value much lower than that of 316 stainless steel or the Co-based alloys to a value about equal to that of annealed 316 stainless steel or the cast CoCrMo alloy. However, with regard to specific strength (strength per density), the titanium alloy exceeds all other materials as shown in Figure 8-15. Nevertheless, titanium has poor *shear* strength, making it unsuitable for bone screws and similar applications. It also tends to gall or seize when in sliding contact with itself or another metal, which makes machining difficult.

Titanium derives its resistance to corrosion by the formation of a solid oxide layer. Under *in vivo* conditions, the oxide (TiO_2) is the only stable reaction product according to the Pourbaix diagram (see Figure 8-25). The oxide layer forms a thin adherent film and passivates the material.

An interesting early study[36] investigated the correlation between the chemical behavior and the electrical potentials for known currents across an electrolytic cell with a calomel half cell as cathode and with a rotating rod of the metal sample as the anode. The chemical test consisted of determining the weight loss of metal samples placed in aerated sterile horse serum for 20

Table 8-11. Mechanical Properties of Ti and Ti Alloys (ASTM F136)

Property[a]	Unalloyed				Alloyed	
	Grade 1	Grade 2	Grade 3	Grade 4	Ti6Al4V	
Tensile strength (MPa)	240	345	450	550	860[b]	895[c]
Yield strength (0.2% offset) (MPa)	170	275	380	485	795	830
Elongation (%)	24	20	18	15	10	10
Reduction of area (%)	30	30	30	25	25	—

[a] All are minimum requirements.
[b] Thickness of 4.75 to 44.4 mm.
[c] Thickness up to 4.75 mm.

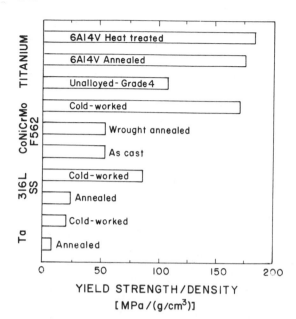

YIELD STRENGTH/DENSITY
[MPa /(g/cm³)]

Figure 8-15. Yield strength to density ratio of some implant materials. (From Ref. 34.)

days. The "anodic back emf," which was obtained by measuring the potential difference for known currents across the electrolytic cell and subtracting the "IR drop" (I, current; R, resistance), corresponded closely to the known inertness of the metals being tested as shown in Figure 8-16. These tests established reference points against which new metals could be screened before proceeding with costly *in vivo* tests and gave another indication of the biocompatibility of titanium.

Although there had never been any incidence of corrosion failures of implants made of titanium materials, a study has shown a rather large count of titanium particles by neutron activation analysis in the dried soft tissues about the implant.[37] The clinical significance, if any, was not given.

8.3.3. Manufacturing of Implants Using Ti-Based Alloys

Titanium is very reactive at high temperature and burns readily in the presence of oxygen; thus, it requires an inert atmosphere or vacuum for melting. Oxygen diffuses readily in titanium and the dissolved oxygen embrittles the metal, reducing its strength. As a result, any hot-working or -forging operation should be carried out below 925°C. Machining at room temperature is not a solution to all the problems since the material also tends to gall or seize the cutting tools as previously mentioned. Very sharp tools with slow speeds and large feeds are used to minimize this effect. Electrochemical machining is an attractive alternate method.[38]

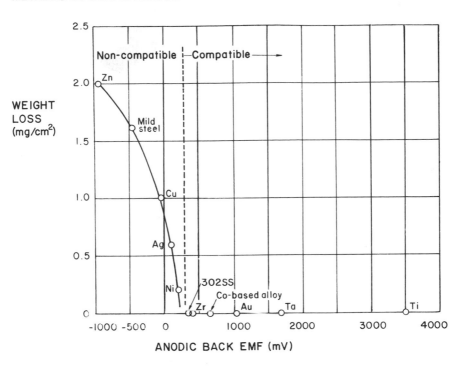

Figure 8-16. Anodic back emf versus weight loss. (From Ref. 36.)

8.4. OTHER METALS

Few other metals have been used for a variety of specialized implant applications. Tantalum has been subjected to animal implant studies and has been shown to be very biocompatible.[39] Due to its poor mechanical properties (Table 8-12) and its high density (16.6 g/ml), it is restricted to a few applications such as wire sutures for plastic and neurosurgery and a radioisotope for bladder tumors.

Table 8-12. Mechanical Properties of Tantalum (ASTM F560) Flat Mil Products

Property[a]	Fully annealed	Cold-worked
Tensile strength (MPa)	205	515
Yield strength (0.2% offset) (MPa)	140	435
Elongation (%)	20–30	2

[a] All are minimum requirements.

Platinum and other noble metals in the platinum group are extremely corrosion resistant but have poor mechanical properties. They are mainly used as alloys for electrodes such as pacemaker tips because of their high resistance to corrosion and more importantly their low threshold potentials.

Gold and silver are also resistant to environmental attack but they have very poor mechanical properties and are of little current interest as implant materials.

8.5. DETERIORATION OF METALLIC IMPLANT MATERIALS

The main cause of metallic deterioration is oxidation and corrosion. Pure metals exist in metastable equilibrium. Therefore, they tend to lower their energy state by electrochemical reaction with oxygen either in the air (oxidation) or in aqueous solution (corrosion). Some basic principles are discussed in this section.

8.5.1. Electrochemical Principles of Corrosion

Because the valence electrons of metals are loosely bound to nuclei, they can be removed easily and become

$$M \rightarrow M^{n+} + ne^- \tag{8-1}$$

where n is the number of valence electrons of the metal. This is an *anodic* or *oxidation* reaction which produces electrons. A *reduction* or *cathodic* reaction that consumes the electrons can be

$$2H_2O + O_2 + 4e^- \rightarrow 4OH^- \tag{8-2}$$

$$4H^+ + O_2 + 4e^- \rightarrow H_2O \tag{8-3}$$

$$2H^+ + 2e^- \rightarrow H_2 \tag{8-4}$$

The reaction represented by equation (8-2) is most relevant to implant corrosion since such corrosion occurs at nearly neutral pH. Equations (8-3) and (8-4) can sometimes occur in confined areas such as pits and crevices where the pH can reach acidic values due to the hydrolysis reactions

$$Fe^{2+} + 2H_2O \rightarrow Fe(OH)_2 + 2H^+ \tag{8-5}$$

which produce hydrogen ions. When they are confined in an area as shown

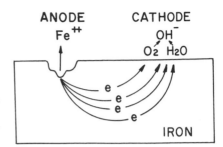

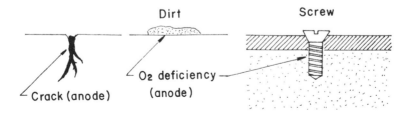

Figure 8-17. Schematic illustration of electrochemical cell set up between anodic and cathodic sites on an iron surface undergoing corrosion. (From Ref. 40.)

in Figure 8-17, they can produce an electrochemical cell with respect to a metal surface.

Variations in oxygen concentration over the surface in the environment can induce the electrochemical cell, sites with lower oxygen concentration become anodes, and corrosion takes place (Figure 8-18). Most corrosion occurs through the interaction of the solution and oxidation process. The dissolved ions and electrons (equation 8-2) build up an electrical potential called an *electrode potential* which depends on the nature of the metal and the solution. The electrode potential with respect to the solution is a measure of the Gibbs free energy of the reaction. If the reaction potential is E_0, then we can write the following relationship:

$$\Delta G = - nE_0 F \tag{8-6}$$

where F is Faraday's constant (amount of electricity associated with the flow of electrons, 96,487 C/mol). For a spontaneous reaction, E_0 will be positive.

The electrode potential of any material can be measured by using a standard hydrogen electrode as shown in Figure 8-19. The equilibrium

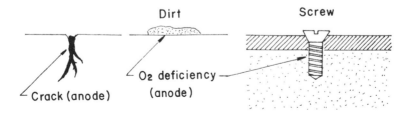

Figure 8-18. Examples of oxygen-deficient corrosion cells.

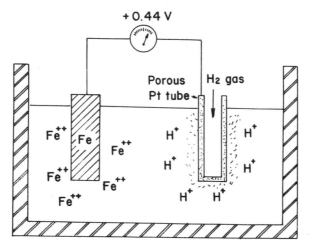

Figure 8-19. Iron–hydrogen corrosion cell. The platinum tube is porous for permeation of the hydrogen gas.

reaction of the hydrogen is the same as equation (8-4). Table 8-13 lists the electrode potentials of various metals with reference to hydrogen.

Equations (8-1) and (8-3) describe the differential hydrogen concentration cell which can be analyzed in terms of thermodynamic equilibrium states. The Gibbs free energy changes of the systems can be obtained from equations (4-5) and (4-12):

$$dG = V dP - S dT \tag{8-7}$$

Assuming the hydrogen gas behaves as an ideal gas,

$$PV = nRT \tag{8-8}$$

therefore at a constant temperature,

$$dG = nRT dP/P \tag{8-9}$$

For n moles of hydrogen from pressure P_1 to P_2,

$$\Delta G = \int_{P_1}^{P_2} nRT \frac{dP}{P} = nRT \ln \frac{P_2}{P_1} \tag{8-10}$$

Table 8-13. Electrode Potentials
of Various Ions

	Potential (V)	
Li^+	+2.96	Anode
K^+	+2.92	
Ca^{2+}	+2.90	
Na^+	+2.71	
Mg^{2+}	+2.40	
Ti^{3+}	+2.00	
Al^{3+}	+1.70	
Zn^{2+}	+0.76	
Cr^{2+}	+0.56	
Fe^{2+}	+0.44	
Ni^{2+}	+0.23	
Sn^{2+}	+0.14	
Pb^{2+}	+0.12	
Fe^{3+}	+0.045	
H	0.000	Reference
Cu^{2+}	−0.34	
Cu^+	−0.47	
Ag^+	−0.80	
Pt^{3+}	−0.86	
Au^+	−1.50	Cathode

If we transport n moles of hydrogen gas, then $2n$ moles of charge flow in the external circuit; thus, from equation (8-6),

$$\Delta G = -2nE_0 F \qquad (8\text{-}11)$$

Thus, from equations (8-10) and (8-11),

$$E_0 = \frac{RT}{2nF} \ln \frac{P_2}{P_1} \qquad (8\text{-}12)$$

The electrode potential of a metal can be altered by its thermomechanical state (cold-worked, annealed, as cast, etc.) and by the concentration of impurities present. For example, the potential energy at grain boundaries or second phase is higher than in the middle of a grain, causing the grain boundaries or second phase to be anodic and resulting in corrosion (Figure 8-20).

Any region of distortion or stress will be anodic with respect to the unstressed region of the same material because the stressed region has a higher energy level. This is the reason why the head or bent portion of a nail is more readily corroded.

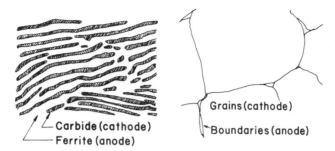

Figure 8-20. Galvanic corrosion caused by energy difference in microstructures. (Left) Galvanic microcell of pearlite. (Right) Galvanic microcell of grain and grain boundary.

Corrosion can be accelerated in the presence of static or dynamic external stress and is called *stress* or *fatigue* corrosion. Most implants are subjected to this type of corrosion. Also the endurance (fatigue) limit of a metal or alloy can be lowered by this process.

So far we have discussed various forms of *galvanic corrosion* as summarized in Table 8-14. As noted, the transfer of electrons occurs due to the differences in composition, energy level, and electrolytic environment. The electrode with higher potential or energy becomes the anode which corrodes.

Most metals and alloys form an *oxide film* in air which will retard the dissolution of the metal ions into the environment. Some metals such as titanium and chromium form a coherent thin oxide film and therefore are highly corrosion resistant despite their relative position of electrode potentials (see Table 8-13). The formation of a protective oxide layer is called *passivation*. Aluminum also forms a stable oxide film in air but it becomes unstable in salt solution. This indicates that the corrosion resistance of metals in saltwater, which is similar to *in vivo* conditions, is different and depends on their passivity as given in Table 8-15.

Table 8-14. Summary of Galvanic Corrosion

Differences in	Examples	
	Anode	Cathode
Composition	Fe	H_2
	Ferrite (α)	Carbide
Energy level	Boundaries	Inside grain
	Stressed region	Unstressed region
Electrolytic Environment	Low pO_2	High pO_2
	Dilute solution	Concentrated solution

Table 8-15. The Galvanic Series for Metals and Alloys, Including Some Stainless Steels[a,b]

Anodic end (electropositive)

Magnesium
Magnesium alloys
Zinc
Aluminum
Cadmium
Aluminum alloy
Carbon steel
Copper steel
Cast iron
4 to 6% Cr steel

A $\begin{cases} \text{12 to 14\% Cr steel} \\ \text{16 to 18\% Cr steel} \\ \text{23 to 30\% Cr steel} \end{cases}$ Active

B $\begin{cases} \text{7\% Ni, 17\% Cr steel} \\ \text{8\% Ni, 18\% Cr steel} \\ \text{14\% Ni, 23\% Cr steel} \\ \text{20\% Ni, 25\% Cr steel} \\ \text{12\% Ni, 18\% Cr, 3\% Mo steel} \end{cases}$ Active

Lead–tin solder
Lead
Tin
Nickel

C $\begin{cases} \text{60\% Ni, 15\% Cr, 20\% Fe} \\ \text{Inconel} \\ \text{80\% Ni, 20\% Cr} \end{cases}$ Active

Brasses
Copper
Bronzes
Nickel–silver (Ni-rich brass)
Copper–nickel
Monel metal
Nickel

C $\begin{cases} \text{60\% Ni, 15\% Cr, 20\% Fe} \\ \text{Inconel} \\ \text{80\% Ni, 20\% Cr} \end{cases}$ Passive

A and B $\begin{cases} \text{12 to 14\% Cr steel} \\ \text{16 to 18\% Cr steel} \\ \text{7\% Ni, 17\% Cr steel} \\ \text{8\% Ni, 18\% Cr steel} \\ \text{14\% Ni, 23\% Cr steel} \\ \text{23 to 30\% Cr steel} \\ \text{20\% Ni, 25\% Cr steel} \\ \text{12\% Ni, 18\% Cr, 3\% Mo Steel} \end{cases}$ Passive

Silver
Graphite

Cathodic end (electronegative)

[a] From Ref. 41.
[b] Note the role of passivation in moving the stainless steels toward the bottom of the table.

If the passivation film breaks down, corrosion will take place at that point which becomes anodic and the rest of the material becomes cathodic. Since the broken-down anodic region has to provide electrons for all the cathodic regions, it will result in accelerated corrosion (*pitting* or *fretting corrosion*) at the anode.

Since oxygen is a necessary element in many corrosion processes, its variation of concentration in an electrolyte can lead to corrosion. Examples are the differential aeration cell and *crevice corrosion* cell. Both are caused by the local oxygen concentration variations.

8.5.2. Rates and Passivity of Corrosion

The traditional corrosion rate test is the measurement of weight change of a sample in a solution with time (Figure 8-21). From the figure one can see that the metal can undergo either corrosion or passivation. When the passivation breaks down, it will corrode rapidly as discussed before. This test does not give information on the mode of corrosion (localized or uniform) and it is hard to predict the corrosion behavior beyond the test time. Similarly, the amount of metallic products (related to the weight loss of the sample) in tissues around implants can be studied by using various techniques such as spectrochemical analysis.[42]

Another method employs a potentiostat to impose an external potential emf (electromotive force) to a specimen which is made anodic under conditions of slowly increasing polarization.[40] A typical potential versus current curve is shown in Figure 8-22 where AB represents the active dissolution of ions by breakdown of passivity; CD, the repassivation process and slow rate of metallic dissolution through a protective passive film; and DE, the breakdown of the (re)passivated film and again rapid dissolution of metals.

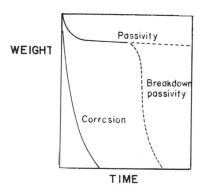

Figure 8-21. Schematic diagram of weight-loss curves for metals that undergo corrosion or passivity.

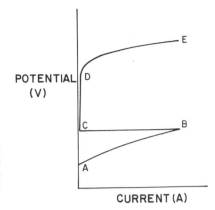

Figure 8-22. Schematic diagrams showing potential versus time after scratches (arrows) in the implant metals at room temperature and Hanks' and 0.17 M NaCl solution. (From Ref. 14.)

Another interesting study was the scratch tests of implant alloys in which the passivated surface was scratched after a steady-state electrical potential was reached (close to the passivation equilibrium) and subsequent determination of the potential changes with time (Figure 8-23). The results show that titanium and Vitallium® (CoCrMo) can be repassivated very rapidly within a few seconds but that the 18Cr10Ni3Mo austenitic stainless steel is not repassivated as rapidly.

The electrical potential can be measured in various ranges of pH as shown for iron in Figure 8-24, which is called a Pourbaix diagram.[43] The diagram may indicate the likelihood of passivation (or corrosion) behavior of an implant *in vivo* since the pH varies from 7.35 to 7.45 in normal extracellular fluid but can reach as low as 3.5 around the wound site. A more detailed Pourbaix diagram for titanium is given in Figure 8-25. The

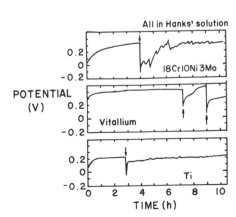

Figure 8-23. Potential versus current curve of a passivated metal. (From Ref. 14.)

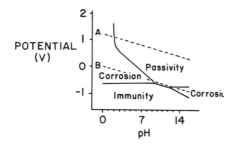

Figure 8-24. Simplified Pourbaix diagram for iron in aqueous solutions. (From Ref. 43.)

diagram shows the stable areas of various species (Ti^{2+}, Ti^{3+}, TiO, TiO_2, Ti_2O_3, and $TiO_3 \cdot 2H_2O$) and the various regions indicate the pH and potential combination for which each phase is stable as in the phase diagram. The stable region of Ti metal is designated as *immunity*, the stable oxide formation as *passivity* and the ion formation or stable corrosion products as *corrosion* ($TiO_3 \cdot 2H_2O$ is a stable corrosion product in that it does not form adherent film as does TiO_2).

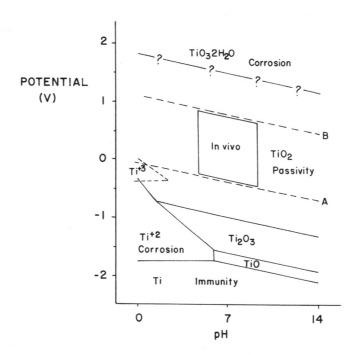

Figure 8-25. Simplified Pourbaix diagram for titanium in aqueous solution. (From Ref. 43.)

Below line B (Figures 8-24 and 8-25) water is reduced and hydrogen is evolved and above line A oxygen is reduced. Therefore, all stable aqueous solutions lie between lines A and B. The boxed area represents the *in vivo* environment. This region coincides with the (passivated) TiO_2. The oxide adheres to the metal base and prevents further oxidation. Actually, the oxide layer forms in air to a thickness of about 0.01 μm, making the titanium stable in aqueous solution.

Table 8-16 shows the rate of corrosion product formation measured by two methods: potentiostatic measurements and spectrochemical analysis.[14] The large discrepancy is mainly attributed to the unavailability of those ions which were carried away from the surrounding tissues near the implants to such tissues as liver or kidney or excreted out of the body.

8.5.3. Corrosion Fatigue

When metal undergoes fatigue in a corrosive environment, one of the following cases may occur[44]:

1. Corrosion has no effect on the fatigue limit and the S–N curve measured in air does not change.
2. Corrosion decreases the fatigue limit but metal repassivation is fast enough to permit definition of a new limit.
3. It is not possible to ascertain a new fatigue limit.

Table 8-16. Comparison of Rate of Formation of Corrosion Products for Alloys in Hanks' Solution during Current–Time Tests in Vivo (Rabbit)[a]

Alloy	Metal converted into compound (ng/cm² per h)	Metal found in tissue (ng/cm² per h)
18Cr–10Ni–3Mo steel		
Mechanically polished	7.8	0.274
Chemically polished	230.0	—
Vitallium®		
Mechanically polished	150	0.249
Commercial finish	20	
Ti		
Mechanically polished	4.1	0.430
Chemically polished	3.5	—
Ti–16Mo	1.5	—
Ti–5Ta	0.26	—

[a] From Ref. 14.

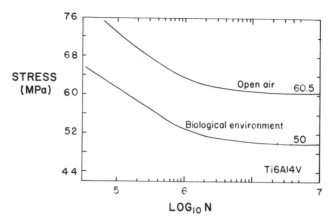

Figure 8-26. Fatigue and corrosion test results in biological environment for annealed Ti alloy. (From Ref. 44.)

A typical corrosion fatigue test result is shown in Figure 8-26 which shows that the endurance limit is lowered from 60 to 50 MPa for Ti6Al4V alloy. Table 8-17 summarizes the fatigue limits given in Tables 8-5 and 8-8 along with data given by Grover.[45] It is noted that the fatigue limit is lowered in all cases. It is not clear why there is a large discrepancy between the fatigue limit from Cornet *et al.*[44] and that from Williams and Roaf[46] on the titanium alloy and wrought titanium of commercial purity. One obvious reason is that cold-working of the wrought titanium increased the fatigue limit.

Table 8-17. Fatigue Limits of Some Metals Used
for Implants

Material	Fatigue limit (MPa)	
	In air	In corrosion medium
316 stainless steel (annealed)	260–280[a]	230–270[b]
Cast CoCrMo alloy	310[b]	240–280[b]
Ti6Al4V	60[c]	50[c]
Wrought commercial pure Ti	300[d]	240[b]

[a] From Tables 8-5 and 8-8.
[b] From Ref. 45.
[c] From Ref. 44.
[d] From Ref. 46.

PROBLEMS

8-1. Calculate the amount of volume change when iron is oxidized to FeO ($\rho = 5.95$ g/ml). The density of Fe is 7.89 g/ml.

8-2. The allotropic phase change of titanium at 882°C is from hcp structure ($a = 0.2956$ nm, $c = 0.4683$ nm) to bcc structure ($a = 0.332$ nm). Calculate the volume change in cm^3/g by heating above the transformation temperature. The density is 4.54 g/ml and the atomic weight is 47.9 g.

8-3. What is the emf potential for the hydrogen reaction if the pressure changes are $P_1 = 3P_2$ at room temperature?

8-4. A 1-mm-diameter eutectoid steel is coated with 1-mm-thick aluminum. Using the following data, answer parts a–f.

Material	Young's modulus (Gpa)	Yield strength (MPa)	Density (g/ml)	Expansion coeff. (°C^{-1})
Eutectoid steel	205	300	7.84	10.8×10^{-6}
Aluminum	70	100	2.7	22.5×10^{-6}

a. If the composite is loaded in tension, which metal will yield first?
b. How much load can the composite carry in tension without plastic deformation?
c. What is Young's modulus of the composite?
d. What is the thermal expansion coefficient of the composite? Hint: $(\Delta L/L)_{st} = (\Delta L/L)_{al}$, $F_{al} = - F_{st}$,

$$\left[\alpha \Delta T + \left(\frac{F/A}{E} \right) \right]_{st} = \left[\alpha \Delta T + \left(\frac{F/A}{E} \right) \right]_{al}$$

e. What is the density of the composite?
f. If the composite is exposed in an electrolytic medium, which one will be anode, aluminum or steel?

8-5. Cu has the following characteristics:

Atomic number	29
Atomic mass	63.54 amu
Melting point	1084.5°C
Density	____ g/ml
Crystal structure	fcc
Atomic radius	0.1278 nm
Ionic radius	0.096 nm

a. Calculate the density of Cu.
b. Calculate the packing efficiency (or factor) of Cu.

c. Would the grain size increase, decrease, or remain the same if the Cu was cooled more slowly than the original Cu after remelting?

d. Would the grain size increase, decrease, or remain the same if the metal of part c was annealed?

8-6. Silver has the following characteristics:

Atomic number	47
Atomic mass	107.87 amu
Melting point	961.9°C
Density	10.5 g/ml
Crystal structure	fcc
Atomic radius	0.1444 nm
Ionic radius	0.126 nm

A bioengineer is trying to construct a binary phase diagram of Cu–Ag and the following additional information was given:

Eutectic temperature	779.4°C
Eutectic composition	28.1 Cu–71.9 Ag
Maximum solubility of Cu in Ag	8.8%
Maximum solubility of Ag in Cu	8.0%

a. Construct the binary phase diagram by knowing the information on Cu in Problem 8-5.

b. Label all the phases in the phase diagram.

8-7. Sterling silver is made of 92.5% Ag and 7.5% Cu.

a. Construct a temperature versus % fraction chart for α, β, and liquid phases. Indicate specific temperatures where the phases start to change.

b. Calculate the density of the sterling silver by using the mixture rule.

c. When the sterling silver was equilibrated at 780°C, what phase(s) exist? What are their compositions?

d. Can the sterling silver be solution-hardened? Give a reason.

8-8. In general, pearlite steels are susceptible to corrosion. Explain why this is to be expected and how it may be prevented.

8-9. A steel pier is partially immersed in water. The likely reduction reaction is

$$H_2O + \tfrac{1}{2}O_2 + 2e^- \rightarrow 2OH^-$$

a. Does this occur above or below the water line? Explain why and describe the pattern of corrosion of the steel post.

b. How can the posts be protected from corrosion?

8-10. Type 316L stainless steel has a maximum carbon content of 0.03%. Welding of finished components is acceptable for this steel but not for type 316. Why? Explain how you would expect the mechanical properties to differ from each other.

8-11. A sheet of copper (Young's modulus $E = 120$ GPa) is stressed elastically by a tensile stress σ. If an electrochemical cell is created using an appropriate electrolyte, the stressed sheet of copper, and a similar unstressed sheet, state which sheet becomes the anode and find the potential developed by the cell.

REFERENCES

1. W. D. Sherman, Vanadium steel plates and screws, *Surg. Gynecol. Obstet. 14*, 629–634, 1912.
2. E. Browning, *Toxicity of Industrial Metals*, 2nd ed., Butterworths, London, 1969.
3. D. C. Ludwigson, Today's prosthetic metals, are they satisfactory for surgical use?, *J. Metals 1964*, 226–229, 1964.
4. D. I. Bardos, Stainless steels in medical devices, in: *Handbook of Stainless Steels*, D. Peckner and I. M. Bernstein (ed.), pp. 1–10, McGraw-Hill, New York, 1977.
5. M. G. Fontana and N. O. Greene, *Corrosion Engineering*, pp. 163–168, McGraw–Hill, New York, 1967.
6. F. B. Puckering (ed.), *The Metallurgical Evolution of Stainless Steels*, pp. 1–42, American Society for Metals and the Metals Society, Metals Park, Ohio, 1979.
7. *Annual Book of ASTM Standards*, Part 46, American Society for Testing and Materials, Philadelphia, 1980.
8. F. H. Keating, *Chromium–Nickel Austenitic Steels*, Butterworths, London, 1956.
9. A. M. Weinstein, W. P. Spires, Jr., J. J. Klawitter, A. J. T. Clemow, and J. O. Edmunds, Orthopedic implant retrieval and analysis study, in: *Corrosion and Degradation of Implant Materials*, ASTM STP 684, B. C. Syrett and A. Acharya (ed.), pp. 212–228, American Society for Testing and Materials, Philadelphia, 1979.
10. H. S. Dobbs and J. T. Scales, Fracture and corrosion in stainless steel hip replacement stems, in: *Corrosion and Degradation of Implant Materials*, ASTM STP 684, B. C. Syrett and A. Acharya (ed.), pp. 245–258, American Society for Testing and Materials, Philadelphia, 1978.
11. L. E. Sloter and H. R. Piehler, Corrosion-fatigue performance of stainless steel hip nails—Jewett type, in: *Corrosion and Degradation of Implant Materials*, ASTM STP 684, B. C. Syrett and A. Acharya (ed.), pp. 173–192, American Society for Testing and Materials, Philadelphia, 1979.
12. D. D. Moyle, *Biomaterials*, lecture notes of BioE801 course, Clemson University, 1979.
13. *Source Book on Industrial Alloy and Engineering Data*, p. 223, American Society for Metals, Metal Park, Ohio, 1978.
14. D. C. Mears, *Materials and Orthopaedic Surgery*, Williams & Wilkins, Baltimore, 1979.
15. C. J. Smithells (ed.), *Metals Reference Book*, p. 549, Butterworths, London, 1976.
16. Micro-grain Zimaloy, Technical Report, Zimmer-USA, Warsaw, Ind., 1978.
17. Vitallium FHS Forged Alloy (High Strength), Technical Monograph, Howmedica Inc., Rutherford, N.J., 1979.
18. G. Smith, Cobalt–nickel base alloys containing chromium and molybdenum, U.S. Patent No. 3,356,542, December 6, 1967.
19. T. M. Devine and J. Wulff, Cast vs. wrought cobalt–chromium surgical implant alloys, *J. Biomed. Mater. Res. 9*, 151–167, 1975.
20. Biophase Implant Material, Technical Information Publication 3846, p. 7, Richards Manufacturing Co., Memphis, Tenn., 1980.

21. M. Semlitch, Properties of wrought CoNiCrMo alloy Protasul-10, a highly corrosion and fatigue resistant implant material for joint endoprostheses, *Eng. Med. 9*, 201–207, 1980.
22. B. C. Syrett and E. E. Davis, Crevice corrosion of implant alloys—A comparison of *in vitro* and *in vivo* studies, in: *Corrosion and Degradation of Implant Materials*, ASTM STP 684, B. C. Syrett and A. Acharya (ed.), pp. 229–244, American Society for Testing and Materials, Philadelphia, 1979.
23. J. H. Dumbleton and J. Black, *An Introduction to Orthopedic Materials*, p. 178, Thomas, Springfield, Ill., 1975.
24. P. Sury and M. Semlitch, Corrosion behavior of cast and forged cobalt-based alloys for double-alloy joint endoprostheses, *J. Biomed. Mater. Res. 12*, 723–741, 1978.
25. J. B. Park, *Biomaterials: An Introduction*, p. 202, Plenum Press, New York, 1979.
26. R. D. Crowninshield, R. A. Brand, and R. C. Johnson, An analysis of femoral stem design in total hip arthroplasty, *Trans. 25th Annv. Orthop. Res. Soc. 4*, 33, 1979.
27. J. J. Klawitter, S. D. Cook, A. M. Weinstein, and S. Das, The effect of elastic modulus and trabecular structure on stress distribution around dental implants, *J. Dent. Res. 58*, 410, 1979.
28. R. R. Tarr, J. L. Lewis, D. Jaycox, A. Sarmiento, J. Schmidt, and L. L. Latta, Effect of materials, stem geometry, and collar-calcar contact on stress distribution in proximal femur with total hip, *Trans. 25th Annv. Orthop. Res. Soc. 4*, 34, 1979.
29. P. Townsend and R. Diamond, Aspects of prosthetic system, *Trans. 25th Annv. Orthop. Res. Soc. 4*, 196, 1979.
30. R. T. Booth, L. E. Beaton, and H. A. Davenport, Reaction of bone to multiple metallic implants, *Surg. Gynecol. Obstet. 71*, 598–602, 1940.
31. R. I. Jaffee and N. E. Promisel (ed.), *The Science, Technology, and Application of Titanium*, Pergamon Press, Elmsford, N.Y., 1968.
32. R. J. Solar, S. R. Pollack, and E. Korostoff, *In vitro* corrosion testing of titanium surgical implant alloys: An approach to understanding titanium release from implants, *J. Biomed. Mater. Res. 13*, 217–250, 1979.
33. G. M. Down, The use of titanium as an implant material, *Eng. Med. 2*, 58–63, 1972.
34. G. H. Hille, Titanium for surgical implants, *J. Mater. 1*, 373–383, 1966.
35. C. J. E. Smith and A. N. Hughes, The corrosion fatigue behavior of a titanium-6 w/o aluminum-4 w/o vanadium alloy, *Eng. Med. 7*, 158–171, 1978.
36. E. G. C. Clarke and J. Hickman, An investigation into the correlation between the electrical potentials of metals and their behavior in biological fluids, *J. Bone Jt. Surg. 35B*, 467–473, 1953.
37. G. Meachim, Histological interpretation of tissue changes adjacent to orthopaedic implants, in: *Biocompatibility of Implant Materials*, D. F. Williams (ed.), pp. 120–127, Sector, London, 1976.
38. L. P. Jahnke, Titanium in jet engines, in: *The Science, Technology, and Application of Titanium*, R. I. Jaffee and N. E. Promisel (ed.), pp. 1099–1115, Pergamon Press, Elmsford, N.Y., 1968.
39. J. T. McFadden, Tissue reactions to standard neurosurgical metallic implants, *J. Neurosurg. 26*, 598–603, 1972.
40. J. Kruger, Fundamental aspects of the corrosion of metallic implants, in: *Corrosion and Degradation of Implant Materials*, ASTM STP 684, B. C. Syrett and A. Acharya (ed.), pp. 107–126, American Society for Testing and Materials, Philadelphia, 1979.
41. C. A. Zapffe, *Stainless Steels*, American Society for Metals, Metals Park, Ohio, 1949.
42. A. B. Ferguson, Jr., P. G. Laing, and E. S. Hodge, Ionization of metal implants in living tissue, *J. Bone Jt. Surg. 42A*, 77–90, 1960.
43. M. Pourbaix, *Atlas of Electrochemical Equilibria in Aqueous Solutions*, Pergamon Press, Elmsford, N.Y., 1966.

44. A. Cornet, D. Muster, and J. H. Jaeger, Fatigue-corrosion of endoprostheses titanium alloys, *Biomat. Med. Devices Artif. Organs 7*, 155–167, 1979.
45. H. J. Grover, Metal fatigue in some orthopedic implants, *J. Mater. 1*, 413–424, 1966.
46. D. F. Williams and R. Roaf, *Implants in Surgery*, Saunders, Philadelphia, 1973.

BIBLIOGRAPHY

L. V. Azaroff, *Introduction to Solids*, Chapters 4 and 5, McGraw–Hill, New York, 1960.
C. O. Bechtol, A. B. Ferguson, and P. G. Laing, *Metals and Engineering in Bone and Joint Surgery*, Ballière, Tindall & Cox, London, 1959.
J. H. Dumbleton and J. Black, *An Introduction to Orthopedic Materials*, Chapter 9, Thomas, Springfield, Ill., 1975.
M. G. Fontana and N. O. Greene, *Corrosion Engineering*, McGraw–Hill, New York, 1967.
A. G. Guy, *Physical Metallurgy for Engineers*, Addison–Wesley, Reading, Mass., 1962.
B. Harris and A. R. Bunsell, *Structure and Properties of Engineering Materials*, Chapters 7–9, Longmans, London, 1977.
S. N. Levine (ed.), *Materials in Biomedical Engineering, Ann. N.Y. Acad. Sci. 146*, 1968.
D. C. Mears, *Materials and Orthopaedic Surgery*, Chapter 5, Williams & Wilkins, Baltimore, 1979.
B. C. Syrett and A. Acharya (ed.), *Corrosion and Degradation of Implant Materials*, ASTM STP 684, American Society for Testing and Materials, Philadelphia, 1979.
L. H. Van Vlack, *A Textbook of Materials Technology*, Chapters 3–6, Addison–Wesley, Reading, Mass., 1973.
L. H. Van Vlack, *Materials Science for Engineers*, Chapters 6 and 22, Addison–Wesley, Reading, Mass., 1970.
D. F. Williams and R. Roaf, *Implants in Surgery*, Chapters 6 and 8, Saunders, Philadelphia, 1973.

CERAMIC IMPLANT MATERIALS

Although the use of ceramic materials is well known in dentistry, their use in medicine as implants is relatively new. The main advantage of ceramics over other implant materials is their "inertness" or "biocompatibility," which is due to their low chemical reactivity. However, certain ceramics are made reactive so as to induce direct bonding to hard tissues. Some ceramics are also made to be absorbed *in vivo* after their original function is fulfilled.

Carbon implants have been found to be especially useful for blood interfacing materials such as heart valves. They are also used as a reinforcing component of composite implant materials due to their specific strength as fibers and their excellent biocompatibility.

9.1. ALUMINUM OXIDES

The main source of high-purity alumina is bauxite and native corundum. The commonly available (α) alumina can be prepared by calcining alumina trihydrate, resulting in calcined alumina.[1] Tabular aluminas are massive low-shrinkage forms that have been sintered without adding permanent binders. The chemical composition and density of commercially available "pure" aluminas are given in Table 9-1. It is also possible to obtain 99.9% pure alumina (Linde Co.) prepared from ammonium alum. However, the ASTM requires only 99.5% pure alumina and less than 0.1% of combined SiO_2 alkali oxides (mostly Na_2O) for implant use.[2]

α-alumina has a rhombohedral crystal structure ($a = 0.4758$ nm, and $c = 1.2991$ nm). Single-crystal alumina, known as sapphire and ruby (depending on the types of impurities: Fe and Ti for sapphire, Cr for ruby), has

Table 9-1. Chemical Composition and Density
of Aluminas[a]

	Calcined A-14	Tabular T-60
Al_2O_3	99.6	99.5+
SiO_2[b]	0.12	0.06
Fe_2O_3	0.03	0.06
Na_2O	0.04	0.20
Density (g/ml)	3.8–3.9	3.65–3.8

[a] From Ref. 1.
[b] ASTM F603 specifies that combined SiO_2 alkali oxides should be less than 0.1% for surgical implant application.[2]

been used successfully to make implants.[3,4] The single crystals can be prepared by feeding fine alumina powders onto the surface of a seed crystal which is slowly withdrawn from the electric arc or oxyhydrogen flame as the fused powder builds up. Alumina crystals up to 10 cm in diameter have been grown by this method.

The bend strength (S_b) of a polycrystalline alumina with constant grain size can be expressed as

$$S_b = S_0 \exp(-bP) \tag{9-1}$$

where S_0 is the bend strength at zero porosity, b is a constant, and P is the

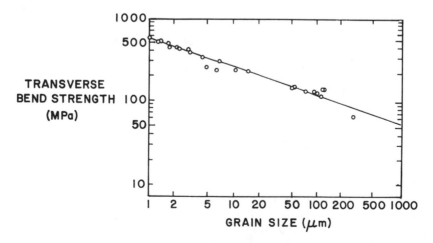

Figure 9-1. Effect of grain size on transverse bend strength. (From Ref. 5.)

porosity. The zero-porosity strength is given by the following equation[5]:

$$S_0 = Kd^{-1/3} \tag{9-2}$$

where K is a constant and d is the grain size as shown in Figure 9-1.

The relationship between grain size and porosity for fully dense alumina is given in Figure 9-2. When the porosity is below 2%, the grains become much larger, which will according to equation (9-2) decrease its strength. The size of the grains can be kept below 2 μm by adding 0.1% MgO. The addition of MgO will make the alumina almost translucent so that it can be used even for housing sodium vapor lamps (Lucalox®). This type of alumina has not as yet been used for implants. Table 9-2 gives mechanical properties of commercially available alumina.

Alumina in general is quite hard (Moh's number is 9 compared to 10 for diamond). Its hardness varies from 2000 kg/mm² (19.6 GPa) to 3000 kg/mm² (29.4 GPa), and hence it is used as an abrasive (emery) and

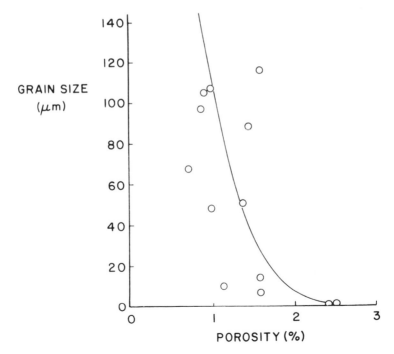

Figure 9-2. Grain size versus porosity of highly pure (99.9 + %) fully dense alumina. (From Ref. 5.)

Table 9-2. Mechanical Properties of Alumina[a, b]

Properties and materials	Values
Bending strength (modulus of rupture, MPa)	
Sapphire	496–703
Ruby	345
Polycrystalline	241–482
Compressive strength (MPa)	
Sapphire	3055–3413
Polycrystalline	2069–3861
Tensile strength (MPa)	
Single crystal	490
Filaments	
Coated	1448
Uncoated	483
Polycrystalline	259
Modulus of elasticity (GPa)	
Single crystal	362.7
Polycrystalline	408.9
Poisson's ratio	
Sapphire	0.257
Polycrystalline	0.32

[a] From Ref. 1.
[b] All measurements were made at 25°C.

bearings for watch movements. This high hardness combined with low friction and low wear are the major advantages of using alumina as a joint replacement material despite its brittleness and fabrication difficulties.

It is of great interest whether the inert ceramics such as alumina can be fatigued in either dynamic or static conditions. In one study[6] it was shown that the fatigue strength of alumina can be decreased by the presence of water above the critical stress. This decrease in fatigue strength is due to the delayed crack growth which is accelerated by the water molecules. However, another study[7] showed that the reduced strength is obtained if water absorption is observed by scanning electron microscopy of the broken specimens, but did not detect any decrease in strength for samples which showed no water marks on the fractured surface (Figure 9-3). It was suggested that the presence of minor amounts of silica in one lot may have contributed to the permeation of water molecules, which is detrimental to the strength of the specimen. It is not clear whether the same static fatigue mechanism operates in single-crystal alumina. It is, however, reasonable to assume that the same static fatigue will occur if the crystals contain flaws or impurities, which will act as the sources of crack initiation and grow under stress.

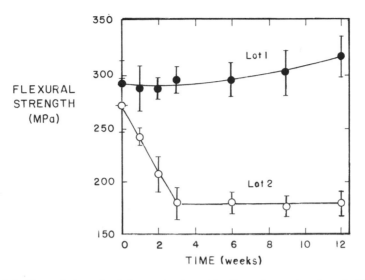

Figure 9-3. Flexural strength of dense alumina rods after aging under stress in Ringer's solution. (From Ref. 7.)

Some investigators[8] have used probability theory to investigate the static and dynamic fatigue behavior of polycrystalline alumina. The cumulative probability of failure under stress as given by them is

$$F = 1 - \exp\left[-\int_A \left(\frac{L_t - L_u}{L_0} \right)^m dA \right] \qquad (9\text{-}3)$$

where

L_t = the logarithm of time to failure

L_u = lower bound of the logarithm of time to fracture

L_0 = scale parameter

dA = surface element under stress

m = Weibull modulus

For constant surface area, equation (9-3) can be reduced to

$$F = 1 - \exp\left[-\left(\frac{L_t - L_u}{L_0} \right)^m A \right] \qquad (9\text{-}4)$$

Figure 9-4 shows the results of static and dynamic fatigue tests of alumina

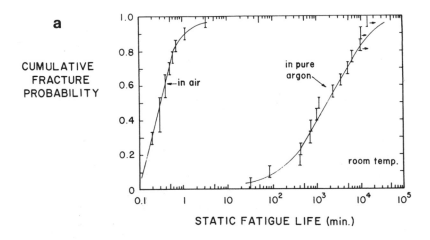

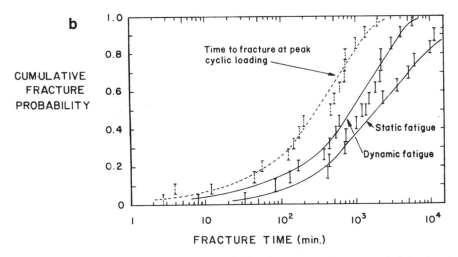

Figure 9-4. (a) Time to fracture under static loading versus fracture probability in air and (50% RH) in pure argon at room temperature. (b) Cumulative fracture probability versus fracture time for static and dynamic fatigue tests at room temperature.

at room temperature, from which two observations can be made: (1) the atmosphere has a drastic effect on the fatigue life, which is again due to stress corrosion with water vapor in the atmosphere; and (2) the resistance of a polycrystalline alumina ceramic to cyclic loading is lower than the static loading at room temperature, which may be significant in implant design.

Some[9] predicted the fatigue life of alumina and Bioglass® (glass-ceramic, see Section 9.3) coated alumina based on fracture mechanics theory, which is based on the assumption that fatigue is controlled by the slow crack growth of preexisting flaws. Generally, the strength distribution of ceramics in an inert atmosphere (S_i) can be correlated to the probability of failure (F) by the Weibull relationship, which according to Trantina[10] is similar to equation (9-3):

$$\ln\ln\left(\frac{1}{1-F}\right) = m\ln\left(\frac{S_i}{S_0}\right) \tag{9-5}$$

where m and S_0 are constants. Figure 9-5 shows a good fit for Bioglass®-coated alumina tested in a tris buffer solution and liquid nitrogen.

A minimum service life (t_{min}) of a sample can be predicted by means of a proof test, wherein a sample is subjected to stresses that are greater than those expected in service in order to eliminate the weak ones.[9] This

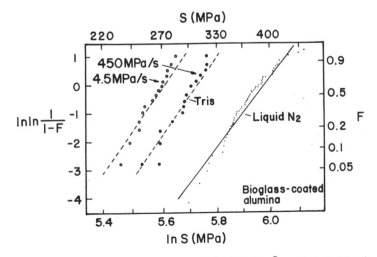

Figure 9-5. Plot of $\ln\ln[1/(1-F)]$ versus $\ln S$ for Bioglass®-coated alumina in a tris hydroxyamino methane buffer solution and liquid nitrogen. (From Ref. 9.)

minimum life can be predicted by the following equation:

$$t_{min} = B\sigma_p^{N-2}\sigma_a^{-N} \tag{9-6}$$

where σ_p is the proof test stress, σ_a is the applied stress, and B and N are constants. Rearranging equation (9-6) will give

$$t_{min}\sigma_a^2 = B\left(\frac{\sigma_p}{\sigma_a}\right)^{N-2} \tag{9-7}$$

which will result in a straight line if we plot $\log t_{min}$ versus $\log(\sigma_p/\sigma_a)$ with a slope of $N-2$ and an intercept equal to B as shown in Figure 9-6 for alumina. These results indicate that to assume a minimum lifetime of 50 years for a constant applied stress of 10,000 psi (68.95 MPa, $\log t_{min}$ $\sigma_a = 17.2$) in a tris buffer solution, the alumina sample would have to be

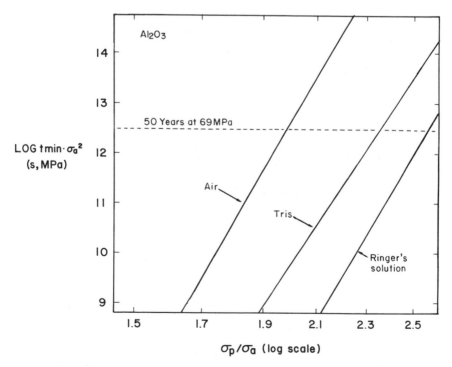

Figure 9-6. Plot of equation (9-7) for alumina after proof testing. $N = 43.85$, $\log B = 3.256$, $m = 13.21$, and $S_0 = 55,728$ (psi). (From Ref. 9.)

proof tested at a level of about 2.3 times the applied stress, that is, 23,000 psi (158.6 MPa).

9.2. HYDROXYAPATITE

Hydroxyapatite as an implant material has been tested many times as an artificial bone since it can be harvested from natural bone by eliminating organic constituents such as collagen and polysaccharides.[11] Recently, hydroxyapatite has been synthesized and used for manufacturing various forms of implants: solid, porous, and as coatings on other implants.

9.2.1. Structure of Hydroxyapatite

As mentioned previously (Section 6.2) the mineral part of bone and teeth is made of an apatite of calcium and phosphate similar to hydroxy-apatite crystals $[Ca_{10}(PO_4)_6(OH)_2]$. The apatite family of minerals, $A_{10}(BO_4)_6X_2$, crystallizes into the hexagonal rhombic prism. Hydroxy-apatite has unit cell dimensions of $a = 0.9432$ nm and $c = 0.6881$ nm.[12] The atomic structure of hydroxyapatite projected down the c-axis on the basal plane is given in Figure 9-7. Note that the hydroxyl ions lie on the corners of the projected basal plane and occur at equidistance intervals [one-half of the cell (0.344 nm)] along columns perpendicular to the basal plane and parallel to the c-axis. Six of the ten calcium ions in the unit cell are associated with the hydroxyls in these perpendicular columns, resulting in strong interactions between them.

The ideal Ca/P ratio of hydroxyapatite is 10/6 and the calculated density is 3.219 g/ml.[14] It is interesting to note that the substitution of OH with F gives greater structural stability, due to the closer coordination of F than the hydroxyl to the nearest calcium. This is one of the reasons for the better resistance of dental caries by fluoridation.

9.2.2. Properties of Hydroxyapatite

There is a wide variation in the reported mechanical properties of hydroxyapatite. Jarcho et al.[15] reported that the hydroxyapatite synthesized by them had average compressive and tensile strengths of 917 and 196 MPa, respectively, for fully densified polycrystalline specimens. Kato et al.[16] reported a compressive strength of 3000 kg/cm^2 (294 MPa), a bending strength of 1500 kg/cm^2 (147 MPa), and a Vickers hardness of 350 kg/mm^2 (3.43 GPa).

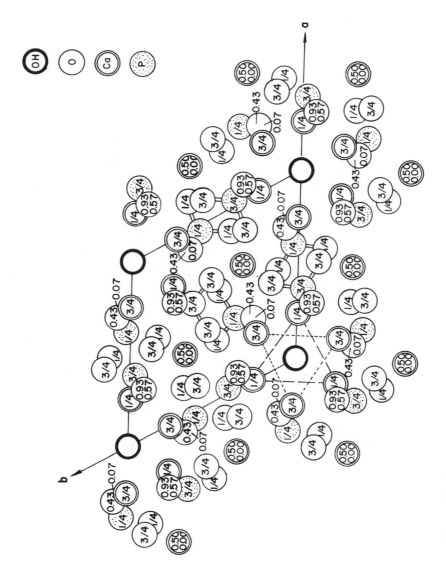

Figure 9-7. Hydroxyapatite structure projected down the c-axis on the basal plane. (From Ref. 12.)

Table 9-3. Elastic Modulus of Hydroxyapatite
and Mineralized Tissues

Test method	Material	Elastic modulus (GPa)
Ultrasonic interference technique	Hydroxyapatite (mineral)	144 (Ref. 17)
	Hydroxyapatite (synthetic)	117 (Ref. 17)
	Dentin	21 (Ref. 18)
	Enamel	74 (Ref. 18)
Destructive technique	Human cortical bone	24.6–35 (Ref. 19)
Resonance frequency technique	Hydroxyapatite (synthetic)	39.4–63 (Ref. 20)
	Canine cortical bone	12–14.6 (Ref. 20)

The elastic modulus of hydroxyapatite measured by ultrasonic interference and resonance frequency techniques are given in Table 9-3. Although there are some variations in the values, depending on the measurement technique, it is clear that hydroxyapatite has a higher elastic modulus than mineralized tissues. Along this line of thought, it is interesting to note that the relatively smaller amount of organic materials (mainly collagen) exists in enamel which has a higher elastic modulus than bone and dentin. This fact is indirect evidence that the mineral portion of the hard tissue is made of hydroxyapatite. Poisson's ratio for the mineral or synthetic hydroxyapatite is about 0.27, which is somewhat close to that of bone (~ 0.3) according to Grenoble.[17]

The most interesting property of hydroxyapatite is its excellent biocompatibility,[21–24] the result of its suspected direct chemical bonding with hard tissues.[16] Hench et al.[25] reported epitaxial hydroxyapatite crystal growth on the surface of Bioglass® wafers (1.23-cm diameter, 0.32-cm thickness) after spreading a 0.254-mm-thick layer of amorphous calcium orthophosphate precipitates on its surface. X-ray diffraction analysis of the crystallization of hydroxyapatite showed an average crystal size of approximately 20 nm, which is in the same range as the observed size for in vivo mineral crystals.[26] A scanning electron micrograph (Figure 9-8) of a fractured section shows the dendritic growth of hydroxyapatite crystals on the glass-ceramic surface.

9.2.3. Manufacture of Hydroxyapatite

Many different methods have been developed to produce precipitates of hydroxyapatite from aqueous solution of $Ca(NO_3)_2$ and NaH_2PO_4.[15,27–29] The precipitates are filtered and dried to form a powder of fine particles.

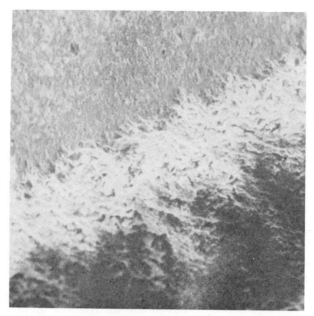

Figure 9-8. Scanning electron micrograph depicting the morphology of hydroxyapatite crystallized on Bioglass® substrate. Note the dendritic crystal growth. (From Ref. 25.)

After calcination for about 3 h at 900°C to promote crystallization, the powder can be pressed into a final form and can be sintered at about 1050 to 1450°C for 3 h.[29] However, according to other groups,[15] the sintering temperature should be between 1100 and 1190°C for 1 h for best results. Above 1250°C, hydroxyapatite shows a second-phase precipitation along the grain boundaries.

9.3. GLASS-CERAMICS

Glass-ceramics are polycrystalline ceramics made by controlled crystallization of glasses developed by S. D. Stookey of Corning Glass Works in the early 1960s. Glass-ceramics were first utilized in photosensitive glasses in which small amounts of copper, silver, and gold are precipitated by ultraviolet light irradiation. These metallic precipitates help to nucleate and crystallize the glass into a fine-grained ceramic which possesses excellent mechanical and thermal properties as seen in everyday life (Corelle® wares). Bioglass® and Ceravital® are two glass-ceramics used in implants.

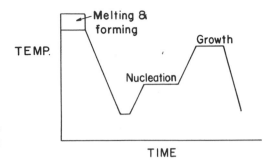

Figure 9-9. Temperature–time cycle for a glass-ceramic. (From Ref. 31.)

9.3.1. Formation of Glass-Ceramics

Important factors in forming glass-ceramics are the nucleation and the growth of crystals of small (<1-μm diameter) uniform size.[30] It is estimated that about 10^{12} to 10^{15} nuclei per cubic centimeter are required to achieve such small crystals. In addition to the metallic agents mentioned (Cu, Ag, and Au), Pt groups, TiO_2, ZrO_2, and P_2O_5 are widely used for the nucleation. The nucleation of glass is carried out at temperatures much lower than the melting temperature when the melting viscosity is in the range of 10^{11} to 10^{12} P for 1 to 2 h.[31] In order to obtain a higher crystalline phase glass, the glass is further heated to a temperature for maximum crystal growth without deformation of the product or phase transformation within the crystalline phases or redissolution of some of the phases. The crystallization is usually more than 90% complete when grain sizes are 0.1 to 1 μm, sizes which are much smaller than the conventional ceramics. Figure 9-9 is a schematic representation of the temperature–time cycle for a glass-ceramic.

Two glass-ceramics have been developed for implantation, SiO_2-CaO-Na_2O-P_2O_5[32-35] and Li_2O-ZnO-SiO_2.[36] In experiments with the former, Hench's group[32-34] varied the composition (except P_2O_5) as given in Table 9-4 in order to obtain maximal induction of direct bonding with bone. The bonding is related to the simultaneous formation of calcium phosphate and a SiO_2-rich film layer on the surface as exhibited by the 46S5.2 system. If a SiO_2-rich layer forms first and a calcium phosphate film develops later (46–55 m/o SiO_2 samples) or no phosphate film is formed (60 m/o SiO_2), then no direct bonding with bone is observed.[34]

The composition of Ceravital® is similar to Bioglass® in SiO_2 content but differs somewhat in the other components (Table 9-4). Also, Al_2O_3, TiO_2, and Ta_2O_5 are present in Ceravital® in order to control the dissolution rate of the ceramic. The mixtures are melted in a platinum crucible at

Table 9-4. Compositions of Bioglass® and Ceravital® Glass-Ceramics^{a,b}

	Code	SiO$_2$	CaO	Na$_2$O	P$_2$O$_5$	MgO	K$_2$O
Bioglass®	42S5.6	42.1	29.0	26.3	2.6	—	—
	46S5.2(45S5)	46.1	26.9	24.4	2.6	—	—
	49S4.9	49.1	25.3	23.0	2.6	—	—
	52S4.6	52.1	23.8	21.5	2.6	—	—
	55S4.3	55.1	22.2	20.1	2.6	—	—
	60S3.8	60.1	19.6	17.7	2.6	—	—
Ceravital®^{c}	Bioactive	40.0–50.0	30.0–35.0	5.0–10.0	10.0–15.0	2.5–5.0	0.5–3.0
	Nonbioactive	30.0–35.0	25.0–30.0	3.5–7.5	7.5–12.0	1.0–2.5	0.5–2.0

^{a} From Refs. 34 and 35.
^{b} Ceravital® compositions are in weight %; Bioglass® compositions are in mole % (m/o).
^{c} In addition, Al$_2$O$_3$ (5.0–15.0), TiO$_2$ (1.0–5.0), and Ta$_2$O$_5$ (5.0–15.0) are present.

1500°C for 3 h, annealed, and then cooled. The nucleation and crystallization temperatures are 680 and 750°C, respectively, each of 24 h. When the crystallites are about 40 nm in size and exhibit no characteristic needle structures, the process is stopped.

The Li$_2$O–ZnO–SiO$_2$ system was originally developed for making laser host material.[36] This glass-ceramic was modified to have better (X-ray) radioopacity as well as better mechanical and thermal properties as filler material for dental restorative composites.

9.3.2. Properties of Glass-Ceramics

Glass-ceramics have several desirable properties compared to glasses and ceramics. The thermal coefficient of expansion is very low, typically 10^{-7} or 10^{-5} per degree and in some cases it can even be made negative. Due to the controlled grain size and improved resistance to surface damage, the mechanical strength of glass-ceramics can be increased at least a factor of two from about 100 MPa to 200 MPa for tensile strength and their resistance to scratching and abrasion are close to those of sapphire.[37] Dissolution of ions into the (aqueous) medium (Figure 9-10) is thought to be the first step in the reaction of glass-ceramics with hard tissues. During this process the pH changes as shown in Figure 9-11. Auger electron spectroscopy (AES) and infrared reflection spectroscopy (IRRS) of the surface of Bioglass® have demonstrated that *in vitro* formation of the calcium phosphate-rich surface film corresponds closely to the capability of the material to bond to living bone as shown in Figures 9-12 and 9-13. The similarity in infrared reflectance spectra between the hydroxyapatite and the glass-ceramics after dissolution is further evidence of the possibility of direct

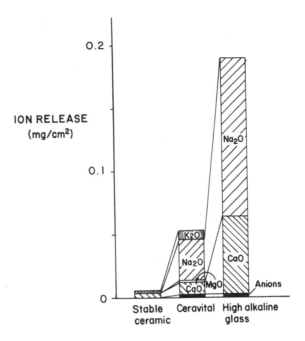

Figure 9-10. Solubility of powder samples of different bioactive glasses and ceramics in a neutral solvent at pH 7.3. (From Ref. 35.)

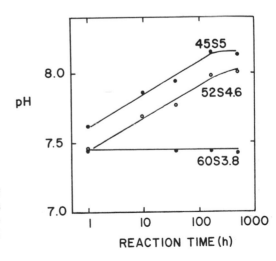

Figure 9-11. Changes in pH of the solution exposed to various glass compositions as a function of reaction time. (From Ref. 34.)

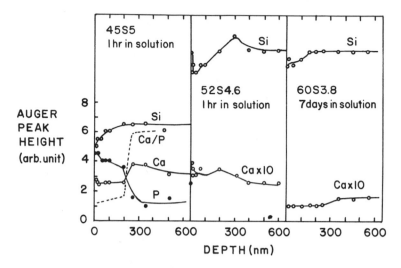

Figure 9-12. Depth profile of 45S5, 52S4.6, and 60S3.8 glasses after reactions by AES using an iron milling technique with an Ar ion beam. (From Ref. 34.)

bonding with bone. Transmission electron microscopy (Figure 9-14) showed an intimate relation between mineralized bone and Bioglass® implanted in the femurs of rats for 6 weeks.[38] The mechanical strength of the interfacial bond between bone and Bioglass® is of the same order of magnitude as the strength of the bulk glass-ceramic (850 kg/cm² or 83.3 MPa) which is about three-fourths that of the host bone strength.[33]

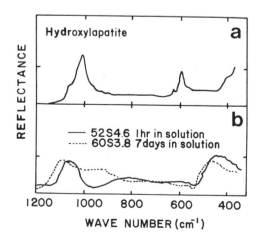

Figure 9-13. (a) IRRS spectrum of hydroxyapatite and (b) difference in IRRS spectra between 1-h-reacted 52S4.6 and 7-day-reacted 60S3.8 glass. (From Ref. 34.)

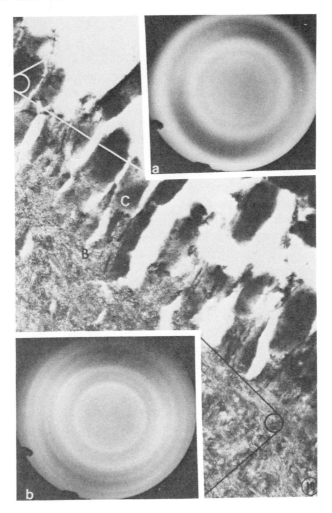

Figure 9-14. Transmission electron micrograph of well-mineralized bone (B) juxtaposed to the glass-ceramic (C), which was fractured during sectioning. 51,500×. Insert (a) is the diffraction pattern from ceramic area and (b) is from bone area. (From Ref. 38.)

The main drawback of the glass-ceramic, as with other glasses and ceramics, is its brittleness. Furthermore, due to the restrictions on its composition for biocompatibility (or osteogenicity), the mechanical strength cannot be substantially improved as it can be done for other glass-ceramics. Therefore, they cannot be used for making major load-bearing implants such as joint implants. However, glass-ceramics can be used as fillers for bone cement,[39] dental restorative composite,[36] and coating material.[9,40]

9.4. OTHER CERAMIC IMPLANTS

In addition to the ceramic materials discussed above, many other ceramics have been tested experimentally,[41] notably titanium oxide (TiO_2), barium titanate ($BaTiO_3$), tricalcium phosphate [$Ca_3(PO_4)_2$], and calcium aluminate ($CaO \cdot Al_2O_3$). TiO_2 was tested for its potential use in a component of the cardiovascular system.[42] $CaO \cdot Al_2O_3$ was used to induce tissue ingrowth for better attachment of an implant by making it porous.[43,44] However, this material loses its strength considerably after *in vivo* and *in vitro* aging[45] as shown in Figure 9-15. $Ca_3(PO_4)_2$ was tried as a biodegradable implant in the hopes of regenerating new bone.[46-48] $BaTiO_3$ was tested with regard to implant attachment by providing a textured surface and a piezoelectric current after the implant was polarized above the Curie temperature and subjected to high voltage.[49,50]

9.5. CARBONS

Carbons can be made in many forms: allotropic, crystalline diamond and graphite, quasi-crystalline glassy, and pyrolytic. Among these only pyrolytic carbon is widely utilized for implant fabrication.[51]

9.5.1. Structure of Carbons[51,52]

The crystalline structure of carbon is similar to that of graphite shown in Figure 9-16. The planar hexagonal structures are formed by strong covalent bonds in which one valence electron per atom is free to move resulting in a high but anisotropic electrical conductivity. Because the bonding between layers is stronger than the van der Waals force, *cross-links* between them were suggested.[51] Indeed, the remarkable lubricating property of graphite cannot be realized unless the cross-links are eliminated.

The poorly crystalline carbons are thought to contain unassociated or unoriented carbons and their hexagonal layers are not perfectly arranged as shown in Figure 9-17. The strong bonding within layers and the weaker bonding between layers cause the properties of the individual crystallites to be highly anisotropic. However, if the crystallites are randomly dispersed, then the aggregate becomes isotropic.

The carbon surfaces have active sites and a model of an oxidized carbon is given in Figure 9-18. The four primary functional groups on the surface are carboxyl, carbonyl, hydroxyl, and lactone. It is also thought that C–H bonds exist within the mass and exposed surfaces of pyrolytic carbons when deposited below 1300°C.

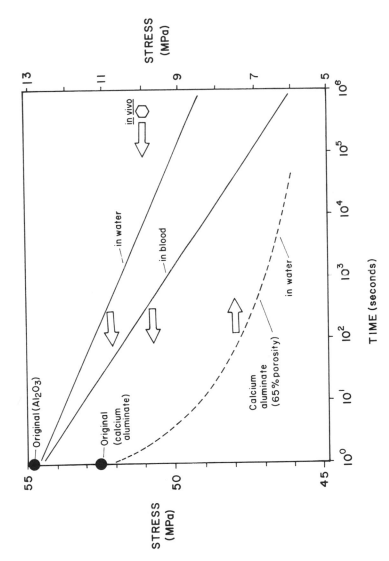

Figure 9-15. Aging effect on the strength of calcium aluminate *in vivo* and *in vitro*. (From Ref. 45.)

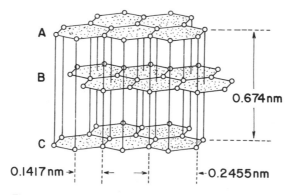

Figure 9-16. Crystal structure of graphite. (From Ref. 53.)

9.5.2. Properties of Carbons

The mechanical properties of carbon, especially the pyrolytic carbons, are largely dependent on the density as shown in Figures 9-19 and 9-20. The increased mechanical properties are directly related to the increased density, which indicates that the properties depend mainly on the aggregate structure of the material.

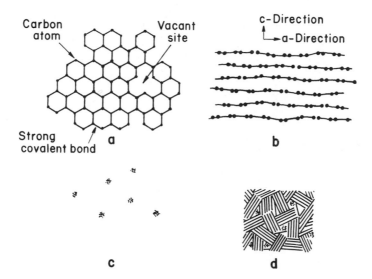

Figure 9-17. Schematic representations of poorly crystalline carbon. (a) Single-layer plane; (b) parallel layers in a crystallite; (c) unassociated carbon; (d) an aggregate of crystallites, single layers, and unassociated carbon. (From Ref. 52.)

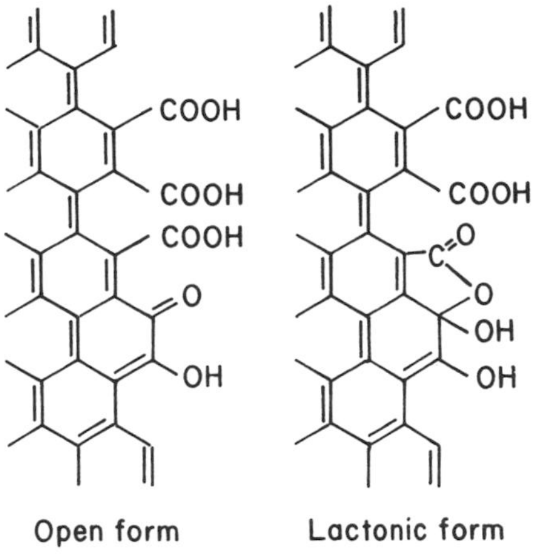

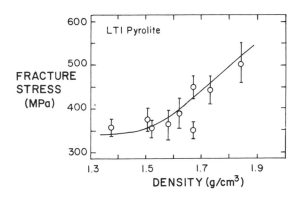

Figure 9-18. Models of oxidized carbon surfaces. (From Ref. 54.)

Open form   **Lactonic form**

Graphite and glassy carbon have much lower mechanical strength than pyrolytic carbon as given in Table 9-5. However, the average modulus of elasticity is almost the same for all carbons. The strength and toughness of pyrolytic carbon are quite high compared to graphite and glassy carbon. This is due to the smaller number of flaws and unassociated carbons in the aggregate.

The fatigue behavior of a vapor-deposited pyrolytic carbon film (400–500 nm thick) onto a stainless steel substrate showed that the film did not break until the substrate underwent plastic deformation at 1.3×10^{-2} strain and up to 10^6 cyclic loading.[51] Therefore, the fatigue behavior is closely related to the substrate as shown in Figure 9-21.

Figure 9-19. Fracture stress versus density for unalloyed LTI pyrolitic carbons. (From Ref. 55.)

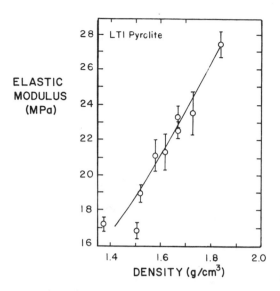

Figure 9-20. Elastic modulus versus density for unalloyed LTI pyrolitic carbons. (From Ref. 55.)

A composite carbon which is reinforced with carbon fibers has been considered for implant fabrication.[57,58] The properties are highly anisotropic as given in Table 9-6. The density is in the range of 1.4–1.45 g/ml with a porosity of 35–38%.

Carbons show excellent compatibility with tissues.[52, 58-64] This compatibility, especially with blood, made the pyrolytic carbon-deposited heart valve and blood vessel walls widely accepted armamentaria for surgeons.

Table 9-5. Properties of Various Types of Carbon

	Type		
	Graphite	Glassy	Pyrolytic
Density (g/ml)	1.5–1.9	1.5	1.5–2.0
Elastic modulus (GPa)	24	24	28
Compressive strength (MPa)	138	172	517 (575[a])
Toughness (m-N/cm^3)[b]	6.3	0.6	4.8

[a] 1.0 w/o Si-alloyed pyrolytic carbon, Pyrolite® (Carbomedics, Austin, Tex.).
[b] 1 m-N/cm^3 = 1.45×10^{-3} in.-lb/in.^{3}

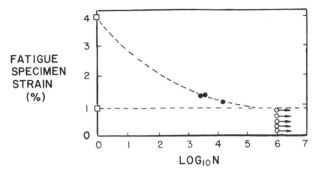

Figure 9-21. Strain versus number of cycles to failure. ○, absence of fatigue cracks in carbon film; ●, fracture of carbon film due to fatigue failure of substrate; □, data for substrate determined in single-cycle tensile test. (From Ref. 56.)

9.5.3. Manufacture of Carbon Implants

Carbons can be deposited onto finished implants from hydrocarbon gas in a fluidized bed at a controlled temperature as shown in Figure 9-22. The anisotropy, density, crystallite size, and structure of the deposited carbon can be controlled by temperature, composition of the fluidizing gas, bed geometry, and residence time of the gas molecules in the bed.[52] The microstructures of deposited carbon should be particularly controlled since the formation of growth features due to the uneven crystallization can result in a weaker material as shown in Figure 9-23. It is also possible to introduce various other elements into the fluidizing gas and codeposit them with

Table 9-6. Mechanical Properties of Carbon Fiber-Reinforced Carbon[a]

	Fiber lay-up	
	Unidirectional	0–90° crossply
Flexural modulus (GPa)		
Longitudinal	140	60
Transverse	7	60
Flexural strength (MPa)		
Longitudinal	1200	500
Transverse	15	500
Interlaminar shear strength (MPa)	18	18

[a] From Ref. 59.

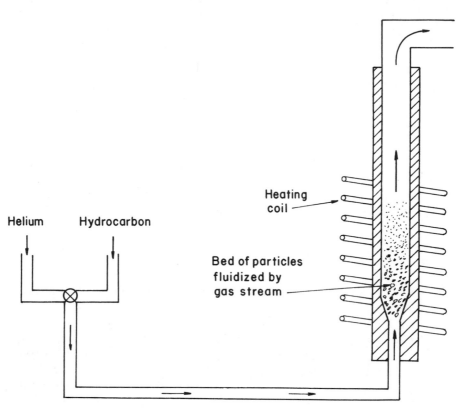

Figure 9-22. Schematic diagram showing particles being coated with carbon in a fluidizing bed. (From Ref. 51.)

carbon. Usually silicon (10–20 w/o) is codeposited (or *alloyed*) to increase hardness for applications requiring resistance to abrasion.

Recently, pyrolytic carbon was successfully deposited onto the surfaces of blood vessel implants made of polymers. This type of carbon is called ultra-low-temperature isotropic (ULTI) carbon instead of LTI (low-temperature isotropic) carbon.[65] The deposited carbon is thin enough not to interfere with the flexibility of the grafts, yet it exhibits excellent blood compatibility.

The vitreous or glassy carbon is made by controlled pyrolysis of polymers such as phenolformaldehyde, rayon (cellulose), and poly-acrylonitrile at high temperature in a controlled environment.[66] This process is particularly useful to make carbon fibers and textiles which can be used by themselves or as components for making composites.[67]

Figure 9-23. Microstructures of carbons deposited in a fluidizing bed. (a) A granular carbon with distinct growth features. (b) An isotropic carbon without growth features. Both under polarized light, 240×. (From Ref. 52.)

PROBLEMS

9-1. A ceramic (Al_2O_3) is used to fabricate a hip joint. Assume a simple ball and socket configuration with a surface contact area of 1.0 cm^2 and continuous static loading in a simulated condition similar to Figure 9-15. (Extrapolate the data if necessary.)

a. How long will it last if the loading is 70 kg (mass), in water and in blood?
b. Will the implant last longer or shorter with dynamic loading? Give reasons.

9-2. What should be the proof stresses of an alumina specimen (see Figure 9-6) if it is to last 20 years at 14 MPa of applied stress in air and in Ringer's solution?

9-3. Express the relationship between transverse bend strength and grain size given in Figure 9-1.

9-4. Consider the following mechanical properties:

	Bone	Vitreous carbon
Ultimate tensile strength (MPa)	100	120
Tensile modulus (GPa)	2	2.8

This comparison suggests that vitreous carbon would be an excellent material for bone replacement. What is wrong with this idea?

REFERENCES

1. W. H. Gitzen (ed.), *Alumina as a Ceramic Material*, American Ceramic Society, Columbus, Ohio, 1970.
2. *Annual Book of ASTM Standards*, Part 46, F603-78, American Society for Testing and Materials, Philadelphia, 1980.
3. H. Kawahara, M. Hirabayashi, and T. Shikita, Single crystal alumina for dental implants and bone screws, *J. Biomed. Mater. Res. 14*, 597–606, 1980.
4. H. Kawahara, A. Yamagami, Y. Koda, J. Yokota, H. Sogawa, Y. Kataoka, H. Kobayashi, S. Maehara, and M. Hirabayashi, Bioceram—A new type of ceramic implant, *Jpn. Soc. Implant Dent*. August 1975.
5. M. Spraggs and T. Vasilos, Effect of grain size on transverse bend strength of alumina and magnesia, *J. Am. Ceram. Soc. 46*, 224–228, 1963.
6. J. T. Frakes, S. D. Brown, and G. H. Kenner, Delayed failure and aging of porous alumina in water and physiological medium, *Am. Ceram. Soc. Bull. 53*, 193–197, 1974.
7. F. E. Krainess and W. J. Knapp, Strength of a dense alumina ceramic after aging *in vitro*, *J. Biomed. Mater. Res. 12*, 241–246, 1978.
8. C. P. Chen and W. J. Knapp, Fatigue fracture of an alumina ceramic at several temperatures, in: *Fracture Mechanics of Ceramics*, Volume 2, R. C. Bradt, D. P. H. Hasselman, and F. F. Lange (ed.), pp. 691–707, Plenum Press, New York, 1974.
9. J. E. Ritter, Jr., D. C. Greenspan, R. A. Palmer, and L. L. Hench, Use of fracture of an alumina and Bioglass-coated alumina, *J. Biomed. Mater. Res. 13*, 251–263, 1979.

10. C. G. Trantina, Brittle fracture and subcritical crack growth in a ceramic structure, in: *Fracture*, Volume 3, D. M. R. Taplin (ed.), pp. 921–927, University of Waterloo, Waterloo, Canada, 1977.

11. M. R. Urist, Bone histogenesis and morphogenesis in implants of demineralized enamel and dentin, *J. Oral Surg. 29*, 88–102, 1971.

12. A. S. Posner, A. Perloff, and A. D. Diorio, Refinement of the hydroxyapatite structure, *Acta Crystallogr. 11*, 308–309, 1958.

13. R. A. Young and J. C. Elliot, Atomic scale bases for several properties of apatites, *Arch. Oral Biol. 11*, 699–707, 1966.

14. D. McConell, *Apatite: Its Crystal Chemistry, Mineralogy, Utilization, and Biologic Occurrence*, Springer-Verlag, Berlin, 1973.

15. M. Jarcho, C. H. Bolen, M. B. Thomas, J. Bobick, J. P. Kay, and H. Doremus, Hydroxyapatite synthesis and characterization in dense polycrystalline form, *J. Mater. Sci. 11*, 2027–2035, 1976.

16. K. Kato, H. Aoki, T. Tabata, and M. Ogiso, Biocompatibility of apatite ceramics in mandibles, *Biomater. Med. Devices Artif. Organs 7*, 291–297, 1979.

17. D. E. Grenoble, The elastic properties of hard tissues and apatites, *J. Biomed. Mater. Res. 6*, 221–233, 1972.

18. R. S. Gilmore, R. P. Pollack, and J. L. Katz, Elastic properties of bovine dentine and enamel, *Arch. Oral Biol. 15*, 787–796, 1970.

19. F. Gaynor Evans, *Mechanical Properties of Bones*, p. 164, Thomas, Springfield, Ill., 1973.

20. A. M. Torgalkar, A resonance frequency technique to determine elastic modulus of hydroxyapatite, *J. Biomed. Mater. Res. 13*, 907–920, 1979.

21. P. Decheyne and K. de Groot, In vivo surface activity of a hydroxyapatite alveolar bone substitute, *J. Biomed. Mater. Res. 15*, 441–445, 1981.

22. R. E. Holmes, Bone regeneration within a coralline hydroxyapatite implant, *Plast. Reconstr. Surg. 63*, 626–633, 1979.

23. E. A. Monroe, W. Votaya, D. B. Bass, and J. McMullen, New calcium phosphate ceramic material for bone and tooth implants, *J. Dent. Res. 50*, 860–861, 1971.

24. S. Niwa, K. Sawai, S. Takahashie, H. Tagai, M. Ono, and Y. Fukuda, Experimental studies on the implantation of hydroxyapatite in the medullary canal of rabbits, *Transactions, First World Biomaterials Congress*, Baden, Austria, April 8–12, 1980.

25. L. L. Hench, R. K. Splinter, and W. C. Allen, Bonding mechanisms at the interface of ceramic prosthetic materials, *J. Biomed. Mater. Symp. 2*, 117–141, 1971.

26. E. D. Eanes and A. S. Posner, Kinetics and mechanisms of conversion of non-crystalline calcium phosphate to crystalline hydroxyapatite, *Trans. N.Y. Acad. Sci. 28*, 233–241, 1965.

27. E. Hayek and H. Newesely, Pentacalcium monohydroxyorthophosphate, *Inorg. Synth. 7*, 63–65, 1963.

28. D. J. Greenfield and E. D. Eanes, Formation chemistry of amorphous calcium phosphates prepared from carbonate-containing solutions, *Calcif. Tissue Res. 9*, 152–162, 1972.

29. T. Kijima and M. Tsutsumi, Preparation and thermal properties of dense polycrystalline oxyhydroxyapatite, *J. Am. Ceram. Soc. 62*, 954–960, 1979.

30. P. W. McMillan, *Glass-Ceramics*, 2nd ed., Academic Press, New York, 1979.

31. W. D. Kingery, H. K. Bowen, and D. R. Uhlmann, *Introduction to Ceramics*, 2nd ed., p. 368, Wiley, New York, 1976.

32. L. L. Hench and H. A. Paschall, Direct chemical bond of bioactive glass-ceramic materials to bone and muscle, *J. Biomed. Mater. Res. Symp. 2*, 5–42, 1973.

33. G. Piotrowski, L. L. Hench, W. C. Allen, and G. J. Miller, Mechanical studies of bone–Bioglass interfacial bond, *J. Biomed. Mater. Symp. 6*, 47–61, 1975.

34. M. Ogino, F. Ohuchi, and L. L. Hench, Compositional dependence of the formation of calcium phosphate film on Bioglass, *J. Biomed. Mater. Res. 14*, 55–64, 1980.
35. B. A. Blencke, H. Bromer, and K. K. Deutscher, Compatibility and long-term stability of glass-ceramic implants, *J. Biomed. Mater. Res. 12*, 307–318, 1978.
36. G. Muller, Glass ceramics as composite fillers, *J. Dent. Res. 53*, 1342–1345, 1974.
37. O. M. Wyatte and D. Dew-Hughes, *Metals, Ceramics, and Polymers*, p. 267, Cambridge University Press, London, 1974.
38. C. A. Beckham, T. K. Greenlee, Jr., and A. R. Crebo, Bone formation at a ceramic implant interface, *Calcif. Tissue Res. 8*, 165–171, 1971.
39. W. Hennig, B. A. Blencke, H. Bromer, K. K. Deutscher, A. Gross, and W. Ege, Investigation with bioactivated polymethacrylates, *J. Biomed. Mater. Res. 13*, 89–99, 1979.
40. P. Griss, D. C. Greenspan, G. Heimke, B. Krenpien, R. Buchinger, L. L. Hench, and G. Jentchura, Evaluation of a Bioglass-coated Al_2O_3 total hip prosthesis in sheep, *J. Biomed. Mater. Res. Symp. 7*, 511–518, 1976.
41. S. F. Hulbert and F. A. Young (ed.), *Use of Ceramics in Surgical Implants*, Gordon & Breach, New York, 1978.
42. T. L. Bridges, A Basic Investigation into the Potential Use of Titanium Dioxide as a Component of the Cardiovascular System, M. S. thesis, Clemson University, 1970.
43. J. J. Klawitter, A Basic Investigation of Bone Growth into a Porous Ceramic Material, Ph.D. thesis, Clemson University, 1970.
44. J. J. Klawitter and S. F. Hulbert, Application of porous ceramics for the attachment of load bearing internal orthopedic applications, *J. Biomed. Mater. Res. Symp. 2*, 161–229, 1972.
45. G. S. Schnittgrund, G. H. Kenner, and S. D. Brown, *In vivo* and *in vitro* changes in strength of orthopedic calcium aluminate, *J. Biomed. Mater. Res. Symp. 4*, 435–452, 1973.
46. T. D. Driskell, C. R. Hassler, and L. McCoy, Significance of resorbable bioceramics in the repair of bone defects, *Annv. Conf. Eng. Med. Biol. 15*, 199, 1973.
47. H. U. Cameron, I. Macnab, and R. M. Pilliar, Evaluation of a biodegradable ceramic, *J. Biomed. Mater. Res. 11*, 179–186, 1977.
48. G. A. Graves and R. L. Hentrich, Resorbable ceramic implants, *J. Biomed. Mater. Res. Symp. 2*, 91–115, 1972.
49. J. B. Park, A. F. von Recum, G. H. Kenner, B. J. Kelly, W. W. Coffeen, and M. F. Grether, Piezoelectric ceramics: A feasibility study, *J. Biomed. Mater. Res. 14*, 269–277, 1980.
50. J. B. Park, B. J. Kelly, A. F. von Recum, G. H. Kenner, W. W. Coffeen, and M. F. Grether, Piezoelectric ceramic implants: *In vivo* results, *J. Biomed. Mater. Res. 15*, 103–110, 1981.
51. J. C. Bokros, Deposition structure and properties of pyrolytic carbon, in: *Chemistry and Physics of Carbon*, P. L. Walker (ed.), Volume 5 pp., 70–81, Dekker, New York, 1969.
52. J. C. Bokros, L. D. LaGrange, and G. J. Schoen, Control of structure of carbon for use in bioengineering, in: *Chemistry and Physics of Carbon*, P. L. Walker (ed.), Volume 9, pp. 103–171, Dekker, New York, 1972.
53. E. I. Shobert, II, *Carbon and Graphite*, Academic Press, New York, 1964.
54. H. P. Boehm, Funktionelle Gruppen an Festkorper-Oberflachen, *Angew. Chem. 78*, 617–652, 1966.
55. J. L. Kaae, Structure and mechanical properties of isotropic pyrolytic carbon deposited below 1600°C, *J. Nucl. Mater. 38*, 42–50, 1971.
56. H. S. Shim and A. D. Haubold, The fatigue behavior of vapor deposited carbon films, *Biomater. Med. Devices Artif. Organs 8*, 333–344, 1980.
57. P. G. Rose, F. Gerstenberger, U. Gruber, W. Loos, D. Wolter, and R. Neugebauer, New aspects of the design and application of carbon fibre reinforced carbon for prosthetic

devices, *Transactions, First World Biomaterials Congress*, p. 1.6, Baden, Austria, April 8–12, 1980.

58. H. Nruckman, H. J. Mauer, K. J. Huttinger, H. Rettig, and U. Weber, New carbon materials for joint prostheses, *Transactions, First World Biomaterials Congress*, p. 1.7, Baden, Austria, April 8–12, 1980.

59. D. Adams and D. F. Williams, Carbon fiber-reinforced carbon as a potential implant material, *J. Biomed. Mater. Res. 12*, 35–42, 1978.

60. J. C. Bokros, R. J. Atkins, H. S. Shim, A. D. Haubold, and N. K. Agarwal, Carbon in prosthetic devices, in: *Petroleum Derived Carbons*, M. L. Deviney and T. M. O'Grady (ed.), pp. 237–265, American Chemical Society, Washington, D.C., 1976.

61. J. L. Nilles and M. Lapitsky, Biomechanical investigations of bone-porous carbon and porous metal interfaces, *J. Biomed. Mater. Res. Symp. 4*, 63–84, 1973.

62. C. L. Stanitski and V. Mooney, Osseous attachment to vitreous carbons, *J. Biomed. Mater. Res. Symp. 4*, 97–108, 1973.

63. V. Mooney, P. K. Predecki, J. Renning, and J. Gray, Skeletal extension of limb prosthetic attachments—Problems in tissue reaction, *J. Biomed. Mater. Res. Symp. 2*, 143–159, 1971.

64. J. Benson, Elemental carbon as a biomaterial, *J. Biomed. Mater. Res. Symp. 2*, 41–47, 1971.

65. A. D. Haubold, H. S. Shim, and J. C. Bokros, Carbon cardiovascular devices, in: *Assisted Circulation*, F. Unger (ed.), pp. 520–532, Academic Press, New York, 1979.

66. F. C. Cowland and J. C. Lewis, Vitreous carbon—A new form of carbon, *J. Mater. Sci. 2*, 507–512, 1967.

67. R. M. Gill, *Carbon Fibres in Composite Materials*, Butterworths, London, 1972.

BIBLIOGRAPHY

J. C. Bokros, R. J. Atkins, H. S. Shim, A. D. Haubold, and N. K. Agarwal, Carbon in prosthetic devices, in: *Petroleum Derived Carbons*, M. L. Deviney and T. M. O'Grady (ed.), American Chemical Society, Washington, D.C., 1976.

J. J. Gilman, The nature of ceramics, in: *Materials*, D. Flanagen *et al.* (ed.), Freeman, San Francisco, 1967.

G. W. Hastings and D. F. Williams (ed.), *Mechanical Properties of Biomaterials*, Part 3, pp. 207–274, Wiley, New York, 1980.

S. F. Hulbert and F. A. Young (ed.), *Use of Ceramics in Surgical Implants*, Gordon & Breach, New York, 1978.

S. F. Hulbert, F. A. Young, and D. D. Moyle (ed.), *J. Biomed. Mater. Res. Symp. 2*, 1972.

W. D. Kingery, H. K. Bowen, and D. R. Uhlmann, *Introduction to Ceramics*, 2nd ed., Wiley, New York, 1976.

F. Norton, *Elements of Ceramics*, 2nd ed., Addison–Wesley, Reading, Mass., 1974.

POLYMERIC IMPLANT MATERIALS

Polymeric materials have a wide variety of applications for implantation since they can be easily fabricated into many forms: fibers, textiles, films, and solids. Polymers bear a close resemblance to natural tissue components such as collagen which allows direct bonding with other substances, e.g., heparin coating on the surface of polymers for the prevention of blood clotting. Adhesive polymers can be used to close wounds or lute orthopedic implants in place.

The fundamentals of polymerization and their properties will be briefly reviewed first. Then we will examine the most widely used polymers. More detailed coverage of each subject can be found in the References and Bibliography.

10.1. POLYMERIZATION

There are many ways of linking basic polymer (*poly*, many; *mer*, unit) molecules together to form long chains which will impart useful physical characteristics. For this brief review we will look at two typical types of polymerization.

10.1.1. Condensation Polymerization

During condensation polymerization a small molecule such as water or alcohol will be condensed out of the chemical reaction:

Methyl terephthalate   Ethylene glycol

$$\text{(10-1)}$$

This particular process is used to make a polyester, e.g., Dacron® (see structure 10-16), which is one of the most widely used implant polymers.

Most natural polymers like cellulose (polysaccharides) and proteins are made by condensation polymerization (see Section 6.1). Cellulose can be polymerized from the common monosaccharide, glucose, by condensing out a water molecule:

Glucose   Cellulose

$$\text{(10-2)}$$

Hyaluronic acid, chondroitin, and chondroitin sulfate are important polysaccharides present in connective tissues. These polysaccharides lubricate the joints and fibrous tissue layers like collagen and elastin.

Collagen and elastin are proteins composed of amino acids which can be considered as monomers. There are about 20 naturally occurring amino acids (Table 6-2) which are polymerized into (poly)peptides by the condensation process:

$$\begin{array}{c}
\text{H} \quad \text{O} \\
| \quad \| \\
\text{H}_2\text{N} - \text{C} - \text{C} - \text{OH} \\
| \\
\text{R}
\end{array}
\xrightarrow{\text{Enzyme}}
\begin{array}{c}
\text{H} \quad \text{O} \\
| \quad \| \\
- \text{C} - \text{C} - \text{N} \\
| \quad \quad | \\
\text{R} \quad \quad \text{H}
\end{array}
\left(
\begin{array}{c}
\text{H} \quad \text{O} \\
| \quad \| \\
\text{C} - \text{C} - \text{N} \\
| \quad \quad | \\
\text{R} \quad \quad \text{H}
\end{array}
\right)_n
\begin{array}{c}
\text{C} - \\
\end{array}
\quad (10\text{-}3)$$

Peptide linkage

Some typical condensation polymers and their chemical reactions are given in Table 10-1. One major drawback of condensation polymerization is the tendency for the reaction to cease before the chains grow to a sufficient length. This is due to the decreased mobility of the chains as polymerization progresses. The result is many short chains. In the case of nylon the chains are polymerized to a sufficient length before this occurs and useful physical properties of the polymer can be obtained.

10.1.2. Addition or Free Radical Polymerization

Addition polymerization can be achieved by rearranging the bond within each monomer. Since each "mer" has to share at least two covalent electrons with other mers, the monomer has to have at least one double bond. For example, in the case of propylene,

$$\begin{array}{c}
\text{H} \quad \text{CH}_3 \\
| \quad | \\
\text{C} = \text{C} \\
| \quad | \\
\text{H} \quad \text{H}
\end{array}
\longrightarrow
\begin{array}{c}
\text{H} \quad \text{CH}_3 \\
| \quad | \\
- \text{C} - \text{C} - \\
| \quad | \\
\text{H} \quad \text{H}
\end{array}
\left(
\begin{array}{c}
\text{H} \quad \text{CH}_3 \\
| \quad | \\
\text{C} - \text{C} - \\
| \quad | \\
\text{H} \quad \text{H}
\end{array}
\right)_n
\begin{array}{c}
\text{H} \quad \text{CH}_3 \\
| \quad | \\
\text{C} - \text{C} - \\
| \quad | \\
\text{H} \quad \text{H}
\end{array}
\quad (10\text{-}4)$$

The breaking of a double bond can be effected by an *initiator*. This is usually a free radical such as benzoyl peroxide:

$$C_6H_5COO-OOC_6H_5 \rightarrow 2C_6H_5COO\cdot \rightarrow 2C_6H_5\cdot + 2CO_2 \uparrow \quad (10\text{-}5)$$
$$(R\cdot)$$

The initiation can be activated by heat, ultraviolet light, and chemicals.

Table 10-1. Typical Condensation Polymers

Type and interunit linkage	Reaction examples
Polyester $$\begin{matrix} & O \\ & \parallel \\ -C&-O- \end{matrix}$$	$HO(CH_2)_nCOOH \rightarrow HO[-(CH_2)_nCOO-]_mH + H_2O$ $HO(CH_2)_nOH + HOO(CH_2)_{n'}COOH \rightarrow$ $$HO[-(CH_2)_n\overset{\overset{O}{\parallel}}{O}C(CH_2)_{n'}-\overset{\overset{O}{\parallel}}{C}O-]_mH + H_2O$$ $$\begin{matrix} CH_2OH \\ | \\ CHOH + HOOC(CH_2)_nCOOH \rightarrow \text{3-dim. network} + H_2O \\ | \\ CH_2OH \end{matrix}$$
Polyamide $$\begin{matrix} & O \\ & \parallel \\ -C&-NH- \end{matrix}$$	$NH_2(CH_2)_nCOOH \rightarrow H[-NH(CH_2)_nCO-]_mOH + H_2O$ $NH_2(CH_2)_nNH_2 + HOOC(CH_2)_{n'}COOH \rightarrow$ $\quad H[-NH(CH_2)_nNHCO(CH_2)_{n'}CO-]_mOH + H_2O$
Polyurethane $$\begin{matrix} & & O \\ & & \parallel \\ -O&-C&-NH- \end{matrix}$$	$HO(CH_2)_nOH + OCN(CH_2)_{n'}CNO \rightarrow$ $\quad [-O(CH_2)_nOCONH(CH_2)_{n'}NHCO-]_m$
Polyurea $$\begin{matrix} & & O \\ & & \parallel \\ -NH&-C&-NH- \end{matrix}$$	$NH_2(CH_2)_nNH_2 + OCN(CH_2)_{n'}CNO \rightarrow$ $\quad [-NH(CH_2)_nNHCONH(CH_2)_{n'}NHCO-]_m$
Silk fibroin $$\begin{matrix} & O \\ & \parallel \\ -C&-NH- \end{matrix}$$	$NH_2CH_2COOH + NH_2CHROOH \rightarrow$ $\quad H[-NHCH_2CONHCHRCO-]_mOH + H_2O$
Polysiloxane $$\begin{matrix} R & & R \\ | & & | \\ -Si&-O-&Si- \\ | & & | \\ R & & R \end{matrix}$$	$$HO-\underset{\underset{CH_3}{|}}{\overset{\overset{CH_3}{|}}{Si}}-OH \rightarrow HO\left(\underset{\underset{CH_3}{|}}{\overset{\overset{CH_3}{|}}{Si}}-O\right)_n + H_2O$$
$$\begin{matrix} R & & R \\ | & & | \\ -Si&-O-&Si- \\ | & & | \\ R & & O \\ & & | \\ & & R-Si-R \end{matrix}$$	$$HO-\underset{\underset{CH_3}{|}}{\overset{\overset{CH_3}{|}}{Si}}-OH + HO-\underset{\underset{CH_3}{|}}{\overset{\overset{CH_3}{|}}{Si}}-OH \rightarrow \text{3-dim. network}$$

The free radicals (initiators) can react with monomers:

$$R\cdot + CH_2{=}CHX \rightarrow RCH_2 - \underset{\underset{X}{|}}{\overset{\overset{H}{|}}{C}}\cdot \qquad (10\text{-}6)$$

and this free radical can react with another monomer:

$$RCH_2 - \underset{\underset{X}{|}}{\overset{\overset{H}{|}}{C}}\cdot + CH_2{=}CHX \rightarrow RCH_2 - CHX - CH_2 - \underset{\underset{X}{|}}{\overset{\overset{H}{|}}{C}}\cdot \quad (10\text{-}7)$$

and the process can continue indefinitely. This process is called *propagation* and can be written in short forms:

$$\begin{aligned} R\cdot \quad + M &\rightarrow RM\cdot \\ RM\cdot \quad + M &\rightarrow RMM\cdot \\ RMM\cdot + M &\rightarrow RMMM\cdot, \text{ etc.} \end{aligned} \qquad (10\text{-}8)$$

where M represents a monomer.

The propagation process can be *terminated* by combining two free radicals, transfer, and/or disproportionate processes:

$$RM_nM\cdot + R\cdot \text{ (or } RM\cdot) \ \rightarrow RM_{n+1}R \text{ (or } RM_{n+2}R) \qquad (10\text{-}9)$$

$$RM_nM\cdot + RH \qquad\qquad \rightarrow RM_{n+1}H + R\cdot \qquad (10\text{-}10)$$

$$RM_nM\cdot + MM_nR \qquad \rightarrow RM_{n+1} + M_{n+1}R \qquad (10\text{-}11)$$

An example of disproportionate termination is given in the following for vinyl polymers:

$$-\underset{\underset{X}{|}}{\overset{\overset{H}{|}}{C}}H_2C\cdot + \cdot\underset{\underset{X}{|}}{\overset{\overset{H}{|}}{C}}-CH_2- \longrightarrow -CH_2\underset{\underset{X}{|}}{\overset{\overset{H}{|}}{C}}H + \underset{\underset{X}{|}}{\overset{\overset{H}{|}}{C}}{=}CH- \quad (10\text{-}12)$$

Some of the commercially important monomers for addition polymers are given in Table 10-2.

The degree of polymerization (DP) is one of the most important parameters in determining physical properties. It is defined as the number of mers per molecule. Each molecule may have small or large numbers of mers depending on the condition of the polymerization. Therefore, we speak of the *average* degree of polymerization, DP, or *average* molecular weight, *M*. The relationship between molecular weight and DP can be expressed as

$$M = DP \times M.W. \text{ of mer} \quad \text{(or repeating unit)} \qquad (10\text{-}13)$$

The number average molecular weight can be calculated according to the number of molecules (X_i) in each molecular weight fraction (MW_i):

$$M_n = \Sigma(X_i \cdot MW_i)/\Sigma X_i \qquad (10\text{-}14)$$

Similarly, the weight average molecular weight can be calculated according to the weight fraction (W_i) in each molecular weight fraction:

$$M_w = \Sigma(W_i \cdot MW_i)/\Sigma W_i \qquad (10\text{-}15)$$

The ratio of the weight average and the number average molecular weight (M_w/M_n) indicates the uniformity of the molecular size distribution. This is why M_w/M_n is called (poly)dispersity. If the ratio is 1, then only molecules of one size exist throughout the polymer. Usually, commercial products vary between 1.5 and 2.5. Since the uniformity of the molecular size distribution is an important factor for physical properties, one should try to obtain a polymer with a low polydispersity.

10.2. SOLID STATE OF POLYMERS

As the molecular chains become longer by polymerization, their relative mobility decreases. The chain mobility is also related to mechanical strength which in turn is directly proportional to the molecular weight, as shown in Figure 5-23.

The polymer chains can be arranged in three ways: linear, branched, and cross-linked or three-dimensional network as shown in Figure 10-1. Linear polymers such as polyethylene and polyamide (nylon) cannot be

Table 10-2. Monomers for Addition Polymerization

Vinyl chloride	($CH_2{=}CHCl$)	Vinyl acetate	($CH_3COOCH{=}CH_2$)
Styrene	($CH_2{=}CH{-}C_6H_5$)	Vinylidene chloride	($CH_2{=}CCl_2$)
Methyl acrylate	($CH_2{=}CH{-}COOCH_3$)	Methyl methacrylate	$\left(CH_2{=}C{<}{\begin{smallmatrix}CH_3\\COOCH_3\end{smallmatrix}}\right)$
Acrylo-nitrile	($CH_2{=}CH{-}CN$)		

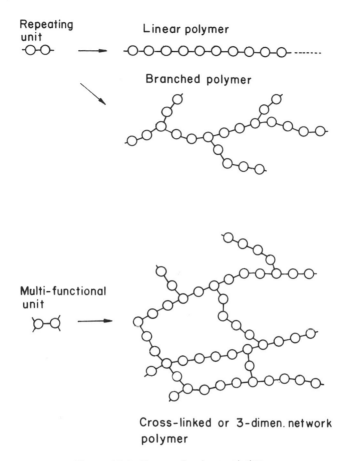

Figure 10-1. Types of polymer chains.

crystallized easily as for metals because the individual chains are long, making the relative motion difficult. Another factor is chain length, which varies within the polymer to create many defect sites because the chain ends act as defects (similar to edge dislocations).

Nevertheless, linear polymers do crystallize from melt or solution. Generally, a complete crystallization is impossible and the resulting structure is a mixture of crystalline and noncrystalline regions as shown in Figure 10-2. The arrangement of chains in crystalline regions is believed to be a combination of folded and extended chains. The seemingly more difficult to form (chain) folds are necessary to explain a single-crystal structure in which the thickness is too small to accommodate the length of the chain (X-ray and electron diffraction patterns of the polymer crystals show that the chains should be aligned in the direction of chain length).

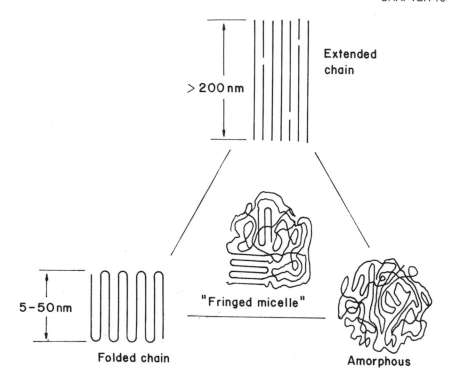

Figure 10-2. Two-dimensional representation of polymer structure. Generally the fibers have more extended chains, single crystals have folded chains, and glassy polymers have the amorphous structure. The semicrystalline polymers are thought to have the "fringed micelle" structure. (From Ref. 1.)

The extended chain configuration can occur by alignment of chains. The orientation of chains can be accomplished easily by drawing through fine holes from polymer melt or solution. Even this process does not prevent chain folds entirely. The ends of folds as in chain ends act as defects, thus lowering strength. To some degree the chain foldings can be prevented by stretching at high temperature (below T_m), which results in higher-strength fibers.

The three-dimensional network polymers such as (poly)phenol-formaldehyde (Bakelite®) do not usually crystallize. This type of structure can be considered as a random cross-linking of chains of amorphous polymers. In fact, solid isoprene rubber is made by cross-linking with sulfur (vulcanization). Too much cross-linking may restrict chain movement severely, making the rubber very hard. Because of this lack of freedom, the chains of the three-dimensional network polymers do not melt but rather burn.

10.3. EFFECT OF STRUCTURAL MODIFICATIONS ON PROPERTIES

The physical properties of polymers can be altered in many ways. In doing so, we can tailor the polymers to meet specific needs. Basically, chemical composition and arrangement of chains will have a great effect on the final properties.

10.3.1. Effect of Molecular Weight and Composition

The molecular weight and its distribution have a great effect on the properties of a polymer since its rigidity is primarily due to the immobilization of the chains by neighboring chains. The chains may be compared to cooked strands of spaghetti in a bowl. With increasing molecular weight, the polymer chains become longer and less mobile, resulting in a more rigid structure. It is equally important that all chains be equal in length (remember polydispersity?). Short chains will act as plasticizers, which in turn decrease the melting and glass transition temperatures, rigidity (modulus of elasticity), density, etc.

Another obvious way of altering the properties is to change the chemical composition of the backbone or side chains. Substituting the backbone carbon of a polyethylene with divalent oxygen or sulfur will decrease the melting and glass transition temperatures because the chain becomes more flexible due to the increased rotational freedom:

(10-16)

Caution should be given that the oxygens in equation (10-16) are randomly distributed rather than regularly distributed. If they are distributed in every other carbon position of the polyethylene, the resulting polyoxymethylene is a high-strength and stiff polymer (see Section 10-9). But the opposite effect can be achieved by substituting the backbone chains with a rigid molecule like benzene:

(10-17)

Polyethylene terephthalate (polyester, Dacron®)

10.3.2. Effect of Side Chain Substitution, Cross-Linking, and Branching

Increasing the size of side groups in linear polymers such as polyethylene will decrease the melting temperature as a result of a lower degree of molecular packing, i.e., decreased crystallinity. This effect prevails until the

Table 10-3. Effect of
Side Chain Substitution on
Melting Temperature in Polyethylene

Side chain	T (°C)
—H	140
—CH_3	165
—CH_2CH_3	124
—$CH_2CH_2CH_3$	75
—$CH_2CH_2CH_2CH_3$	−55
—$CH_2CH-CH_2CH_3$ 　　　$\vert$ 　　CH_3	196
CH_3 　　$\vert$ —$CH_2C-CH_2CH_3$ 　　$\vert$ 　CH_3	350

side group itself becomes large enough to hinder the movement of the main chain, as shown in Table 10-3. Very long side groups can be thought of as branches.

Cross-linking of the main chains is in effect similar to chain substitution with a small molecule, i.e., it lowers the melting temperature. This is due to the interference of the cross-linking, which can decrease the mobility of the chains resulting in further retardation of the crystallization rate. In fact, a large degree of cross-linking can completely prevent crystallization. However, opposite results can be obtained when elastomers or rubbers are *cross-linked*. The more cross-linking, the stiffer and harder the rubber becomes. The strength and strengthening mechanisms of polymers have been discussed (Section 5.4). The reader is advised to review this material before studying the following sections which will deal with individual polymers used for implantation.

10.4. POLYOLEFINS

Polyethylene and polypropylene and their copolymers are called polyolefins. These are linear thermoplastics. Polyethylene is available commercially in three major grades: low and high density and ultrahigh molecular weight (UHMWPE).

10.4.1. Structure and Properties of Polyethylene

Polyethylene has the repeating unit structure

$$
\begin{array}{cc}
\text{H} \quad \text{H} \\
| \quad | \\
\text{C}=\text{C} \\
| \quad | \\
\text{H} \quad \text{H}
\end{array}
\longrightarrow
\begin{array}{cc}
\text{H} \quad \text{H} \\
| \quad | \\
-\text{C}-\text{C} \\
| \quad | \\
\text{H} \quad \text{H}
\end{array}
\left(
\begin{array}{cc}
\text{H} \quad \text{H} \\
| \quad | \\
\text{C}-\text{C} \\
| \quad | \\
\text{H} \quad \text{H}
\end{array}
\right)_n
\begin{array}{cc}
\text{H} \quad \text{H} \\
| \quad | \\
\text{C}-\text{C}- \\
| \quad | \\
\text{H} \quad \text{H}
\end{array}
\qquad (10\text{-}18)
$$

which can be readily crystallized. In fact, it is almost impossible to produce amorphous polyethylene because of the small hydrogen side groups which cause high mobility of chains.

The first polyethylene was made by reacting ethylene gas at high pressure (100–300 MPa) in the presence of a catalyst (peroxide) to initiate polymerization. The process yields *low-density* polyethylene. By using a Ziegler catalyst, *high-density* polyethylene can be produced at low pressure (10 MPa). Unlike the former, high-density polyethylene does not contain branches. The result is better packing of the chains, which increases density and crystallinity. The crystallinity is usually 50–70% and 70–80% for low-

Table 10-4. Properties of Polyethylene

	Low density	High density	UHMWPE[a]
Molecular weight (g/mol)	$3 \sim 4 \times 10^3$	5×10^5	2×10^6
Density (g/ml)	0.90–0.92	0.92–0.96	0.93–0.944
Tensile strength (MPa)	7.6	23–40	30
Elongation (%)	150	400–500	200–250
Modulus of elasticity (MPa)	96–260	410–1240	—

[a] Data from ASTM F 648.

and high-density polyethylene, respectively. Some important physical properties of polyethylenes are given in Table 10-4.

The ultrahigh-molecular-weight polyethylene ($> 2 \times 10^6$ g/mol) has been used extensively for orthopedic implants, especially for load-bearing surfaces such as total hip and knee joints. This material has no known solvent at room temperature; therefore, only high temperature and pressure sintering may be used to produce desired products. Conventional extrusion or molding processes are difficult to use.

10.4.2. Polypropylene

Polypropylene can be synthesized by using a Ziegler-type stereospecific catalyst which controls the position of each monomer unit as it is being polymerized to allow the formation of a regular chain structure from the asymmetric repeating unit,

$$\left(\begin{array}{cc} H & CH_3 \\ | & | \\ -C & -C- \\ | & | \\ H & H \end{array} \right)_n \qquad (10-4)$$

Three types of structures can exist, depending on the position of the methyl group along the polymer chain, as shown in Figure 10-3. The random distribution of methyl groups in the *atactic* polymer prevents close packing of chains in comparison with the *isotactic* and *syndiotactic* structures. This results in less crystallinity for the atactic material. However, the presence of the methyl side groups restricts the movement of the polymer chains, and crystallization rarely exceeds 50–70% even for materials with isotacticity over 95%.

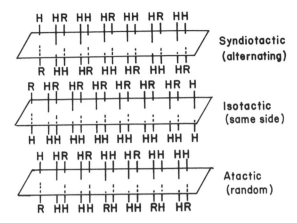

Figure 10-3. Syndiotactic, isotactic, and atactic polypropylene chain structure.

The properties of polypropylene depend largely on the percentage of isotactic materials present, crystallinity, and molecular weight. An increased (iso)tacticity increases crystallinity and enhances properties such as density, softening temperature, and chemical resistance. Table 10-5 lists typical properties of commercial polypropylenes.

Because polypropylene has an exceptionally high flex life, it is used to make integrally molded hinges for finger joint prostheses. It also has excellent environmental stress-cracking resistance. The permeability of polypropylene to gases and water vapor is between that of low- and high-density polyethylene.

10.5. POLYAMIDES (NYLONS)

The polyamides are known as nylons and are designated by the number of carbon atoms in the parent diamine and diacid. Nylons can be polymer-

Table 10-5. Properties of Polypropylene

	Value
Density (g/ml)	0.90–0.91
Tensile strength (MPa)	28–36
Elongation (%)	400–900
Modulus of elasticity (GPa)	1.1–1.55
Softening temperature (°C)	150

ized by step-reaction (or condensation) and ring-scission polymerization. They have excellent fiber-forming abilities due to interchain hydrogen bonding and a high degree of crystallinity, which increases strength in the fiber direction. However, the hydrogen bonds can be destroyed under *in vivo* conditions; therefore, they can be utilized for biodegradable applications, such as absorbable sutures.

The basic chemical structure of the repeating unit of polyamides can be written in two ways:

$$-\left[NH(CH_2)_x NHCO(CH_2)_y CO\right]_n- \qquad (10\text{-}19)$$

and

$$-\left[NH(CH_2)_x CO\right]_n- \qquad (10\text{-}20)$$

Formula (10-19) represents polymers made from diamine and diacids such as type 6/6 and 6/10. Polyamides made from ω-amino acids (formula 10-20) are designated as nylon 6, 11, and 12. These types of polyamides are produced by ring-scission polymerization:

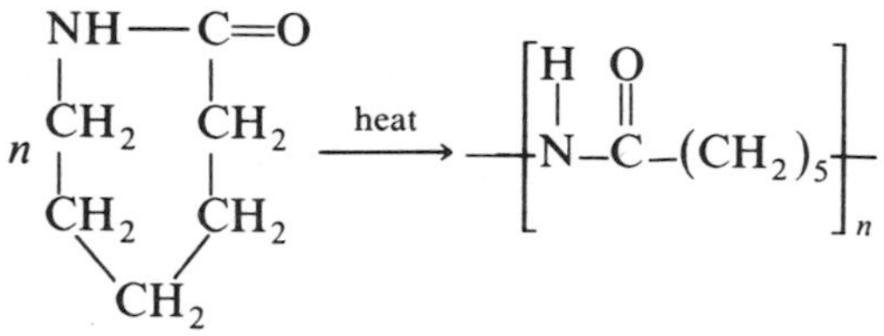

$$(10\text{-}21)$$

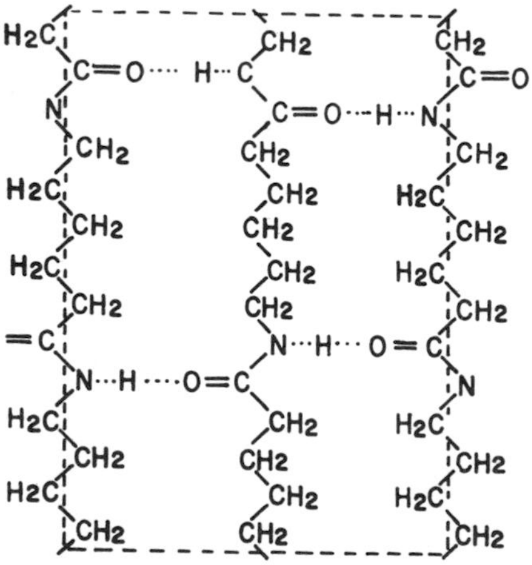

Figure 10-4. Hydrogen bonding in polyamide chains (nylon 6).

Table 10-6. Softening Temperature
of Polyamides

Type	Softening temperature (°C)
Nylon 6/6	265
6/10	220
6	215
11	185

The presence of –CONH– groups in polyamides results in strong attraction between chains via hydrogen bonding as shown in Figure 10-4. Since the hydrogen bond plays a major role in determining properties, the number and distribution of CONH groups are important factors. For example, the softening temperature (T_g) can be decreased by decreasing the number of CONH groups as given in Table 10-6. On the other hand, an increase in the number of CONH groups improves physical properties such as strength: Table 10-7 shows that 6/6 is stronger than 6/10 and 6 is stronger than 11.

As mentioned previously, the nylons are hygroscopic and lose their strength *in vivo* when implanted.[2] The water molecules serve as *plasticizers* which attack the amorphous region. Proteolytic enzymes also aid hydrolysis by attacking the amide group.

10.6. ACRYLIC POLYMERS

These polymers are used extensively in medical applications as (hard) contact lenses, implantable ocular lenses, and bone cement for joint fixation. Dentures and maxillofacial prostheses are also made from acrylics because they have excellent physical and coloring properties and are easy to fabricate.

Table 10-7. Properties of Polyamides

	Type			
	6/6	6/10	6	11
Density (g/ml)	1.14	1.09	1.13	1.05
Tensile strength (MPa)	76	55	83	59
Elongation (%)	90	100	300	120
Modulus of elasticity (GPa)	2.8	1.8	2.1	1.2

10.6.1. Structure and Properties of Acrylics and Hydrogels

The basic chemical structure of repeating units of acrylics can be represented by

$$\left(\!\!\begin{array}{c} R_1 \\ | \\ CH_2 - C - \\ | \\ COOR_2 \end{array}\!\!\right)_n \qquad (10\text{-}22)$$

The only difference between polymethylacrylate (PMA) and polymethyl-methacrylate (PMMA) is the R group, R_1 and R_2 being respectively H and CH_3 for PMA and CH_3 for PMMA. These polymers are addition (or free radical) polymerized by various methods: bulk, solution, emulsion, suspension, and granulation polymerization. These polymers can be obtained in liquid monomer or fully polymerized beads, sheets, rods, etc. The monomers have been used to embed hard tissue samples for histological evaluation of implants.[3]

Because of the bulky side groups, these polymers are usually obtained in clear amorphous state. For the same reason, PMMA has a higher tensile strength (60 MPa) and softening temperature (125°C) than PMA (7 MPa and 33°C, respectively). PMMA has an excellent light transparency (92% transmission), a high index of refraction (1.49), and excellent weathering properties. This material can be cast, molded, or machined with conventional tools. It has an excellent chemical resistivity and is highly biocompatible in pure form. The material is somewhat brittle in comparison with other polymers (Table 10-8).

The first hydrogel polymer (i.e., absorption of water > 30% of its weight) for soft lens applications was polyhydroxyethylmethacrylate (poly-

Table 10-8. Properties of PMMA

Properties	Cast	Molded
Refractive index	1.48–1.50	1.49
Density (g/ml)	1.17–1.20	1.17–1.20
Tensile strength (MPa)	55–80	48–70
Elongation (%)	3–7	3–10
Modulus of elasticity (GPa)	2.4–3.1	3.1
Water absorption (%, 24 h)	0.3–0.4	0.3–0.4

HEMA).[4] The chemical formula is similar to formula (10-22):

$$\left(CH_2 - \underset{\underset{COOCH_2OH}{|}}{\overset{\overset{CH_3}{|}}{C}} \right)_n \qquad (10\text{-}23)$$

where the OH group is the hydrophilic group responsible for hydration of the polymer. Generally, hydrogels for contact lenses are made by the polymerization or copolymerization of certain hydrophilic monomers with small amounts of a cross-linking agent such as ethylene glycol dimethacrylate (EGDM):

$$\begin{array}{c} CH_3 \\ | \\ CH_2 = C - COOCH_2 \\ | \\ CH_2 = C - COOCH_2 \\ | \\ CH_3 \end{array} \qquad (10\text{-}24)$$

Another type of hydrogel has the following chemical formula[5]:

$$\begin{array}{c} H \\ | \\ CH_2 = CCONH_2 \end{array} \qquad (10\text{-}25)$$

The acrylamide can be cross-linked by N,N-methylene-bis-acrylamide:

$$\begin{array}{c} H \\ | \\ CH_2 = CCONH \\ | \\ CH_2 \\ | \\ CH_2 = CCONH \\ | \\ H \end{array} \qquad (10\text{-}26)$$

In order to maintain the water content of polyHEMA, it is usually copolymerized with its higher homologs:

$$\begin{array}{c} CH_3 \\ | \\ CH_2 = C - COO(CH_2CH_2O)_n CH_2CH_2OH \end{array} \qquad (10\text{-}27)$$

Table 10-9. Oxygen Permeability Coefficient
of Contact Lens Materials[a]

	$P_g \times 10^4$ (μl-cm/cm^2-h kPa)[b]	Comments
Polymethylmethacrylate	0.27	Hard contact lens
Polydimethylsiloxane	1750	Flexible
Polyhydroxyethylmethacrylate	24	39% H$_2$O, soft lens

[a] From Ref. 6.
[b] At STP, to convert μl-cm/cm^2-h kPa to μl-cm/cm^2-h mm Hg, divide by 7.5.

($n = 1$, diethylene glycol monomethacrylate; $n = 2$, triethylene glycol mono-methacrylate). More conveniently, polyHEMA can be copolymerized with other hydrophilic monomers such as vinyl pyrrolidinone:

$$
\begin{array}{c}
CH_2\!\!-\!\!CH_2 \\
| \quad\quad | \\
CH_2 \quad C\!\!=\!\!O \\
\diagdown \;/ \\
N \\
| \\
CH\!\!=\!\!CH_2
\end{array}
\qquad (10\text{-}28)
$$

The water content of the copolymer can be increased to over 60% (the normal water content for polyHEMA is 40%).

The hydrogels have a relatively low oxygen permeability in comparison with silicone rubber (see Table 10-9). However, the permeability can be increased with increased hydration (water content) or decreased (lens) thickness. Silicone rubber is not a hydrophilic material but its high oxygen permeability and transparency make it an attractive lens material. It is usually used after coating with hydrophilic hydrogels, such as vinyl pyrro-lidinone, by grafting.[7]

10.6.2. Bone Cement (PMMA)

Recently, bone cement has been used in a greater number of clinical applications to secure a firm fixation of joint prostheses such as hip and knee joints. Bone cement is primarily made of PMMA powder and mono-mer methylmethacrylate liquid. One commercial product, Surgical Simplex® Radiopaque Bone Cement, gives the following description for its use.[8]

Surgical Simplex® P Radiopaque Bone Cement (methylmethacrylate; mixture of PMMA, methylmethacrylate-styrene-copolymer, and barium sulfate, U.S.P.) is a drug packaged in two sterile components.

One component is an ampule containing a 20-ml full dose of a colorless, flammable liquid monomer that has a sweet, slightly acrid odor and has the following composition:

Methyl methacrylate (monomer)	97.4v/o (volume %)
N,N-dimethyl-p-toluidine	2.6v/o
Hydroquinone	75 + 15ppm

Formula:

$$CH_2 = C - COOCH_3$$
$$|$$
$$CH_3$$

(10-29)

Methylmethacrylate (monomer)

Hydroquinone is added to prevent premature polymerization which may occur under certain conditions; e.g., exposure to light, elevated temperatures. N,N-dimethyl-p-toluidine is added to promote or accelerate cold-curing of the finished compound. The liquid component is sterilized by membrane filtration.

The other component is a packet of 40 g of finely divided white powder (mixture of polymethyl methacrylate, methyl methacrylate-styrene-copolymer, and barium sulfate, U.S.P.) of the following composition:

Polymethyl methacrylate	15.0w/o (weight %)
Methyl methacrylate-styrene-copolymer	75.0w/o
Barium sulfate (Ba SO$_4$), U.S.P.	10.0w/o

Formula:

$$\begin{array}{ccc} CH_3 & & CH_3 \\ | & & | \\ -CH_2-C-CH_2-C- \\ | & & | \\ COOCH_3 & COOCH_3 \end{array}$$

(10-30)

PMMA

$$\begin{array}{cccc} CH_3 & & O & & CH_3 \\ | & & | & & | \\ -CH_2-C-CH_2-CH-CH_2-C- \\ | & & & & | \\ COOCH_3 & & & & COOCH_3 \end{array}$$

(10-31)

Methylmethacrylate-styrene copolymer

When the two components are mixed together, the monomer liquid is polymerized by the free radical (addition) polymerization process. An

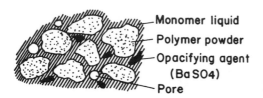

- Monomer liquid
- Polymer powder
- Opacifying agent (BaSO4)
- Pore

Figure 10-5. Two-dimensional representation of bone cement structure after curing. (From Ref. 11.)

activator, dibenzoyl peroxide (see reaction 10-5), will react with a monomer to form a monomer radical which will then attack another monomer to produce a dimer. The process will continue until long-chain molecules are produced. The monomer liquid will wet the polymer powder particle surfaces and link them together after polymerization, as shown in Figure 10-5. The following formulas give a concise view of free radical polymerization.

Initiation:

$$\langle\bigcirc\rangle-COOOOC-\langle\bigcirc\rangle \longrightarrow 2\langle\bigcirc\rangle\cdot + 2CO_2 \tag{10-32}$$

$$\langle\bigcirc\rangle\cdot + CH_2{=}\overset{\displaystyle COOCH_3}{\underset{\displaystyle CH_3}{C}} \longrightarrow \langle\bigcirc\rangle{-}CH_2{-}\overset{\displaystyle COOCH_3}{\underset{\displaystyle CH_3}{C}}\cdot \ + heat \tag{10-33}$$

$$(R\cdot) \qquad (M) \qquad\qquad\qquad (RM\cdot)$$

Propagation:

$$RM\cdot\ + M \longrightarrow RMM\cdot\ + heat$$
$$RMM\cdot + M \longrightarrow RMMM\cdot + heat$$
$$\vdots \qquad \vdots \qquad\qquad \vdots \qquad \vdots \tag{10-34}$$
$$RM_n\cdot\ + M \longrightarrow RM\cdot_{n+1}\ + heat$$

Termination:

Combining two free radicals or addition of a terminator

The free radicals associated with the above polymerization process can be monitored by *electron paramagnetic resonance* (EPR) spectrometry[9,10] as shown in Figure 10-6. Polymerization during curing obviously increases the degree of polymerization, that is, increases the molecular weight as given in Table 10-10. However, the molecular weight distribution does not change

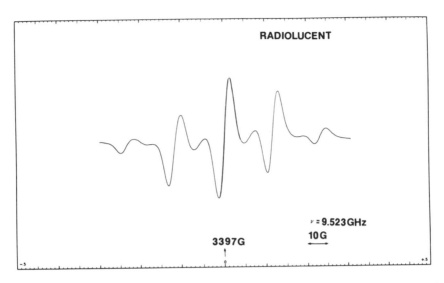

Figure 10-6. Typical EPR spectrum of bone cement during polymerization. (From Ref. 9.)

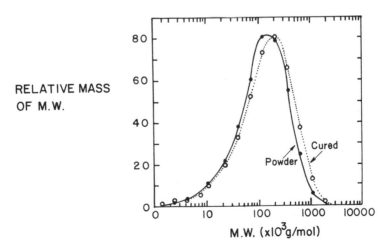

Figure 10-7. Molecular weight distribution of bone cement powder and after curing. (From Ref. 5.)

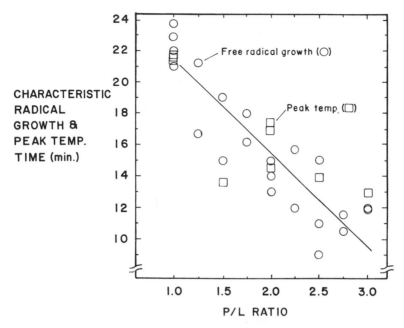

Figure 10-8. EPR characteristic free radical growth and peak temperature time versus powder/liquid (P/L) ratios of bone cement. (From Ref. 10.)

significantly after curing as shown in Figure 10-7. The properties of cured bone cement are compared with those of commercial acrylic resins in Table 10-11. These studies show that bone cement properties can be affected by the intrinsic and extrinsic factors listed in Table 10-12.

Figure 10-8 shows the relationship between the characteristic radical growth time (defined as the time to half-maximum EPR peak) and the powder/liquid (P/L) ratio. The characteristic growth time decreases with increasing P/L ratio. Similar results were obtained for the peak sample temperature time, which is plotted in the same figure. The decrease in both

Table 10-10. Molecular Weight of Bone Cement[a]

Types of M.W. (g/mol)	Powder	Cured
M_n (number average)	44,000	51,000
M_w (weight average)	198,000	242,000

[a] From Ref. 11.

Table 10-11. Physical Properties of Bone Cement
and Commercial Acrylic Resins

	Radiopaque bone cement[a]	Commercial acrylic resins[b]
Tensile strength (MPa)	28.9 ± 1.6	55–76
Compressive strength (MPa)	91.7 ± 2.5	76–131
Young's modulus (compressive loading, MPa)	2200 ± 60	2960–3280
Endurance limit (ultimate tensile strength)[c]	0.3	0.3
Density (g/ml)	1.10–1.23	1.18
Water adsorption (%)	0.5	0.3–0.4
Shrinkage after setting (%)	2.7–5	—

[a] From Ref. 11
[b] F-

rowth time and peak sample temperature time
due to the greater amount of initiator present in
n initiator, which is caused by the increased
ng the initiator, allows the polymerization reac-
The increased surface area of powder available
sing P/L ratio may also promote the reaction.
reached by the sample *decreases* with increasing
re 10-9. This is because of the lesser amount of
merization with increasing P/L ratio, resulting
ion reactions, and consequently reaching lower
her studies can be conducted using the EPR
radicals are present in the system.

)-12. Factors Affecting Bone Cement Properties

Intrinsic factors
  Composition of monomer and powder
  Powder particle size, shape, and distribution: degree
    of polymerization
  Liquid/powder ratio
Extrinsic factors
  Mixing environment, especially temperature
  Mixing technique
  Curing environment: temperature, pressure, contacting
    surface (tissue, air, water, etc.)

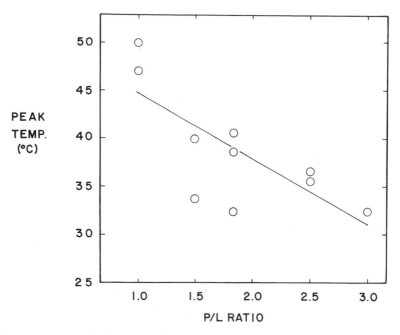

Figure 10-9. Peak temperature of bone cement setting versus P/L ratios. (From Ref. 10.)

Another interesting observation is the growth and decay of free radicals with time (i.e., polymerization and curing or aging), as shown in Figure 10-10. Although the bone cement "cures" and "hardens" in less than 30 min, it is obvious that complete curing takes much longer.

10.7. FLUOROCARBON POLYMERS

The best known fluorocarbon polymer is polytetrafluoroethylene (PTFE), commonly known as Teflon®. Other polymers containing fluorine are polytrifluorochloroethylene (PTFCE), polyvinylfluoride (PVF), and fluorinated ethylene propylene (FEP). Only PTFE will be discussed here since the others have rather inferior chemical and physical properties and are rarely used in implant fabrication.

PTFE is made from tetrafluoroethylene under pressure with a peroxide catalyst in the presence of excess water for removal of heat. The repeating unit is similar to that of polyethylene, except that the hydrogen atoms are

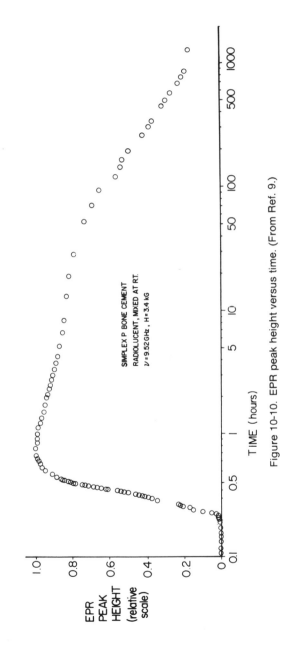

SIMPLEX P BONE CEMENT
RADIOLUCENT, MIXED AT R.T.
$\nu = 9.52\,GHz$, H=3.4 kG

TIME (hours)

Figure 10-10. EPR peak height versus time. (From Ref. 9.)

replaced by fluorine atoms:

$$\left(\begin{array}{cc} F & F \\ | & | \\ -C-C- \\ | & | \\ F & F \end{array}\right)_n \tag{10-35}$$

The polymer is highly crystalline ($> 94\%$ crystallinity) with an average molecular weight of 5×10^5 to 5×10^6 g/mol. This polymer has a very high density (2.15–2.2 g/ml), low tensile strength (17–28 MPa), and low modulus of elasticity (0.5 GPa). It also has a very low surface tension (18.5 ergs/cm^2) and friction coefficient (0.1).

PTFE cannot be injection molded or melt extruded because of its very high melt viscosity and it cannot be plasticized. Usually, the powders are sintered to above 327°C under pressure to produce implants.

10.8. RUBBERS

Three types of rubbers—silicone, natural, and synthetic rubbers—have been used in implant fabrication. Rubbers are defined by the ASTM as "a material which at room temperature can be stretched repeatedly to at least twice its original length and upon release of the stress, returns immediately with force to its approximate original length." Rubbers are stretchable because of the kinks of the individual chains as seen in *cis*-1,4-polyisoprene:

$$\tag{10-36}$$

On the other hand, the *trans* form cannot be stretched since the chains are not kinked. Instead, they crystallize, resulting in brittle gutta-percha:

$$\tag{10-37}$$

The *repeated* stretchability is due to the cross-links between chains which hold the chains together. The amount of cross-linking for natural rubber controls the flexibility of the rubber: the addition of 2–3% sulfur results in a flexible rubber, while adding as much as 30% sulfur makes a hard rubber.

Rubbers contain *antioxidants* to protect them against decomposition by oxidation, hence improving aging properties. *Fillers* of various types are also used to improve their physical properties. These include carbon black, hydrated silica, and quartz.

10.8.1. Natural and Synthetic Rubbers

Natural rubber is made mostly from the latex of the *Hevea brasiliensis* tree and its chemical formula is the same as that given in formula (10-36). In its pure form, rubber is compatible with blood.[15] Cross-linking by X-rays and organic peroxides produces a rubber with superior blood compatibility compared with rubbers made by conventional sulfur vulcanization.[16]

Synthetic rubbers were developed to substitute for natural rubber. There are many types available, as listed in Table 10-13. The Natta and Ziegler types of stereospecific polymerization techniques have made this variety possible. Polyisobutylene, butyl rubber (isobutylene/isoprene copolymer), SBR (styrene/butadiene copolymer), and polychloroprene (neoprene) are some of the synthetic rubbers produced in commercial quantity.

10.8.2. Silicone Rubbers

Silicone rubber is one of the few polymers developed for medical use.[17,18] The basic polymer is made by reacting silicon and methyl chloride:

$$Si + 2CH_3Cl \longrightarrow Cl-\underset{\underset{CH_3}{|}}{\overset{\overset{CH_3}{|}}{Si}}-Cl \qquad (10\text{-}38)$$

Methylchloride   Dimethyldichlorosilane

Table 10-13. Properties of Natural and Synthetic Rubbers

	Natural	SBR	Butyl	Neoprene
Tensile strength (MPa)	7–30	7–14	7–20	20
Elongation (%)	100–700	300–1000	100–700	—
Hardness (Shore A Durometer)	30–90	40–90	40–75	40–95
Density (g/ml)	0.92	0.92	0.92	1.23

Reacting further with water:

$$\underset{\underset{CH_3}{|}}{\overset{\overset{CH_3}{|}}{Cl-Si-Cl}} + H_2O \longrightarrow \underset{\underset{CH_3}{|}}{\overset{\overset{CH_3}{|}}{HO-Si-OH}} \qquad (10\text{-}39)$$

The reaction product is unstable and condenses, resulting in polymers:

$$n\ \underset{\underset{CH_3}{|}}{\overset{\overset{CH_3}{|}}{HO-Si-OH}} \longrightarrow \left(\underset{\underset{CH_3}{|}}{\overset{\overset{CH_3}{|}}{Si-O}}\right)_n + nH_2O \qquad (10\text{-}40)$$

Low-molecular-weight polymers have low viscosity and can be cross-linked to make a rubberlike material. Two types of vulcanization (cross-linking) processes can be used: heat vulcanization and room temperature vulcanization (RTV). The heat vulcanization process uses dichlorobenzoyl peroxide as the cross-linking agent. The peroxide breaks down on heating into a free radical which reacts with the polymer chain, causing a hydrogen to break away from the methyl groups and allowing cross-links to form. After cross-linking, the dichlorobenzene is dissipated by heat during curing.

Two types of RTV silicone rubbers are available, one-component and two-component types. The one-component silicone rubber uses a cross-linking agent, methyltriacetoxysilane [$(CH_3-Si-CO-CO-CH_3)_3$], which can be activated by water molecules. Acetic acid is a by-product. The two-component system uses a catalyst added at the time of vulcanization. For instance, medical-grade silicone rubber uses stannous octate as a catalyst.

Table 10-14. Properties of Silicone Rubbers

	Types of silicone rubber		
	Heat-vulcanizing		RTV Type 382
	Soft (MDX4-4515)	Hard (MDX4-4516)	
Density (g/ml)	1.12	1.23	1.13
Tensile strength (MPa)	6	7	2.8
Elongation (%)	600	350	160
Brittle point (°C)	−115	−75	−75
Elastic modulus	———— Varies up to 10 MPa ————		

These rubbers use silica (SiO_2) powder as fillers to improve their mechanical properties. The more fillers used, the higher the density of the rubber since silica has a higher density. As given in Table 10-14, this produces a harder rubber.

10.9. HIGH-STRENGTH THERMOPLASTICS

Recently, new polymeric materials have been developed to match the properties of light metals. These polymers have excellent mechanical, thermal, and chemical properties due to their stiffened main backbone chains. Polyacetals and polysulfones are being tested as implant materials; polycarbonates have found applications in heart/lung assist devices, food packaging, etc.

Polyacetals are produced by reacting formaldehyde

$$n \; \underset{\underset{H}{|}}{\overset{\overset{H}{|}}{C}}{=}O \longrightarrow O \underset{CH_2}{\diagup} O \underset{CH_2}{\diagup} O \left(\underset{CH_2}{\diagup} O \right)_n \qquad (10\text{-}41)$$

The polyformaldehyde (polyoxymethylene) is known widely as Derlin®. These polymers have a reasonably high molecular weight ($> 20,000$ g/mol) and excellent mechanical properties, as given in Table 10-15. More importantly, they display an excellent resistance to most chemicals and to water over wide temperature ranges. This material is being evaluated for the acetabular cup of hip joint replacement.[19]

Of the *polysulfones*, the most useful one is made from the reaction of disodium salt of bisphenol A [sodium salt of 2,2-(4-hydroxyphenol)propane]

Table 10-15. Properties of Polyacetal, Polysulfone, and Polycarbonate

	Polyacetal (Derlin®)	Polysulfone (Udel®)	Polycarbonate (Lexan®)
Density (g/ml)	1.425	1.24	1.20
Tensile strength (MPa)	70	70	63
Elongation (%)	15–75	50–100	60–100
Tensile modulus (GPa)	3.65	2.52	2.45
Water absorption (%, 24 h)	0.25	0.3	0.3

with 4,4'-dichlorodiphenyl sulfone in dimethylsulfoxide:

$$\text{NaO}-\langle\bigcirc\rangle-\underset{\underset{CH_3}{|}}{\overset{\overset{CH_3}{|}}{C}}-\langle\bigcirc\rangle-\text{ONa}+\text{Cl}-\langle\bigcirc\rangle-\text{SO}_2-\langle\bigcirc\rangle-\text{Cl}$$

$$(10\text{-}42)$$

$$\xrightarrow{\text{NaCl}} \left(\langle\bigcirc\rangle-\underset{\underset{CH_3}{|}}{\overset{\overset{CH_3}{|}}{C}}-\langle\bigcirc\rangle-\text{O}-\langle\bigcirc\rangle-\text{SO}_2-\langle\bigcirc\rangle-\text{O}\right)_n$$

These polymers have high thermal stability due to the bulky side groups (therefore amorphous) and rigid main backbone chains. They are also highly stable to most chemicals but weak to polar organic solvents such as ketones and chlorinated hydrocarbons. Some properties of polysulfone (Udel® P1700) are given in Table 10-15. These polymers are being tested for porous coating applications for orthopedic implants.[20]

Polycarbonates are tough, amorphous, and transparent polymers made by reacting bisphenol A and diphenyl carbonate:

$$n\ \text{HO}-\langle\bigcirc\rangle-\underset{\underset{CH_3}{|}}{\overset{\overset{CH_3}{|}}{C}}-\langle\bigcirc\rangle-\text{OH}+n\ \underset{\underset{OC_6H_5}{|}}{\overset{\overset{OC_6H_5}{|}}{C}}{=}\text{O} \xrightarrow{\text{NaOH}}$$

$$(10\text{-}43)$$

$$\left(\text{O}-\langle\bigcirc\rangle-\underset{\underset{CH_3}{|}}{\overset{\overset{CH_3}{|}}{C}}-\langle\bigcirc\rangle-\text{O}-\overset{\overset{O}{\|}}{C}\right)_n$$

The best known commercial polycarbonate is Lexan®. It has excellent mechanical and thermal properties as given in Table 10-15.

10.10. DETERIORATION OF POLYMERS

Polymers deteriorate due to chemical, thermal, and physical factors. These factors may act synergistically, accelerating the deterioration process.

The deteriorations affect the main backbone chain, side groups, cross-links, and their original molecular arrangement.

10.10.1. Chemical Effects

If a linear polymer is undergoing deterioration, the main chain will usually be *randomly scissioned*. Sometimes depolymerization occurs, which differs from random chain scission. This process is the inverse of the disproportionate chain termination of addition polymerization (cf. reactions 10-11 and 10-12).

Cross-linking of a linear polymer may result in deterioration. An example of this is low-density polyethylene. On the other hand, if cross-linking is broken by the oxygen or ozone attack on (poly)isoprene rubber, the rubber becomes brittle.

It is also undesirable to change the nature of bonds, as in the case of polyvinyl chloride and polyvinyl acetate:

$$
\begin{array}{c}
\text{H H H H H} \\
-\text{C}-\text{C}-\text{C}-\text{C}-\text{C}- \\
\text{H Cl H Cl H}
\end{array}
\longrightarrow
\begin{array}{c}
\text{H H H H} \\
-\text{C}-\text{C}-\text{C}-\text{C}-\text{C}- \ + \text{HCl} \\
\text{H H Cl H}
\end{array}
$$

Polyvinyl chloride

$$(10\text{-}44)$$

$$
\begin{array}{c}
\text{H H} \\
-\text{C}-\text{C}- \\
\text{H COOCH}_3
\end{array}
+ \text{H}_2\text{O} \longrightarrow
\begin{array}{c}
\text{H H} \\
-\text{C}-\text{C}- \\
\text{H OH}
\end{array}
+ \text{CH}_3\text{COOH}
\qquad (10\text{-}45)
$$

Polyvinyl
acetate

Polyvinyl alcohol
(water soluble)

The by-products of degradation can be irritable to tissues since they are, in this case, acids.

10.10.2. Sterilization Effects

In conjunction with sterilization, thermal effects play an important role in polymer deterioration.[21] In dry heat sterilization, the temperature varies between 160 and 190°C. This is above the melting and softening temperatures of many linear polymers like polyethylene and PMMA. In the case of

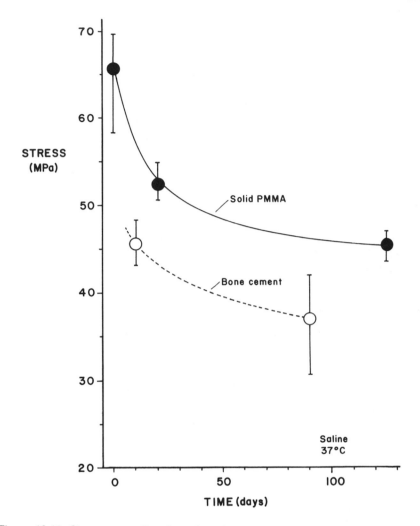

Figure 10-11. Stress versus time for polymethylmethacrylate in saline solution at 37°C. Note the large decrease in tensile strengths for both solid and porous bone cement. (Unpublished data of T. Parchinski, G. Cipoletti, and F. W. Cooke, Clemson University, 1977.)

polyamide (nylon), oxidation will occur at the dry sterilization temperature although this is below its melting temperature. The only polymers which can safely be dry sterilized are PTFE and silicone rubber.

Steam sterilization (autoclaving) is performed under high steam pressure at relatively low temperature (120–135°C). However, if the polymer is subjected to attack by water vapor, this method cannot be employed.

Polyvinyl chloride, polyacetals, polyethylenes (low-density variety), and polyamides belong to this category.

Chemical agents such as ethylene and propylene oxide gases, and phenolic and hypochloride solutions are widely used for sterilizing polymers since they can be used at low temperatures. Chemical sterilization takes a longer time and is more costly. Chemical agents sometimes cause polymer deterioration even when sterilization takes place at room temperature. However, the time of exposure is relatively short (overnight), and most polymeric implants can be sterilized with this method.

Radiation sterilization using the isotope ^{60}Co can also deteriorate polymers since at high dosage the polymer chains can be broken and recombined. In the case of polyethylene, at high dosage (above 10^6 Gy) it becomes a brittle, hard material. This is due to a combination of random chain scission and cross-linking.

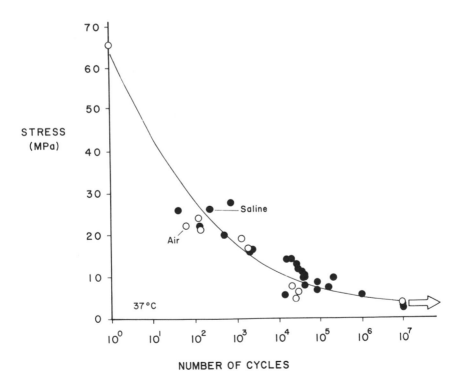

Figure 10-12. Fatigue test of solid polymethylmethacrylate. Note the S–N curve is the same both for nontreated samples and for samples soaked in saline solution at 37°C. (After T. Parchinski and F. W. Cooke, Clemson University, 1977.)

10.10.3. Mechanochemical Effects

It is well known that cyclic or constant loading deteriorates polymers. This effect can be accelerated if the polymer is simultaneously subjected to chemical and mechanical activation processes. Thus, if the polymer is stored in water or in saline solution, its strength will decrease as shown in Figure 10-11. Another reason for the decrease in strength is the plasticizing effect of the water molecules at higher temperature. However, the plasticizing effect compensates for the deleterious effect of the saline solution under cyclic loading, so that there is no difference between samples stored in saline versus in air as shown in Figure 10-12.

10.10.4. *In Vivo* Environmental Effects

Even though the materials implanted inside the body are not subjected to light, radiation, oxygen, ozone, or temperature variations, the body environment is very hostile, and all polymers start to deteriorate as soon as

Table 10-16. Effect of Implantation on Polymers

	Effects
Polyethylene	Low-density ones absorb some lipids and lose tensile strength; high-density ones are inert and no deterioration occurs
Polypropylene	Generally no deterioration
Polyvinyl chloride (rigid)	Tissue reaction, plasticizers may leach out and become brittle
Polyethylene terephthalate (polyester)	Susceptible to hydrolysis and loss of tensile strength
Polyamides (nylon)	Absorb water and irritate tissue, lose tensile strength rapidly
Silicone rubber	No tissue reaction, very little deterioration
Polytetrafluoroethylene	Solid specimens are inert; if it is fragmented into pieces, irritation will occur
Polymethylmethacrylate	Rigid form: crazing, abrasion, and loss of strength by heat sterilization Cement form: High heat generation during polymerization may damage tissues

they are implanted. The most probable cause of deterioration is ionic attack (especially OH^-) and dissolved oxygen. Enzymatic degradation[22] may also play a significant role if the implant is made from natural polymeric materials like reconstituted collagen.

It is safe to predict that if a polymer deteriorates in physiological solution *in vitro*, the same will be true *in vivo*. Most hydrophilic polymers such as polyamides and polyvinyl alcohol will react with body water and undergo rapid deterioration. The hydrophobic polymers like PTFE and polypropylene are less prone to deteriorate *in vivo*.

The deterioration products may induce tissue reactions. In the case of *in vivo* deterioration, the original physical properties will be changed if the implant deteriorates. For example, polyolefins (polyethylene and polypropylene) will lose their flexibility and become brittle. For polyamides, the amorphous region is selectively attacked by water molecules which act as plasticizers, making polyamides more flexible. Table 10-16 shows the effects of implantation on several polymeric materials.

PROBLEMS

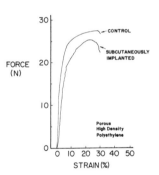

10-1. Porous polyethylene (diameter 110–500 μm, interconnecting pores) is tested in the form of rods with a diameter 3.4 mm *in vitro* and *in vivo* for 2 months. Typical force–elongation curves are obtained with a 1-cm gauge length and as shown in Figure P10-1.
a. Calculate the modulus of elasticity and tensile strength for both curves.
b. Explain why the curve for the *in vivo* sample is not as smooth as the control sample and why the strength is lower than in the control.

Figure P10-1.

10-2. A sample of methacrylate ($CH_2\text{=}CHCOOCH_3$) is polymerized. The resulting polymer has a DP of 1000. Draw the structure for the repeating unit of the polymer and calculate the polymer molecular weight.

10-3. An applied strain of 0.4 produces an immediate stress of 10 MPa in a piece of rubber, but after 42 days the stress is only 5 MPa.

a. What is the relaxation time?
b. What is the stress after 90 days?

10-4. What is the viscosity and shear modulus for a polymer which behaves as a Voigt model if the shear strains are:

Time (h)	Strain (%)
1	0.6
2	0.84
10	1.00
20	1.00

The shear stress is 1 MPa.

10-5. Polymers containing chemical bonds similar to those found in the body (e.g., amide and ester groups) are less biodegradable than polymers containing C–C and C–F bonds. Is this statement true?

10-6. Match the following with the lettered areas of Figure P10-6: liquids; greases; soft waxes; brittle waxes; tough waxes; and hard and soft plastics.

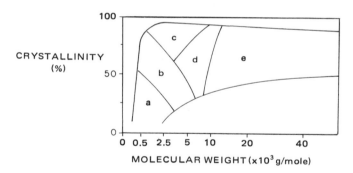

Figure P10-6.

10-7. The average end-to-end distance ($\bar{L}$) of a chain of an amorphous polymer can be expressed as $\bar{L} = lm^{1/2}$, where l is the interatomic distance (0.154 nm for C–C) and m is the number of bonds. If the average molecular weight of a polystyrene is 20,800 g/mol, what is the average end-to-end distance of a chain?

10-8. Name the following polymers.

a.
$$\begin{array}{cc} CH_3 & H \\ \diagdown & \diagup \\ C=C \\ \diagup & \diagdown \\ CH_2 & CH_2 \end{array}$$

b.
$$\begin{array}{c} CH_3 \\ | \\ -Si-O- \\ | \\ CH_3 \end{array}$$

c. $-CH_2CH_2-$

d. 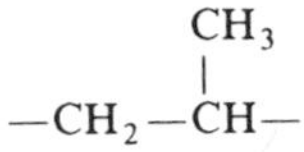 $-CH_2-\overset{\overset{\displaystyle CH_3}{|}}{CH}-$

e. 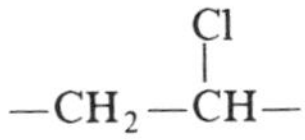 $-CH_2-\overset{\overset{\displaystyle Cl}{|}}{CH}-$

f. 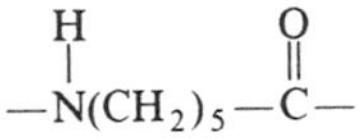 $-\overset{\overset{\displaystyle H}{|}}{N}(CH_2)_5-\overset{\overset{\displaystyle O}{\|}}{C}-$

g. 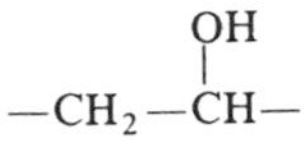 $-CH_2-\overset{\overset{\displaystyle OH}{|}}{CH}-$

h. $-CH_2-O-$

i. 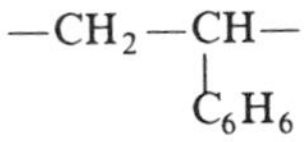 $-CH_2-\underset{\underset{\displaystyle C_6H_6}{|}}{CH}-$

j. 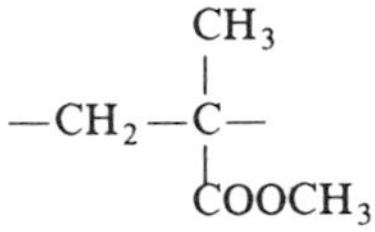 $-CH_2-\overset{\overset{\displaystyle CH_3}{|}}{\underset{\underset{\displaystyle COOCH_3}{|}}{C}}-$

k. $-CF_2-CF_2-$

l.

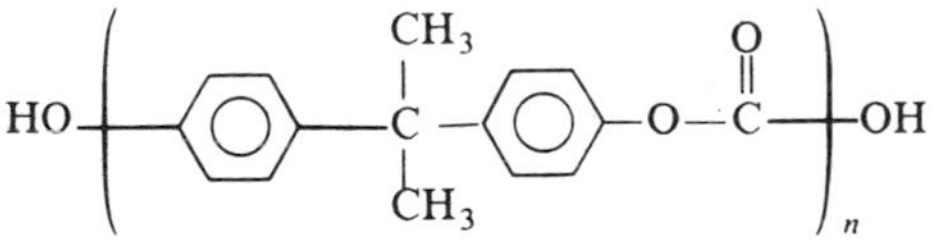

m.

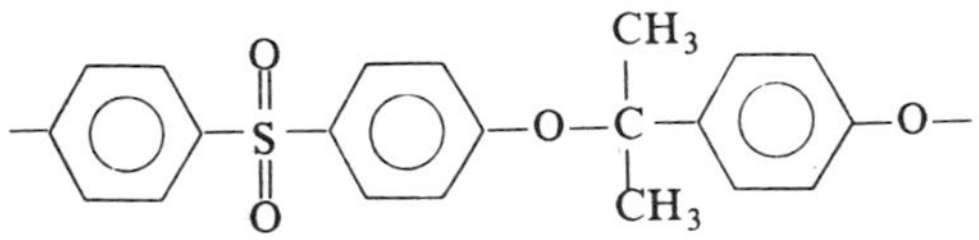

n. 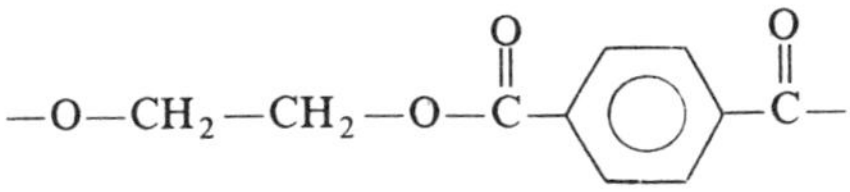 $-O-CH_2-CH_2-O-\overset{\overset{\displaystyle O}{\|}}{C}-$

o. $-CH_2-\overset{\overset{\displaystyle CH_3}{|}}{\underset{\underset{\displaystyle \underset{\underset{\displaystyle CH_2CH_2OH}{|}}{O}}{\underset{\underset{\displaystyle |}{C=O}}{|}}}{C}}-$

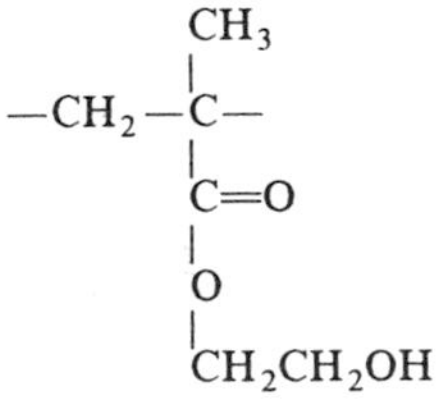

REFERENCES

1. B. Wunderlich, *Crystals of Linear Macromolecules*, ACS Audio Course, American Chemical Society, Washington, D.C., 1973.
2. J. H. Harrison and R. H. Adler, Nylon as a vascular prosthesis in experimental animals with tensile strength studies, *Surg. Gynecol. Obstet. 103*, 813–818, 1956.
3. G. H. Kenner, L. Hendricks, W. Barb, G. Gimenez, and J. B. Park, Bone embedding techniques with inhibited PMMA monomer, *Stain Technol. 57*, 121–128, 1982.
4. O. Wichterle and D. Lim, Hydrophilic gels for biological use, *Nature (London) 185*, 117–118, 1960.
5. W. M. Thomas, Acrylamide polymers, in: *Encyclopedia of Polymer Science and Technology*, N. M. Bikales (ed.), Volume 1, pp. 177–197, Interscience, New York, 1964.
6. M. F. Refojo, Contact lenses, in: *Encyclopedia of Chemical Technology*, 3rd ed., Volume 16, pp. 720–742, Wiley, New York, 1979.
7. A. S. Hoffman, A review of the use of radiation plus chemical and biochemical processing treatments to prepare novel biomaterials, *Radiat. Phys. Chem. 18*, 323–342, 1981.
8. Surgical Simplex P Bone Cement Technical Monograph, Howmedica Inc., Rutherford, N.J., 1977.
9. J. B. Park, R. C. Turner, and P. E. Atkins, EPR studies of free radicals in PMMA bone cement: A feasibility study, *Biomater. Med. Devices Artif. Organs 8*, 23–33, 1979.
10. R. C. Turner, P. E. Atkins, M. A. Ackley, and J. B. Park, Molecular and macroscopic properties of PMMA bone cement: Free radical generation and temperature change versus mixing ratio, *J. Biomed. Mater. Res. 15*, 425–432, 1981.
11. S. S. Haas, G. M. Brauer, and G. Dickson, A characterization of PMMA bone cement, *J. Bone Jt. Surg. 57A*, 380–391, 1975.
12. *Modern Plastics Encyclopedia*, Volume 57, p. 533, McGraw–Hill, New York, 1980.
13. R. P. Kusy, Characterization of self-curing acrylic bone cements, *J. Biomed. Mater. Res. 12*, 271–305, 1978.
14. P. R. Meyer, Jr., E. P. Lautenschlager, and B. K. Moore, On the setting properties of acrylic bone cement, *J. Bone Jt. Surg. 55A*, 149–156, 1973.
15. Y. Nose, J. Wright, M. Mathis, and W. J. Kolff, Natural rubber artificial heart, *Dig. 7th Int. Conf. Med. Biol. Eng.* p. 379, 1967.
16. K. Atsumi, Y. Sakurai, E. Atsumi, S. Narausawa, S. Kunisawa, M. Okikura, and S. Kimoto, Application of specially cross-linked natural rubber for artificial internal organs, *Trans. Am. Soc. Artif. Intern. Organs 9*, 324–331, 1965.
17. S. Braley, Acceptable plastic implants, in: *Modern Trends in Biomechanics*, D. C. Simpson (ed.), pp. 25–51, Butterworths, London, 1970.
18. Bulletin of the Dow Corning Center for Aid to Medical Research, Dow Corning Corp., Midland, Mich., 1970.
19. J. H. Dumbleton, Derlin as a material for joint prosthesis—A review, in: *Corrosion and Degradation of Implant Materials*, ASTM STP 684, B. C. Syrett and A. Acharya (ed.), pp. 41–60, American Society for Testing and Materials, Philadelphia, 1979.
20. M. Spector, M. J. Michon, W. H. Smarook, and G. T. Kwiatrowski, A high-modulus polymer for porous orthopedic implants, *J. Biomed. Mater. Res. 12*, 665–677, 1978.
21. B. Bloch and G. W. Hastings, *Plastics Materials in Surgery*, 2nd ed., Thomas, Springfield, Ill., 1972.
22. D. F. Williams, Some observations on the role of cellular enzymes in the *in vivo* degradation of polymers, in: *Corrosion and Degradation of Implant Materials*, ASTM STP 684, B. C. Syrett and A. Acharya (ed.), pp. 61–75, American Society for Testing and Materials, Philadelphia, 1979.

BIBLIOGRAPHY

B. Bloch and G. W. Hastings, *Plastic Materials in Surgery*, 2nd ed., Thomas, Springfield, Ill., 1972.

S. D. Bruck, *Blood Compatible Synthetic Polymers: An Introduction*, Thomas, Springfield, Ill., 1974.

S. D. Bruck, *Properties of Biomaterials in the Physiological Environment*, CRC Press, Boca Raton, Fla., 1980.

Guidelines for Physiochemical Characterization of Biomaterials, Report of the National Heart, Lung, and Blood Institute Work Group, Devices and Technology Branch, NIH Publication 80-2186, 1980.

Guidelines for Blood – Material Interactions, Report of the National Heart, Lung, and Blood Institute Working Group, Devices and Technology Branch, NIH Publication 80-2185, 1980.

E. P. Goldberg and A. Nakajima (ed.), *Biomedical Polymers, Polymeric Materials and Pharmaceuticals for Biomedical Use*, Academic Press, New York, 1980.

H. Lee and K. Neville, *Handbook of Biomedical Plastics*, Pasadena Technology Press, Pasadena, Calif., 1971.

S. N. Levine (ed.), *Polymers and Tissue Adhesives*, Ann. N.Y. Acad. Sci. *146*, 1968.

R. L. Kronenthal and Z. Oser (ed.), *Polymers in Medicine and Surgery*, Plenum Press, New York, 1975.

R. I. Leinninger, Polymers as surgical implants, *CRC Crit. Rev. Bioeng. 2*, 333–360, 1972.

M. F. Refojo, Contact lenses, in: *Encyclopedia of Chemical Technology*, 3rd ed., Volume 16, pp. 720–742, Wiley, New York, 1979.

M. F. Refojo, The chemistry of soft hydrogel lens materials, in: *Soft Contact Lenses*, M. Ruben (ed.), Chapter 3, Wiley, New York, 1978.

M. Szycher and W. J. Robinson (ed.), *Synthetic Biomedical Polymers, Concepts and Applications*, Technomic, Westport, Conn., 1980.

W. M. Thomas, Acrylamide polymers, in: *Encyclopedia of Polymer Science and Technology*, N. M. Bikales (ed.), Volume 1, pp. 177–197, Interscience, New York, 1964.

L. Vroman and F. Leonard (ed.), *The Behavior of Blood and Its Components at Interfaces*, Ann. N.Y. Acad. Sci. *238*, 1978.

O. Wichterle, Hydrogels, in: *Encyclopedia of Polymer Science and Technology*, N. M. Bikales (ed.), Volume 15, pp. 273–291, Interscience, New York, 1971.

O. Wichterle and D. Lim, Hydrophilic gels for biological use, *Nature (London) 185*, 117–118, 1960.

D. F. Williams, Biodegradation of surgical polymers, *J. Mater. Sci. 17*, 1233–1246, 1982.

11

SOFT TISSUE REPLACEMENT IMPLANTS

The success of soft tissue implants has primarily been due to the development of synthetic polymers. This is mainly because the polymers can be tailor-made to match the physical and chemical properties of soft tissues. In addition, polymers can be made into various physical forms, such as liquid for filling spaces, fibers for suture materials, films for catheter balloons, knitted fabrics for blood vessel prostheses, and solid forms for cosmetic and weight-bearing applications.

It should be recognized that different applications demand different materials with specific properties. The following are minimal requirements for all soft tissue replacements:

1. Reasonably close approximation of physical properties especially flexibility and texture.
2. The implants should not deteriorate.
3. The implants should not cause severe tissue reaction.
4. The implants should not induce thick fibrous tissue encapsulation or ingrowth.
5. The implants should be noncarcinogenic, nontoxic, nonallergenic, and nonimmunogenic.

Other important factors include sterilizability, feasibility of mass production, cost, fatigue life, and esthetic quality.

11.1. SUTURES, SURGICAL TAPES, AND ADHESIVES

The most common implants are sutures. In recent years surgical tapes and tissue adhesives have added to the surgeon's armamentarium. Although their use in actual surgery is limited, for some surgical procedures they are indispensable.

11.1.1. Sutures

There are two types of sutures according to their physical *in vivo* integrity, that is, absorbable and nonabsorbable. They may be distinguished according to their source of raw materials, that is, natural sutures (catgut, silk, and cotton) and synthetic sutures (nylon, polyethylene, polypropylene, stainless steel, and tantalum). Sutures may also be classified according to their physical form, that is, monofilament and multifilament.

The absorbable suture, catgut, is made of collagen and is derived from sheep intestinal submucosa. It is usually treated with a chromic salt to increase its strength and retard resorption by cross-linking. Such treatment extends the life of catgut suture from 3–7 days up to 20–40 days. Table 11-1 gives the strengths of British-made catgut.

It is interesting to note that the stress concentration at a surgical knot decreases the suture strength of catgut by half, no matter what kind of knotting technique is used. It has been suggested that the most effective knotting technique is the square knot with three ties to prevent loosening.[2] According to one study there is no measurable difference in the rate of wound healing whether the suture is tied loosely or tightly. Therefore, loose suturing is recommended because it lessens pain and reduces cutting soft tissues.[3]

Table 11-1. Minimum Breaking Loads for British-Made Catgut[a]

| Size | Diameter (mm) | | Minimum breaking load (lb) | |
	Min.	Max.	Straight pull	Over knot
7/0	0.025	0.064	0.25	0.125
6/0	0.064	0.113	0.5	0.25
5/0	0.113	0.179	1	0.5
4/0	0.179	0.241	2	1
3/0	0.241	0.318	3	1.5
2/0	0.318	0.406	5	2.5
0	0.406	0.495	7	3.5
1	0.495	0.584	10	5
2	0.584	0.673	13	6.5
3	0.673	0.762	16	8
4	0.762	0.864	20	10
5	0.864	0.978	25	12.5
6	0.978	1.105	30	15
7	0.105	1.219	35	17.5

[a] From Ref. 1.

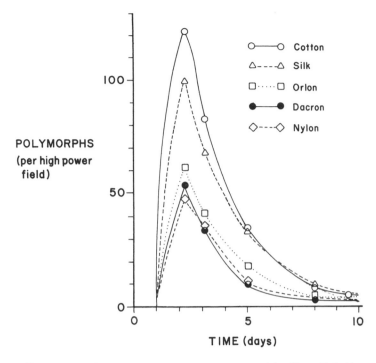

Figure 11-1. Cellular response to sutured materials. (From Ref. 4.)

Catgut and other absorbable sutures [nylon, polyglycolic acid (PGA)] invoke tissue reactions although the effect diminishes as they are being absorbed. This is true of other natural, nonabsorbable sutures like silk and cotton, which show more reaction than synthetic sutures like polyester, nylon, and polyacrylonitrile as shown in Figure 11-1. As is the case in the would-healing process (discussed in Chapter 7), the cellular response is most active 1 day after suturing and subsides in about a week.[4]

It is interesting to note that if the suture is contaminated even slightly, the incidence of infection increases manyfold.[5] The most significant factor is that the chemical structure and the geometric configuration of the suture seem to have no influence on the infection. Polypropylene, nylon, and PGA sutures cause less infection than other sutures, e.g., stainless steel, plain and chromic catgut, or polyester sutures.[6]

11.1.2. Surgical Tapes

The use of surgical tapes is supposed to offer a means of avoiding pressure necrosis, scar tissue formation, problems of stitch abscesses, and weakened tissues.[7] The problems with surgical tapes are similar to those experienced with Band-Aids, that is, (1) misaligned wound edges, (2) poor adhesion due to moisture or dirty wounds, (3) separation of tapes when hematoma, wound drainage, etc. occur.

The wound strength and scar formation in the skin may depend on the type of incision made. If the subcutaneous muscles in the fatty tissue were cut and the overlying skin was closed with tape, then the muscles retract. This in turn increases the scar area, resulting in poor cosmetic appearance when compared to a suture closure. However, due to the higher strength of scar tissue, the taped wound has higher wound strength than the sutured wound only if the muscle was not cut. Because of this, tapes have not enjoyed the success which was anticipated when they were first introduced, although early studies recommended their use enthusiastically.[8]

Tapes have been used successfully for assembling scraps of donor skin for skin graft, correcting nerve tissues for neural regrowth, etc.

11.1.3. Tissue Adhesives

The special environment of tissues and their regenerative capacity make the development of an ideal tissue adhesive difficult. Through past experience, the ideal tissue adhesive should be able to be wet and bind to tissues, be capable of rapid polymerization without producing excessive heat or toxic by-products, be resorbable as the wounds heal without interfering with the normal healing process, have ease of preparation for use in the operating theater, be sterilizable, have adequate shelf life and ease of large-scale production.[9,10]

The main strength of tissue adhesion comes from the covalent bonding between amine, carboxylic acid, and hydroxyl groups of tissues, and functional groups such as

$$R—\underset{\underset{O}{\diagdown\diagup}}{C{-}C}—, \qquad —\underset{\underset{NH}{\diagdown\diagup}}{C{-}C}—, \qquad RCNO$$

There are several adhesives available of which alkyl-α-cyanoacrylate is best known. Among the homologs of alkyl-cyanoacrylate, the methyl- and ethyl-2-cyanoacrylate are most promising. With the addition of some plasticizers and fillers, they are commercially known as Eastman 910 and Alpha S-2, respectively. An interesting comparison is illustrated in Figure 11-2, which shows that the bond strength of adhesive-treated wounds is about half that of the sutured wound after 10 days.[11] Because of the lower strength and lesser predictability of *in vivo* performance of adhesives, their application is limited to use after trauma on fragile tissues such as spleen,

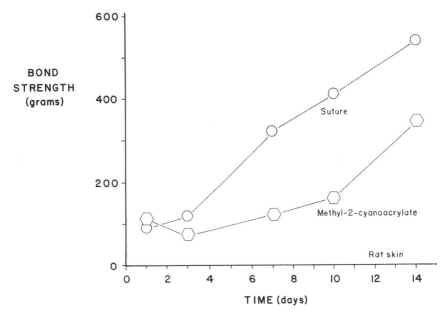

Figure 11-2. Bond strength of wounds with different closure materials. (From Ref. 11.)

liver, and kidney or after an extensive surgery on soft tissues such as lung. The topical use of adhesives in plastic surgery and fractured teeth has been moderately successful. As with any other adhesive, the end results of the bond depend on many variables such as thickness, porosity, and flexibility of the adhesive film, as well as the rate of degradation.

A polyurethane prepolymer type of adhesive was first used to bond bone fractures in 1959[12] and has since been studied for use in soft tissue repair.[13] Beech[14] experimented with three types of adhesives for dental application. The results are given in Table 11-2. These results show that bonding to tooth surfaces by the carboxylate cement is due to the ionized carboxyl group complexed with calcium ions. Therefore, bond strength is decreased when used in dentin, which has less calcium bonding sites than enamel. If the dentin is hypercalcified by a saturated solution of calcium hydrogen phosphate ("Brushite") and 100 ppm fluoride ion at pH 6.5, the bond strength increases.

The alkyl-2-cyanoacrylates have been shown to form covalent bonds with proteins. The TBB-MMA (tributyl, boron-initiated methylmethacry-late) system bonds to protein by graft copolymerization. The P/A-MMA (peroxide, amine-treated methylmethacrylate) system shows almost negligi-ble bonding capacity to any of the substrates when no primary bonding

Table 11-2. Tensile Strength of Various Adhesives[a]

	Substrate (MPa)			
	Enamel	Etched enamel	Dentin	Etched dentin
Polycarboxylate	7[b]	7[b]	3	0
Alkyl-2-cyanoacrylate				
Butyl	0	6	6	6
Ethyl	0	6	11[c]	11[c]
Methylmethacrylate				
Tributyl, boron-initiated	0	—	1	1
Peroxide, amine-treated	0	—	0	0

[a] From Ref. 14.
[b] No bond failure—cohesive fracture in cement.
[c] Values up to 25 MPa observed.

mechanism is present. Others have demonstrated that the interdiffusion and precipitation of polyelectrolytes on a dentin surface can lead to a strong adhesion of the applied, cured polymer layer.[15]

Adhesives have been prepared from fibrinogen, one of the clotting elements of blood.[16] This material has sufficient strength (0.1 MPa) and elastic modulus (0.15 MPa) to sustain the adhesiveness for the anastomoses of nerve, microvascular surgery, dural closing, bone graft fixation, skin graft fixation, and other soft tissue fixation.[17] This material is available commercially in Europe and will be in the U.S. pending FDA approval.

11.2. PERCUTANEOUS AND SKIN IMPLANTS

The need for percutaneous (trans or through the skin) implants has been accelerated by the advent of artificial kidneys and hearts, and by the prolonged injection of drugs and nutrients.

Artificial skin (or dressing) is urgently needed to maintain the body temperature of severely burned patients. Actual replacement of skin by biomaterials is beyond the capability of today's technology.

11.2.1. PERCUTANEOUS DEVICES

The problem of obtaining a functional and a viable interface between the tissue (skin) and an implant (percutaneous) device is primarily due to the following factors[18]:

1. Although initial attachment of the tissue into the interstices of the implant surface occurs, it cannot be maintained for a long period of time, since the dermal tissue cells turn over continuously and dynamically. Furthermore, down-growth of epithelium around the implant or overgrowth of implant (invagination) occurs.
2. Any openings large enough for bacteria to penetrate will result in *infection* even though initially there is complete sealing between skin and implant.

Many variables and factors are involved in the development of percutaneous devices. These are:

1. End-use factors
   a. Transmission of information (biopotentials, temperature, pressure, blood flow rate), energy (electrical stimulation, power for heart assist devices), matter (cannula for blood), and load (attachment of prosthesis)
2. Engineering factors
   a. Materials selection: polymers, ceramics, metals, and composites
   b. Design variation: button, tube with and without skirt, porous or smooth surface, etc.
   c. Mechanical stresses (soft or hard interface, porous or smooth interface)
3. Biological factors
   a. Implant host: man, dog, hog, rabbit, sheep, etc.
   b. Implant location: abdominal, dorsal, forearm, etc.
4. Human factors
   a. Postsurgical care
   b. Implantation technique
   c. Esthetic outlook

Figure 11-3 shows a simplified cross-sectional view of a generalized percutaneous device (PD) which can be broken down into five regions:

A. Interface between epidermis and PD should be completely sealed against invasion by foreign organisms. Dynamic growth provision at this interface should be made.
B. Interface between dermis and PD should reinforce the sealing of (A), as well as resist mechanical stresses. Due to the relatively large thickness of the dermis, the mechanical aspect is more important at this interface.

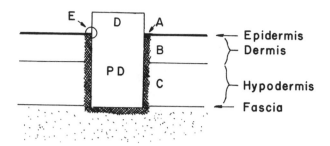

Figure 11-3. Simplified cross-sectional view of PD–skin interfaces. (From Ref. 18.)

C. Interface between hypodermis and PD should reinforce the function of (B). The immobilization of the PD against piston action is a primary function of (C).

D. Implant material per se should meet all the requirements of an implant for soft tissue replacement.

E. The line where epidermis, air, and PD meet is called a three-phase line which is similar to (A).

The stresses generated between a cylindrical percutaneous device and skin tissue can be simplified as shown in Figure 11-4. The relative motion of the skin and implant results in shear stresses which can be avoided if the implant floats (or moves) freely with the movement of the skin. For this reason, PDs without connected leads or catheters function longer. There have been many different PD designs to minimize shear stresses. All designs

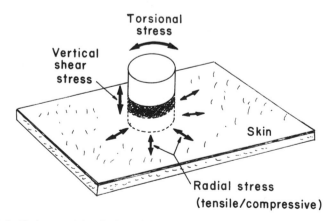

Figure 11-4. Various mechanical stresses acting at the PD–skin interface. (From Ref. 18.)

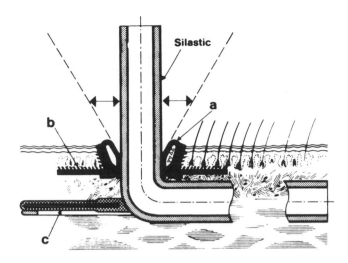

Figure 11-5. Schematic drawing of a Grosse-Siestrup PD. (From Ref. 19.)

have centered around creating a good skin tissue/implant attachment in order to *stabilize* the implant. This is done by providing felts, velours, and other porous materials at the interface. Figure 11-5 shows a design to minimize the transfer of stresses and strains to the skin. These include making an air chamber interposed between skin and PD, firmer fixation of the cannula by providing a large surface for tissue ingrowth (Figure 11-6) as well as minimizing the trauma imposed by the external tubes and wires by providing a pin connector with a good provision for firm tissue attachment subcutaneously (Figure 11-7).

A microporous surface modification (Figure 11-8) of the conventional segmented polyurethane blood access device was claimed to have several advantages over other designs: (1) the avoidance of interconnecting pores prevents open pathways through the interface structure and thus minimizes

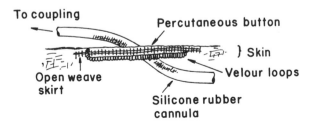

Figure 11-6. Simple PD as proposed by Cooke et al.[20]

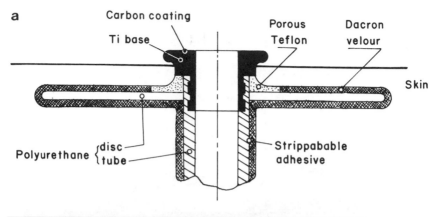

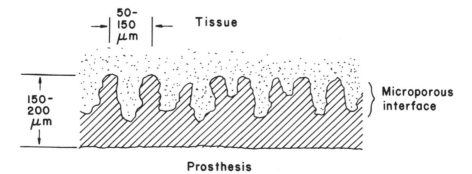

Figure 11-7. PD interfaces as suggested by Daly *et al.*[21] Schematic view; (bottom) finished product.

Figure 11-8. Cross-sectional representation of the microporous interface as suggested by Robinson *et al.*[22]

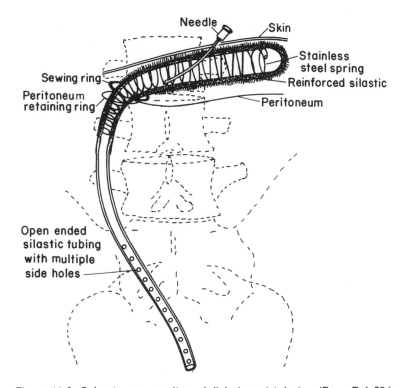

Figure 11-9. Subcutaneous peritoneal dialysis assist device. (From Ref. 23.)

infection and inadequate tissue adhesion; (2) there is a mechanical adhesion between the interconnecting "fingers" (protrusions and cavities) and between the microporous interface and ingrown tissues; and (3) the characteristics of the interface structure contribute to chronic compatibility, comparable to that of connective tissue.[22]

There have been, however, no PDs that are completely satisfactory.[1] Some researchers have switched to subcutaneous implants which can be accessed by a needle as shown in Figure 11-9 for peritoneal dialysis.[23]

11.2.2. Artificial Skins

Because artificial skin is another example of a percutaneous implant, the problems are similar to those described in the previous section. Most needed for this application is a material which can adhere to a large (burned) surface and thus prevent the loss of fluids, electrolytes, and other biomolecules until the wound has healed. Although a permanent skin implant is needed, it is a long way from being realized for the same reasons given in the case of percutaneous implants proper. Presently, autografting and homografting are the only methods available as a permanent solution.

Recently, Yannas and Burke's group[24-26] designed an artificial skin which was well conceived and was developed over several years. The wound closure was achieved by controlling physicochemical properties of wound-covering material (membrane). They determined six ways to improve certain physicochemical and mechanical requirements necessary in the design of artificial skin. These are shown schematically in Figure 11-10. Their bio-mechanical–chemical analysis has led to the design of a cross-linked collagen–polysaccharide (chondroitin 6-sulfate) composite membrane chosen for the ease in controlling porosity (5- to 150-μm diameter), flexibility (by varying cross-link density), and moisture flux rate.[26]

Several polymeric materials including reconstituted collagen have also been tested as burn dressing.[27,28] Among them are the copolymers of vinyl chloride and acetate and methyl-2-cyanoacrylate. The latter was found to be too brittle and histotoxic for use as a burn dressing. The ingrowth of tissue into the pores of sponge (Ivalon®, polyvinyl alcohol), and woven fabric (nylon and silicone rubber velour) was also attempted without much success.

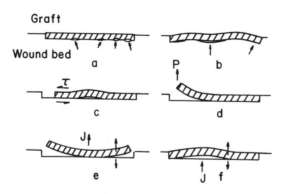

Figure 11-10. Schematic representation (not drawn according to scale) of certain physicochemical and mechanical requirements in the design of an effective wound closure. (a) Skin graft (cross-hatched) does not displace air pockets (arrows) efficiently from graft–wound-bed interface. (b) Flexural rigidity of graft is excessive; graft does not deform sufficiently under its own weight to make contact with depressions in wound-bed surface; as a result, air pockets form (arrows). (c) Shear stresses τ (arrows) cause buckling of graft, ruptures of graft–wound-bed bond, and formation of air pocket. (d) Peeling force P lifts graft away from wound-bed. (e) Excessively high moisture flux rate through graft causes dehydration and development of shrinkage stresses at edges (arrows) which cause lift-off away from wound bed. (f) Very low moisture flux J causes accumulation (edema) at graft–wound-bed interface and peeling off (arrows). (From Ref. 24.)

Plastic tapes have been used to hold skin graft during microtoming and grafting procedures. For severe burns the immersion into silcone fluid was found to be beneficial for prevention of early fluid loss, decubitus ulcers, and reduction of pain.

11.3. MAXILLOFACIAL AND OTHER SOFT TISSUE AUGMENTATION

In the previous section we have dealt with problems associated with wound closing and wound/tissue interfacial implants. In this section we will study (cosmetic) reconstructive implants. Although soft tissue implants can be divided into (1) space filler, (2) mechanical support, and (3) fluid carrier or storer, most have two or more combined functions. For example, breast implants fill space and provide mechanical support.

11.3.1. Maxillofacial Implants

There are two types of maxillofacial implant (often called prosthetics which implies extracorporeal attachment) materials: extraoral and intraoral. The latter is defined as "the art and science of anatomic, functional or cosmetic reconstruction by means of artificial substitutes of those regions in the maxilla, mandible, and face that are missing or defective because of surgical intervention, trauma, etc."

There are many polymeric materials available for the extraoral implant which requires: (1) color and texture should be matched with that of patients; (2) it should be mechanically and chemically stable, i.e., it should not creep and change colors or irritate skin; and (3) it should be easily fabricated. Polyvinyl chloride and acetate (5–20%) copolymers, PMMA, silicone, and polyurethane rubbers are currently used.[29]

The requirements for the intraoral implants are the same as for other implant materials since they are in fact implanted. For the maxillary, mandibular, and facial bone defects, a metallic material such as tantalum or Vitallium® is used.[30] For soft tissues like gum and chin, a polymer such as silicone rubber or PMMA is used for the augmentation. Table 11-3 gives most of the alloplasts used for maxillofacial surgery.[31] Figure 11-11 shows a dramatic improvement of facial features after a chin implant made of silicone rubber with Dacron® felt, used to help seat the implant in place, thus preventing its migration.

The use of injectable silicones which polymerize *in situ* has proven partially successful for correcting facial deformities. Although this is obviously a better approach in terms of the minimal initial surgical damage, this procedure was not accepted due to tissue reaction and the eventual displacement of the implant.

Table 11-3. Implant Materials Used in
Maxillofacial Surgery[a]

Facial contour correction materials
    Silicone rubber—fluid, gel, and solid form
        Chin, zygomatic, nasal maxillary, lateral
            mandibular, and orbital-frontal correction
        More extensive and longer use (20 years) than any
            other facial contour alloplast
        Preformed or trimmed to proper anatomical contour
        Bone resorption with and without loss of
            soft tissue profile reported by many authors
        Difficult in carving to host defect dimensions
        Supraperiosteal application with occasional
            mobility in soft tissues
        May be combined with other porous polymers (felts,
            velours) for tissue ingrowth stabilization
    Teflon[®]—solid or woven form with tissue tolerances
        nearly equal to silicone rubber
        Orbital floor sheeting replacement most successful
            application
        Occasional reports of replacing mandibular segments
            with limited success
        Tissue adherence is poor
    Other polymers
        Polymethylmethacrylate—chronic tissue reactions
            from monomer release and bone resorption have now
            excluded its use
        Polyethylene terephthalate (Dacron[®]) mesh
            impregnated with polyether urethane elastomer
            recently developed for facial and mandibular
            discontinuity defects serve as a framework for
            bone grafting—has had some limited use
    Proplast[®]–graphite-fiber–Teflon[®] composite with porous
        low-modulus characteristics
        Suitable for all facial applications as silicone
            rubber
        Produces tissue ingrowth with subperiosteal
            applications thereby mimicking contour rigidity of
            bone graft
        More easily carved and molded to anatomical shape
            than other polymers
        Low resilience and low modulus minimize tissue
            extrusion and bone resorption
        Antibiotic infusion necessary to avoid occasional
            contaminated implant

Mandibular reconstruction materials
    Ticonium[®] (NiCrCo alloy), Vitallium[®] (CoCrMo
        alloy)
        Not subjected to electrolytic activity of body
            fluids
        Use as tray with cancellous bone graft

Table 11-3 (*Continued*)

Mandibular reconstruction materials (*cont.*)
    Ticonium® (NiCrCo alloy), Vitallium® (CoCrNi alloy) (*cont.*)
        Not workable at surgical table, only able to cut
        Used in large reconstruction for its high strength
    Tantalum
        Not subjected to electrolytic activity of body fluids
        Not very workable at the surgical table
        Not ductile enough to drill with a high-speed drill
        Brittle—postoperative fractures
    Stainless steel mesh (316 type)
        Well tolerated
        Allows ingrowth of fibrous tissue
        Hard but can be cut or bent at table
        After it is cut, filed, or worked, particles will
            be left on the surface
    Titanium mesh
        With cancellous bone
        Used in large reconstruction
        More workable than other metallic mesh
        Metal implants serve to control and fix
            mandibular fragments thus decreasing
            immobilization period
    Kirschner wire and Steinman pin
        Skewered through a rib graft or polymer
        Usually used as a temporary space maintainer
        Will migrate through bone
        Cannot prevent rotation
        Alone lacks bulk to create a bed for the graft,
            but with Silastic®, bulk is added
    Silicone rubber blocks
        Do not absorb protein; stable up to 500°F
        Produce the least tissue reactivity
        Do not produce the degree of local tissue reaction
            necessary for their incorporation into host tissue
        Used as short-term space maintainer in mandibular
            and condyle reconstruction
        Custom-made wax model of mandible is invested and
            the resultant mold is packed with raw silicone
            rubber, flask and heat vulcanized
        Wire-reinforced silicone implant has been used to
            replace more than 80% of mandible
    Teflon® block
        Does not absorb protein; stable up to 500°F;
            inert to acid, alkali, bacteria
        Easily cut or bent without breaking
        Suitable for replacement of large defects
    Polyurethane—polyester (Dacron®) mesh cloth
        Implant prepared by saturating polyethylene
            terephthalate cloth mesh in the catalyzed

(*cont.*)

Table 11-3 (*Continued*)

Mandibular reconstruction materials (*cont.*)
  Polyurethane—polyester (Dacron®) (*cont.*)
    urethane; the wet mesh is applied tightly to a
    solid model of the section to be reconstructed and
    heat cured
  Used with cancellous bone
    Advantages
      Molded easily to fit a variety of shapes
      Does not have sharpness of metal
      Use with bone
    Disadvantages
      Cannot be reshaped at surgery
      Lack of strength
  Polymethylmethacrylate bond cement
    Has polarity and attracts water molecules, absorbs
    protein and cellular debris to cause foreign body
    reaction
    No evidence it produces malignant changes in
    humans, although fibrosarcomas have resulted from
    subcutaneous implantation in mice
    Easily molded to conform to bone contour
    Placement of burr holes in the implant will allow
    for tissue ingrowth

Temporomandibular joint reconstruction
  Deformities
  Chronic hypomobility—ankylosis
  Chronic hypermobility—dislocation
  Arthritis of the TMJ
    Degenerative joint disease or osteoarthritis
    Inflammatory joint disease or rheumatoid
    arthritis
    Infectious and metabolic diseases
  Deformities secondary to trauma—including
  ankylosis and dislocation
  Growth deformities, hyper- and hypoplasia
  Materials
    Fascia, fascia lata, dermis are used primarily in
    ankylosis patients following arthroplasty
    Autogenous costochondral graft and metatarsal
    joints are used in reconstruction of totally
    excised or absent condyles of children with
    ankylosis or agenesis
    Silicone rubber, acrylic, Teflon®, and metallic
    condylar stump prosthesis are advocated as
    interpositional articulation materials
    (pseudoarthrosis) in ankylosis patients
    following gap osteoarthroplasty
    Stainless steel, Co–Cr alloy, and others are
    occasionally used to line the glenoid fossa as
    opposed to capping the mobile mandibular stump

Table 11-3 (*Continued*)

Temporomandibular joint reconstruction (*cont.*)
   Materials (*cont.*)
      Metallic Proplast® condyles with ramus shanks and
        titanium condyles with ramus mesh are advocated
        in ankylosis following condylectomy; mandibular
        advancement to correct malocclusion is feasible
        with these prostheses
      Vitallium® or titanium mesh can be bent to a
        tubular shape for condylar blockage in
        hypermobility patients
      Silastic® and Proplast® composite are used in
        painful arthritis patients following condylar
        shave or high condylectomy—serves as meniscus
        or fibrous substitute to aid in normal joint
        function

---

[a] From Ref. 31.

11.3.2. Ear and Eye Implants

The use of implants can restore the conductive hearing loss from otosclerosis (a hereditary defect which involves a change in the bony tissue of the ear) and chronic otitis media (the inflammation of the middle ear which may cause partial or complete impairment of the ossicular chain; see Figure 11-12). Many different prostheses are available to correct the defects, some of which are shown in Figure 11-13. The porous polyethylene total ossicular replacement implant is used to obtain a firm fixation of the implant by tissue ingrowth. The tilt-top implant is designed to retard tissue ingrowth into the section of the shaft which may diminish sound conduction.[35] It is interesting to note that whereas the early design of the stapes implant was intended to replicate the natural shape, the recent designs have evolved into a much simpler shape as shown in Figure 11-14.

Many different materials have been tested for the fabrication of implants including PMMA, PTFE, polyethylene, silicone rubber, stainless steel, and tantalum. More recently, PTFE–carbon composite (Proplast®), porous polyethylene (Plastipore®), and pyrolytic carbon (Pyrolite®) have been shown to be suitable materials for otology implants.[36]

Eye implants are used to restore the functionality of the cornea and lens (see Figure 11-15) when they are damaged or diseased. Usually the cornea is transplanted from a suitable donor rather than implanted since the longevity of the cornea implant is uncertain due to fixation problems and infection. Figure 11-16 shows some of the eye implants tried clinically. They are made from "transparent" acrylics, especially PMMA, which has a high refractive index (1.49). Recently, intraocular lenses have become somewhat

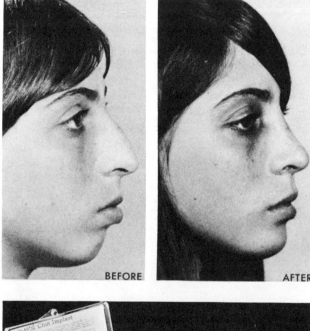

BEFORE   AFTER

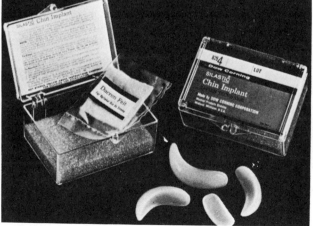

Figure 11-11. Chin implant in actual use and various sizes of chin implants made of silicone rubber. (Courtesy of Dow Corning Co., Midland, Mich.)

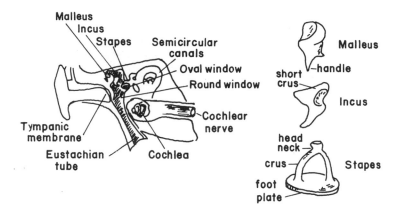

Figure 11-12. (Left) Anatomy of the ear; (right) details of malleus, incus, and stapes.

more popular than the usual thick-lensed glasses required after cataract surgery. The problems of infection and fixation are the major drawbacks of these implants as for the corneal implants.

Recently, there has been an attempt to develop an artificial eye for people who have lost all the conductive functions of the optic nerve by providing stimulation to the brain cells as shown in Figure 11-17.[39] One of the major problems with this type of total organ is the development of suitable electrode materials that will last a long time *in vivo* without changing their characteristics electrochemically.

11.3.3. Fluid Transfer Implants

Fluid transfer implants are required for cases such as hydrocephalus and urinary incontinence. Hydrocephalus, caused by abnormally high pressure of the cerebrospinal fluid in the brain, can be treated by draining the fluid (essentially an ultrafiltrate of blood) through a cannula as shown in Figure 11-18. The earlier shunt had two one-way valves at either ends while the Ames shunt has simple slits at the discharging end which opens when enough fluid pressure is exerted.[42] The Ames shunt empties the fluid in the peritoneum while others drain into the bloodstream through the right internal jugular vein or right atrium of the heart. The simpler peritoneal shunt showed a lower incidence of infection.[43]

The use of implants for correcting the urinary system has not been successful because of the difficulty of joining a fluid-tight prosthesis to the living system. In addition, blockage of the passage by deposits from urine and constant danger of infection have proven to be difficult to overcome. Many materials have been tested including glass, rubber, silver, tantalum,

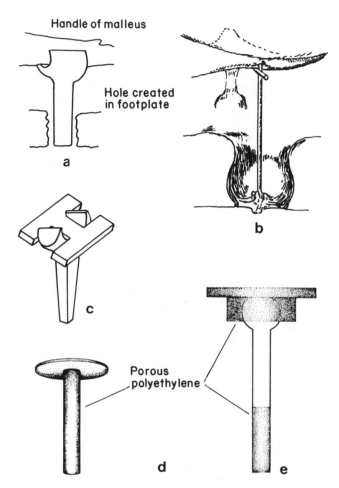

Figure 11-13. Prostheses for the reconstruction of the ossicles. (a) PTFE "piston" stapes prosthesis of Shea et al.[32] (b) Incus replacement prosthesis of Sheehy.[33] (c) Tabor prosthesis for replacement of whole ossicular chain.[34] (d) Porous polyethylene total ossicular replacement prosthesis.[35] (e) Same as (d) except the stem can be tilted.

Vitallium®, polyethylene, Dacron®, Teflon®, and polyvinyl alcohol without much long-term success.[44]

11.3.4. Space-Filling Soft Tissue Implants

Breast implants are quite common space-filling implants. In the early stages the enlargement of breasts was done with various materials such as

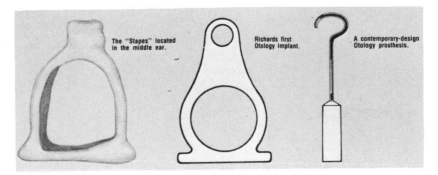

Figure 11-14. Early and contemporary designs of the otology implant. (Courtesy of Richards Manufacturing Co., Memphis, Tenn.)

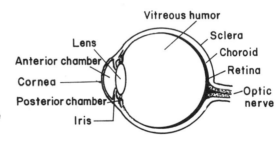

Figure 11-15. Anatomy of the eye.

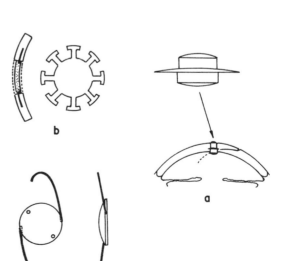

Figure 11-16. (a) Corneal implant of McPherson and Anderson.[37] (b) Corneal implant of Cardona.[38] (c) Intraocular lens. (Courtesy of Intra-Intermedics, Inc., Pasadena, Calif.)

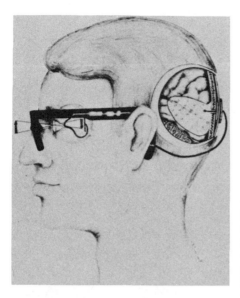

Figure 11-17. Diagram of concept of artificial eye. Television cameras in the glasses relate the message via micro-computers with radiowaves to the array of electrodes on the visual cortex of the brain. (From Ref. 55.)

paraffin wax, beeswax, or silicone fluids by direct injection or by enclosure in a rubber balloon. There have been several problems associated with directly injected implants, including progressive instability and ultimate loss of original shape and texture, as well as infection, pain, etc. In the 1960s the FDA banned such practice by classifying injectable implants such as silicone gel as drugs.

One of the early efforts of augmenting the breast was to implant a sponge made of polyvinyl alcohol. However, the ingrowth of soft tissues calcified with time and the so-called marble breast resulted. Although breast enlargement or replacement for cosmetic reason alone is not recommended, prostheses have been developed for the patient who has undergone radical mastectomy or who has nonsymmetrical deformities. They are probably beneficial for psychological reasons. In this case a silicone rubber bag filled with silicone gel and backed with polyester mesh to permit tissue ingrowth for fixation, is a widely accepted prosthesis. The artificial penis, testicles, and vagina fall into the same category as breast implants.

11.4. BLOOD INTERFACING IMPLANTS

Blood interfacing implants can be divided into two categories i.e., short-term extracorporeal implants such as membranes for artificial organs (kidney and heart/lung machine) and tubes and catheters for the transport

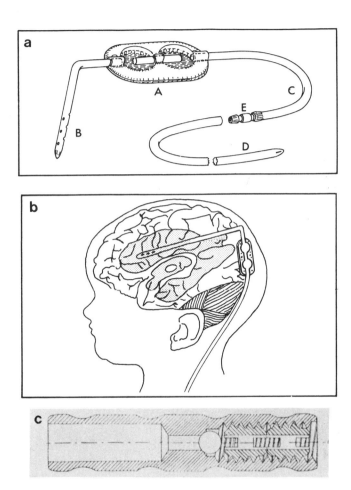

Figure 11-18. Ames design hydrocephalus shunt (a) [*in situ* (b)] and valve for another shunt (c). The shunt in (a) is made of silicone rubber (Silastic®) and consists of: (A) translucent double-chamber flushing device; (B) radiopaque ventricular catheter; (C) radiopaque connector tubing; (D) radiopaque peritoneal catheter; (E) stainless steel connector. (From Refs. 40 and 41.)

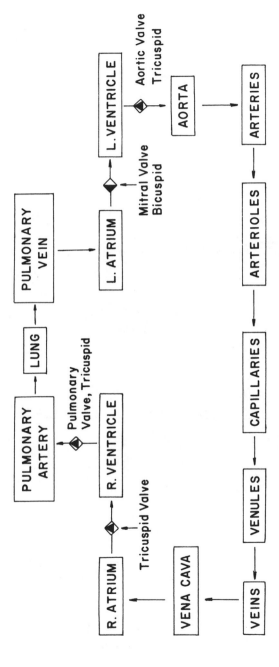

Figure 11-19. Schematic diagram of blood circulation in the body.

of blood, and long-term *in situ* implants such as vascular implants and implantable artificial organs. Although pacemakers for the heart are not interfaced with blood, they are considered here since they are devices that help to circulate blood throughout the body.

The single most important requirement for blood interfacing implants is blood compatibility.[46] Although blood coagulation is most important for blood compatibility, the implants should not damage proteins, enzymes, and formed elements of blood (red blood cells, white blood cells, and platelets). The latter includes hemolysis (red blood cell rupture) and initiation of the platelet release reaction.

Blood is circulated throughout the body according to the sequence shown in Figure 11-19. Implants are usually used to replace or patch large arteries and veins including the heart and its valves although surgical remedy without using implants is usually perferred. However, there are many unavoidable occasions when it is necessary to anastomose a large segment of vital organs with implants.

The basic requirements for blood interfacing implants are the same as for other soft tissue implants except that the surface exposed to blood should be made nonthrombogenic or at least thromboresistant. Most materials used for this type of application are made of polymers due to their flexibility and ease of fabrication.

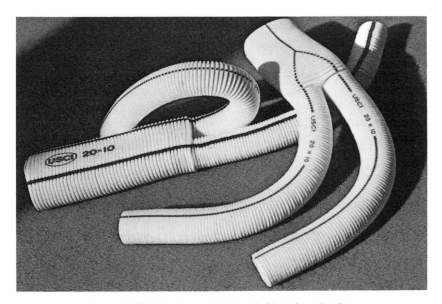

Figure 11-20. Modern arterial graft. Note the crimping.

11.4.1. Vascular Implants

Implants have been used in various circumstances of vascular maladies ranging from simple sutures for anastomosis after removal of vessel segments to vessel patches for aneurysms. Vein implants have encountered some difficulties as the result of the collapse of an adjacent vein or clot formation due to low pressure and stagnant flow. Vein replacements have not been a major concern since autografting can be performed in the majority of cases. Nonetheless, many materials including nylon, PTFE, and polyester were fabricated for clinical applications.

Early designs for material implants were solid tubes made of glass, aluminum, gold, silver, and PMMA. All of the implants developed clots. In the early 1950s porous implants were introduced which allowed tissue growth into the interstices (see Chapter 1). The new tissues interface blood, thus minimizing clotting. Ironically, for this type of application, thrombogenic materials were found more satisfactory. Another advantage of tissue ingrowth is the fixation of the implant by the ingrown tissue making a viable anchor. The initial leakage through pores is disadvantageous but this can be prevented by preclotting the outside surface of the implant prior to placement. Crimping of the prosthesis, as shown in Figure 11-20, is done to prevent kinking when the implant is flexed.[47]

Although the exact sequence of tissue formation in the implant in humans is not fully documented, quite a bit is known about reactions in animals. Generally, soon after implantation the inner and outer surface of the implant are covered with fibrin and fibrous tissue, respectively. A layer of fibroblasts replaces the fibrin, becoming neointima (sometimes called pseudointima). The long-term fate of the neointima varies in animals; in dogs it stabilizes into a constant thickness, whereas in pigs it will grow until it occludes the vessel. In man the initial phase of the healing is the same as for animals but in the later stage the inner surface is covered by both fibrin and a cellular layer of fibroblasts. The sequence for healing of arterial implants in pigs is given in Figure 11-21.

The types of materials and the geometry of the implant influence the rate and nature of tissue ingrowth. A number of polymeric materials are currently used to fabricate implants, including nylon, polyester, PTFE, polypropylene, and polyacrylonitrile. However, PTFE, polyester, and polypropylene are the most favorable materials due to the minimal deterioration of their physical properties *in vivo* as discussed in Chapter 10. Polyester (particularly polyethylene terephthalate, Dacron®) is usually preferred because of its superior handling properties.

Recently a pyrolytic carbon-coated arterial graft has been developed by the technique of ultralow-temperature isotropic (ULTI) deposition.[49] The

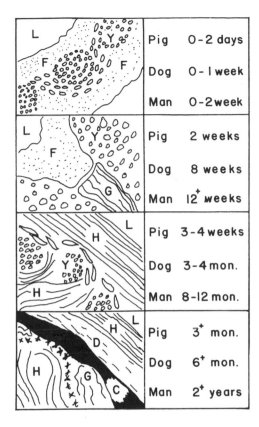

	Pig	0-2 days
	Dog	0-1 week
	Man	0-2 week
	Pig	2 weeks
	Dog	8 weeks
	Man	12^+ weeks
	Pig	3-4 weeks
	Dog	3-4 mon.
	Man	8-12 mon.
	Pig	3^+ mon.
	Dog	6^+ mon.
	Man	2^+ years

Figure 11-21. Basic healing pattern of arterial prosthesis. L, lumen of prosthesis; F, fibrin; Y, yarn bundle; G, organizing granulation tissue; H, healed fibrous capsular tissue; D, degenerative fibrous capsular tissue; C, calcified capsular tissue. (From Ref. 48.)

nonthrombogenic properties of pyrolytic carbon may enhance the patency of the graft made from this material and decrease use of anticoagulant drugs postsurgery.

The geometry of fabrics and porosity have a great influence on healing characteristics. The preferred porosity is such that 5000 to 10000 ml of water is passed per 1 cm^2 of fabric per minute at 120 mm Hg.[48] The lower limit will prevent excessive leakage of blood and the higher limit is better for tissue ingrowth and healing characteristics. The thickness of the implant is directly related to the amount of thrombus formation: the thinner the fabric the smaller, or the thinner, the thrombus deposit, resulting in faster organization of the neointima.

The long-term testing of vascular prostheses is as important as it is with any other implants. Botzko et al.[50] developed a simple in vivo testing machine as shown in Figure 11-22 in which the pseudoextracellular fluid is drawn through valve A and pushed out through valve B of the graft at 96

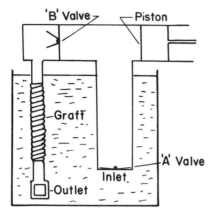

Figure 11-22. Schematic diagram of arterial graft life tester. (From Ref. 50.)

cycles per minute with a peak pressure of 150 mm Hg at 37°C. They have compared various grafts with *in vivo* implant samples as shown in Figure 11-23. It can be seen that the Teflon® knit graft did not lose its tenacity (the initial values for Teflon® and Dacron® knit prostheses are about 1.3 and 3.0 g/denier, respectively), while the Dacron® grafts showed initial decreases and stabilized after 6 months both *in vivo* and *in vitro*.[50]

11.4.2. Heart Valve Implants

There are four valves in the ventricles of the heart as shown in Figure 11-24 (cf. Figure 11-19). In the majority of cases, the left ventricular valves (mitral and aortic) become incompetent more frequently than the right

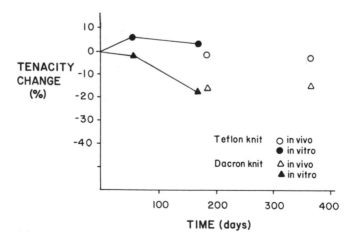

Figure 11-23. Percent change in tenacity for two types of prostheses after life testing of canine implant. (From Ref. 50.)

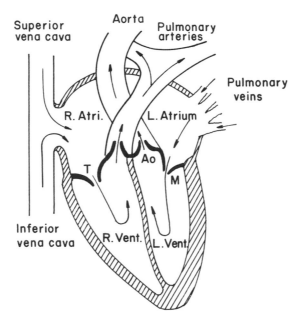

Figure 11-24. Circulation of blood in the heart. Compare with Figure 11-19.

ventricular valves as the result of higher left ventricular pressure. Most important and frequently critical is the aortic valve since it is the last gate the blood has to go through before being circulated in the body.

There have been many different types of valve implants. The early ones in the 1960s were made of leaflets which mimicked the natural valves. Invariably the leaflets could not withstand fatigue for more than 3 years. In addition to hemolysis, regurgitation and incompetence were major problems. Later, butterfly leaflets and ball- or disk-in-the cage were introduced. Some are shown in Figure 11-25. The material requirements for valve implants are the same as for vascular implants. Some additional requirements are related to blood flow and pressure, i.e., the formed elements of blood should not be damaged and should not drop the blood pressure below a clinically significant value. Also, valve noise should be minimal.

Figure 11-26 shows a porcine valve made from collagen-rich material such as pericardial tissues. Basically the pericardium is made up of three layers of collagen fibers oriented 60° from each layer and about 0.5 mm thick for bovine pericardium. It can be cross-linked by formaldehyde. During the process, the cell viability is destroyed which in turn does not provoke immunological reactions. Also, porcine xenograft valves have been used.[51]

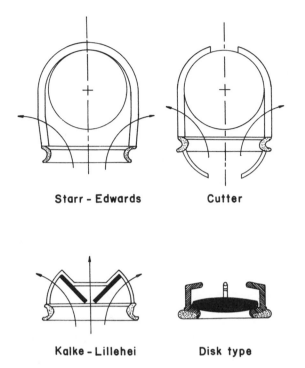

Starr - Edwards   Cutter

Kalke - Lillehei   Disk type

Figure 11-25. Schematic diagram of various types of heart valves.

All valves have a sewing ring which is covered with various polymeric fabrics. This helps during initial fixation of the implant. Later, the tissue ingrowth will render the fixation viable which is similar to the porous vascular implants. The cage itself is usually made of metals and covered with fabrics, to reduce noise, or with pyrolytic carbon for a nonthrombogenic surface (the disk or ball is also coated with pyrolytic carbon at the same time). The practice of covering the struts with fabrics has been abandoned since the fabric fatigued and broke into pieces.

The ball (or disk) is usually made of hollowed solid polymers (e.g., polypropylene, polyoxymethylene, polychlorotrifluoroethylene), metals (titanium, Co–Cr alloy), or pyrolytic carbon deposited on a graphite substrate. The early use of a silicone rubber poppet was found undesirable due to lipid adsorption and subsequent swelling and dimensional changes. Although this was an unfortunate episode (some were fatal), it helped to reinforce that the *in vitro* experiment alone is not sufficient to predict all circumstances that may arise during *in vivo* use, no matter how carefully one tries to predict.[45] This is true of any implant research even with very simple devices.

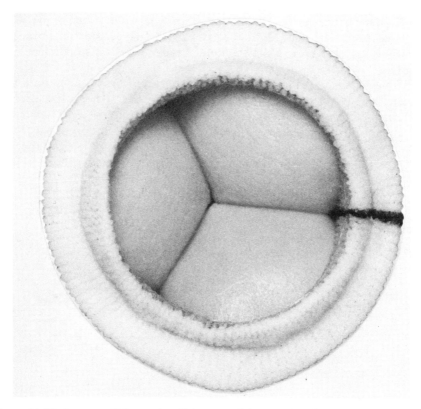

Figure 11-26. Ionescu–Shiley pericardial xenograft heart valve. (Courtesy of Shiley, Inc., Irvine, Calif.)

11.4.3. Heart Assist Devices

Heart assist devices are aimed at sustaining blood circulation when the natural heart cannot function normally or during cardiac surgery. While the blood is circulated by a pump, it can be oxygenated by either the patient's own lung or through an artificial oxygenator. Often the latter method is preferred to simplify surgery.[52]

There are basically three types of oxygenators as shown in Figure 11-27. In all cases, oxygen gas is contacted with blood and simultaneously waste gas (CO_2) is removed. In order to increase the rate of gas exchanges at the blood/gas surface of the bubble oxygenator, the gas is broken into small bubbles (about 1-mm diameter; if smaller, it is hard to remove them from blood) to increase the surface contact area. Sometimes the blood is spread thinly as a film and exposed to the oxygen. This is called a film oxygenator.

A membrane oxygenator is similar to the artificial kidney membrane to be discussed later. The main difference is that the oxygenator membrane is permeable to gases only, while the kidney membrane is also permeable to liquids.

Some of the mechanical and chemical characteristics of the natural and artificial lung (oxygenator) are compared in Table 11-4. The surface area of the artificial membrane is about 10 times larger than the natural lung since the amount of oxygen transfer through a membrane is proportional to the surface area, pressure, and transit time but inversely proportional to the film thickness (blood). It is noted that the blood film thickness of the artificial membrane is about 30 times larger than in the natural lung. This has to be compensated for by increased transit time (16.5 s) and higher pressure (650 mm Hg) to achieve the same amount of oxygen transfer as the lung.

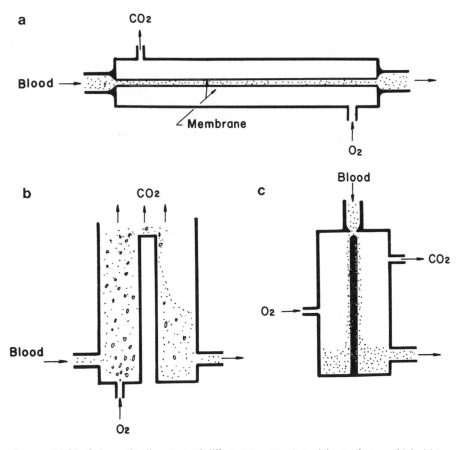

Figure 11-27. Schematic diagrams of different oxygenators: (a) membrane; (b) bubble; (c) film.

Table 11-4. Mechanical and Chemical Characteristics of Natural versus Artificial Lung[a]

	Natural lung	Artificial lung
Pulmonary flow (liters/min)	5	5
Head of pressure (mm Hg)	12	0–200
Pulmonary blood volume (liters)	1	1–4
Blood transit time (s)	0.1–0.3	3–30
Blood film thickness (mm)	0.005–0.010	0.1–0.3
Length of capillary (mm)	0.1	20–200
Pulmonary ventilation (liters/mm)	7	2–10
Exchange surface (m^2)	50–100	2–10
Veno-alveolar O_2 gradient (mm Hg)	40–50	650
Veno-alveolar CO_2 gradient (mm Hg)	3–5	30–50

[a] From Ref. 53.

The membranes are usually made of silicone rubber or PTFE. The gas permeability of these materials is given in Table 11-5. It is noted that silicone rubber is 40 and 80 times more permeable to O_2 and CO_2 than PTFE but the latter can be made about 20 times thinner. Therefore, silicone rubber is only 2 and 4 times better than PTFE for O_2 and CO_2 transfer.

Table 11-5. Gas Permeability of Teflon® and Silicone Rubber Membranes[a,b]

Membrane	Thickness (mil)	Oxygen	Carbon dioxide	Nitrogen	Helium
Teflon®	1/8	239	645	106	1425
	1/4	117	302	56	730
	3/8	77	181	35	430
	1/2	61	126	30	345
	3/4	41	86	23	240
	1	29			
Silicone	3	391	2072	184	224
rubber	4	306	1605	159	187
	5	206	1112	105	133
	7	159	802	81	94
	12	93	425	48	51
	20	59	279	31	43

[a] From Ref. 53.
[b] Permeation rates of oxygen, carbon, dioxide, nitrogen, and helium across Teflon® and silicone rubber membranes of a given thickness, in ml/min-m^2-atm (STP).

Polyurethane, natural and silicone rubbers have been used for constructing balloon-type assist devices as well as for coating the inner surfaces of the total artificial heart. This is because these materials are thromboresistant. Sometimes the surfaces are coated with heparin and other nonthrombogenic molecules. The felt or velour surface was not successful as an imitation of natural tissues on the inner surface of an artificial heart wall.

11.4.4. Artificial Organs

The ultimate triumph of biomaterials science would be to make implants behave or function the same way as the organs or tissues they replace. As mentioned in Chapter 1, most implants are designed to substitute mechanical functions. The electrical functions can be taken over by some implants (pacemakers) and some primitive yet vital chemical functions can also be delegated to implants (kidney machine and oxygenator). Most of the *artificial heart* and *heart assist devices* use a simple balloon and valve system. Figure 11-28 shows some typical heart assist devices. In all cases a balloon is used to displace blood. A simpler heart assist device is the

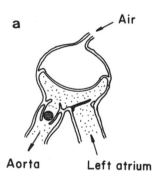

a

Air

Aorta   Left atrium

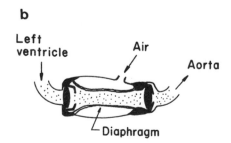

b

Left
ventricle   Air   Aorta

Diaphragm

Figure 11-28. Schematic diagram of heart assist devices. (a) De Bakey left ventricular bypass. (b) Bernard–Teco assist pump. (From Ref. 45.)

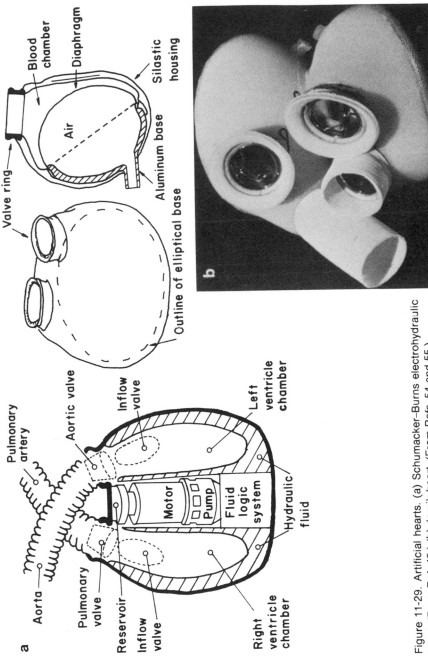

Figure 11-29. Artificial hearts. (a) Schumacker–Burns electrohydraulic heart. (From Ref. 45.) (b) Jarvik heart. (From Refs. 54 and 55.)

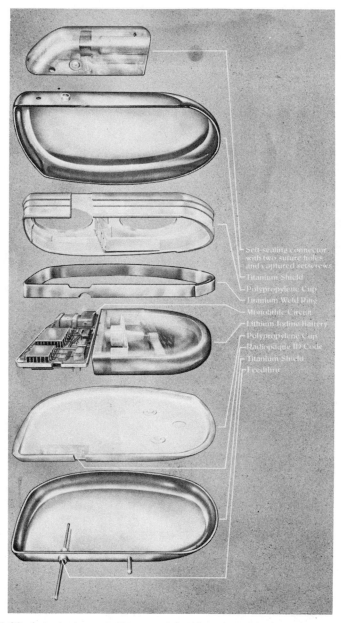

Figure 11-30. A typical pacemaker consists of a power source and an electronic circuitry encased in solid plastic while the electrical wires are coated with a flexible polymer, usually a silicone rubber.

intraaortic balloon which is placed in the descending aorta. During the diastolic phase of the heart, the balloon is inflated to prevent backflow.

Two artificial hearts are shown in Figure 11-29. Although the design principle and material requirements are the same as those for assist devices, the power consumption (~ 6 W) is too high to be completely implanted at this time. A miniature, totally implantable, nuclear-powered, artificial heart is being developed.[54] So far the power is introduced through a percutaneous device (Section 11.2.1) in the form of compressed air or electricity as was the case with the first heart replacement done at the University of Utah.

A *cardiac pacemaker* is used to assist the regular contraction rhythm of heart muscles. The sinoatrial (SA) node of the heart originates the electrical impulses which pass through the bundle of His to the atrioventricular (AV) node. In the majority of cases, the pacemakers are used to correct the conduction problem in the bundle of His. Basically the pacemakers should deliver an exact amount of electrical stimulation to the heart at varying heart rates. The pacemaker consists of conducting electrodes attached to a stimulator as shown in Figure 11-30.

The electrodes are well insulated with rubber (usually silicone) except the tips which are sutured or directly embedded into the cardiac wall as shown in Figure 11-31. The tip is usually made of a noncorrosive noble metal with reasonable mechanical strength such as Pt–10% Ir alloy. The

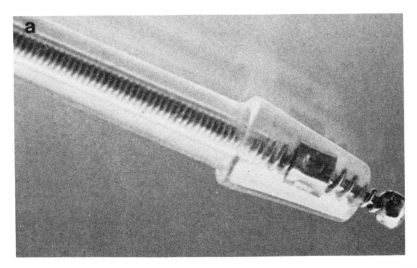

Figure 11-31. Different types of pacemaker electrodes. (a) Ball-point electrode (Cordis). The ball has a 1-mm diameter and the surface area is 8 mm. (b) Screw-in electrode (Medtronic). (c) Details of an arterial electrode (Medtronic 6991). (From Ref. 56.)

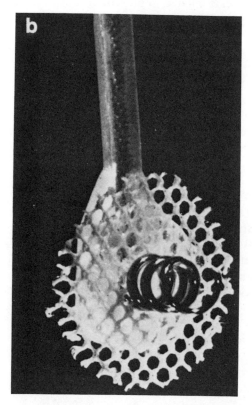

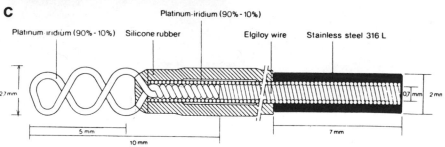

Figure 11-31. (*Continued*)

most significant problems are the fatigue of the electrodes (they are coiled like springs to prevent this) and the formation of collagenous scar tissue at the tip which increases electrical resistance at the point of tissue contact.[56] The battery and electronic components are insulated by casting in a clear polymer resin.[57]

Pacemakers are usually changed after 2–5 years due to the limitation of the power source. A nuclear energy-powered pacemaker is commercially available. Although this and other new power packs (such as the lithium battery) may lengthen the life of the power source, the fatigue of the wires and diminishing conductivity due to tissue thickening limit the maximum life time of the pacemaker to less than 10 years. A porous electrode at the tip of the wires may be fixed to the cardiac muscles by tissue ingrowth as in the case of a vascular prosthesis. This may solve the interfacial problems as shown in Figure 11-32.

The primary function of the kidneys is to remove metabolic waste products. This is accomplished by passing blood through a glomerulus under a pressure of about 75 mm Hg. The glomerulus contains up to 10 primary branches and 50 secondary loops to filter the blood. The glomeruli are contained in Bowman's capsule which in turn is a part of the nephron (Figure 11-33). The main filtrates are urea (70 times the urea content of normal blood), sodium, chloride, bicarbonate, potassium, glucose, creatinine, and uronic acid.

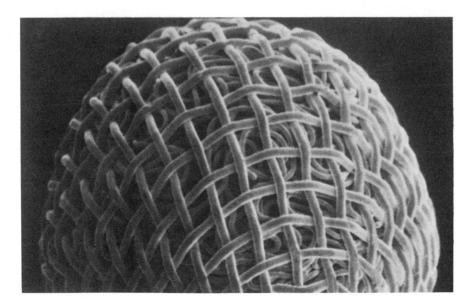

Figure 11-32. Amundsen/CPI porous electrodes. (From Ref. 58.)

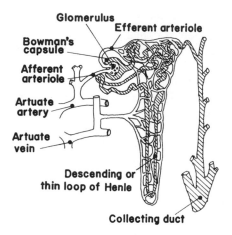

Figure 11-33. Schematic diagram of a kidney nephron.

The membrane is the key component of an artificial kidney machine. In fact, the first attempt to filter or dialyze blood with a machine failed due to an inadequate membrane. Besides a membrane filter the kidney dialyzer consists of a bath and a pump to circulate blood from the artery and return the cleansed blood to the vein as shown in Figure 11-34.

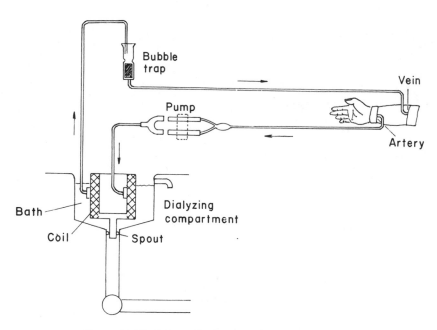

Figure 11-34. Schematic diagram of a typical dialyzer.

There are basically three types of membranes for the kidney dialyzer, shown in Figure 11-35. Historically, the flat plate-type membrane was developed first and can be two or four layers. The blood passes through the spaces between the membrane layers while the dialysate is passed through the spaces between the membrane and the restraining boards. The second and most widely used type is the coil membrane in which two cellophane tubes (each with 9-cm circumference and 108-cm length) are flattened and coiled with an open-mesh spacer material made of nylon. The newest addition is made of hollow fibers. Each fiber has dimensions of 255- and 285-μm inside and outside diameter, respectively, and is 13.5 cm long. Each unit contains up to 11,000 hollow fibers. The blood flows through the fibers while the dialysate is passed through the outside of the fibers. The operational characteristics of the various dialyzers are given in Table 10-6.

Recently there have been some efforts to improve the dialyzers using charcoals. The blood can be circulated directly over the charcoal or the charcoal can be made into microcapsulates incorporating enzymes or other drugs. One drawback of activated carbon filtering is that it ineffectively absorbs urea, which is one of the major by-products to be eliminated by the dialysis.

The majority of dialysis membranes are made from cellophane which is derived from cellulose. Ideally the membrane should selectively remove all the metabolic wastes as does the normal kidney. Specifically the membrane should not selectively sequester materials from dialyzing fluid, should be blood compatible so that an anticoagulant is not needed, and should have sufficient wet strength to permit ultrafiltration without significant dimensional changes. It should allow passages of low molecular weight waste product while preventing passage of plasma proteins.

There are two clinical-grade cellophanes available, Cuprophane® (Bemberg Co., Wuppertal, Germany) and Visking® (American Viscose Co., Fredricksberg, Va.). The cellophane films contain 2.5-nm-diameter pores which can filter molecules smaller than 4000 g/mol. A typical curve for flow rate versus clearance is shown in Figure 11-36. The clearance curve is flattened at a high flow rate, the so-called square meter concept which is an important design factor of a membrane.

There have been many attempts to improve the cellophane membrane wet strength by cross-linking, copolymerization, and reinforcement with other polymers such as nylon fibers. Also, the surface has been coated with heparin to prevent clotting. Other membranes such as copolymer of polyethyleneglycol and polyethylene terephthalate can filter selectively due to the alternate hydrophilic and hydrophobic segments. Besides improving the membrane for better dialysis, the main thrust of kidney research is to make the kidney machine more compact (portable or wearable kidney) and less costly (home dialysis, reusable or disposable filters, etc.)

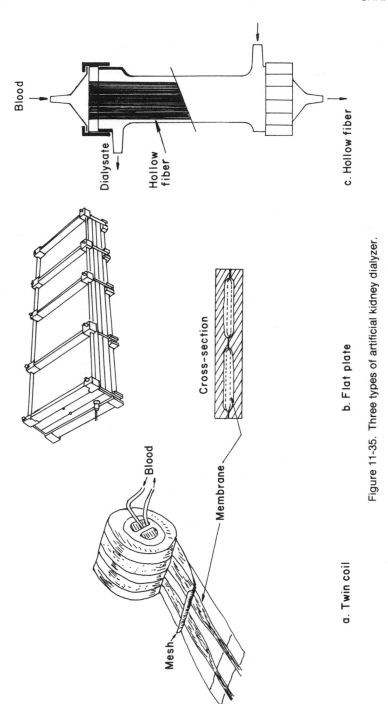

Figure 11-35. Three types of artificial kidney dialyzer.

Table 11-6. Comparison of the Plate and Coil
Artificial Kidneys[a]

	Flat plate (2 layers)	Coil (twin)
Membrane area (m²)	1.15	1.9
Priming volume (ml)	130	1000
Pumped needed?	No	Yes
Blood flow rate (ml/min)	140–200	200–300
Dialysate flow rate (liters/min)	2.0	20–30
Blood channel thickness (mm)	0.2	1.2
Treatment time (h)	6–8	6–8

[a] From Ref. 53.

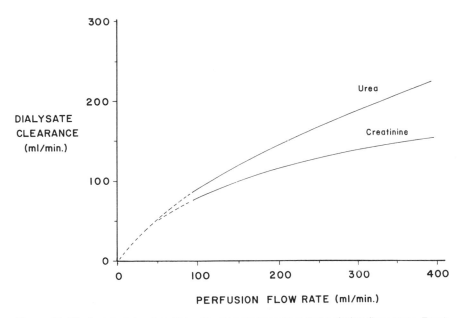

Figure 11-36. Average *in vitro* dialysate clearance versus test solution flow rates. Fresh dialysis bath addition rate in a circulating single pass (RSP) dialyzer is 600 ml/min with bubble trap pressure of 60 mm Hg at 37°C. (Courtesy of Travenol Laboratories, Inc., Morton Grove, III.)

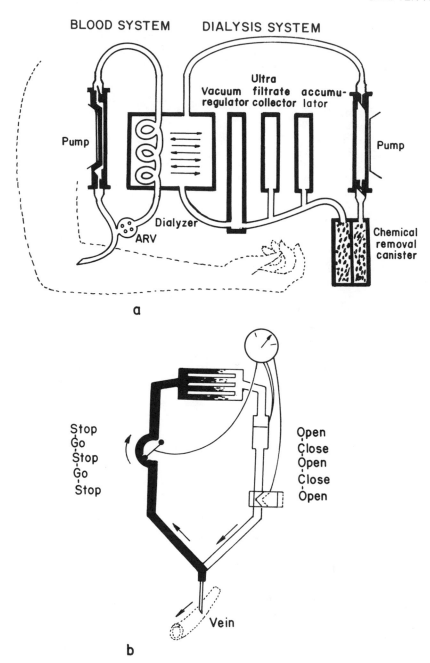

Figure 11-37. (a) Diagram of wearable artificial kidney and (b) schematic arrangement of the single-needle dialysis technique of Dr. Klaus Kopp. The pump operates continuously or intermittently synchronized with the inflow and outflow of blood. [(a) from Ref. 55; (b) from Ref. 54.]

The other important factor in dialysis is the use of a cannula which needs to be connected to the blood vessels as discussed in Section 11.2.1. In order to minimize the repeated trauma on the blood vessels, the cannula may be implanted for a long period in chronic kidney patients. Also, a single-needle dialysis technique has been developed for the same reason as shown in Figure 11-37.

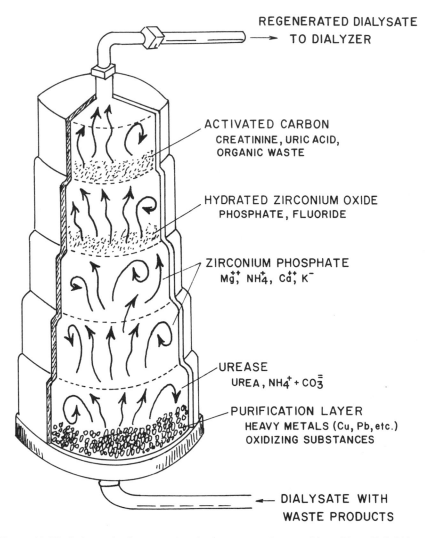

REGENERATED DIALYSATE
→ TO DIALYZER

ACTIVATED CARBON
CREATININE, URIC ACID,
ORGANIC WASTE

HYDRATED ZIRCONIUM OXIDE
PHOSPHATE, FLUORIDE

ZIRCONIUM PHOSPHATE
Mg^{++}, NH_4^+, Cd^{++}, K^-

UREASE
UREA, $NH_4^+ + CO_3^=$

PURIFICATION LAYER
HEAVY METALS (Cu, Pb, etc.)
OXIDIZING SUBSTANCES

← DIALYSATE WITH
WASTE PRODUCTS

Figure 11-38. Schematic diagram of sorbent regeneration cartridge. (From Ref. 61.)

In addition to hemodialysis, dialysis can be carried out by using the patient's own peritoneum which is a semipermeable membrane. The blood is brought to the membrane through the microcirculation of the peritoneum while the dialysate is introduced into the peritoneal cavity through a catheter implanted in the abdominal wall, one of which is shown in Figure 11-9. The dialysate is drained through the same catheter after solute exchange takes place and replaced by a fresh bottle. Glucose is added to the dialysate to increase its osmotic pressure gradient for ultrafiltration since it is impossible to obtain a high hydrostatic pressure gradient.

Recently a sorbent cartridge to regenerate the dialysis for reuse has been advanced.[59] The cartridge was originally developed for hemodialysis and could reduce the water requirement substantially, making dialysis more portable.[60] A schematic diagram of the sorbent-regenerated cartridge is shown in Figure 11-38. It is interesting to note that in order to remove urea, enzymolysis is carried out by urease since carbon cannot absorb urea effectively as mentioned before.

PROBLEMS

11-1. What will the blood urea nitrogen concentration be after 5 and 10 h of dialysis if the initial concentration is 100 mg%? The concentration after dialysis can be expressed exponentially:

$$C_t = C_o \exp\left[\frac{Q_b(b-1)t}{V}\right]$$

where C_o is the original dialysate concentration, Q_b is the blood flow rate, t is time, V is the volume of body fluid (60% of body weight), and b is a constant determined by mass transfer coefficient (K), Q_b, and surface area (A) of the membrane by $\exp(KA/Q_b)$ according to Cooney.[53]

11-2. In each of the following three groups, select the most related number–letter combinations.

(a) Low coefficient of friction   1. Bone ( )
(b) Tensile force transmission   2. Skin ( )
(c) La place equation   3. Tendon ( )
(d) Haversian system   4. Artery ( )
(e) Elastin   5. Ligamentum nuchae ( )
(f) Langer's line   6. Cartilage ( )

(a) Al_2O_3 1. Fluidized bed ( )
(b) Hydroxyapatite   2. Resorbable ceramic ( )
(c) Pyrolitic carbon   3. Sapphire ( )
(d) Calcium phosphate   4. Mineral phase of bone ( )

(e) Bioglass®   5. Piezoelectric ( )
(f) Barium titanate   6. $SiO_2-CaO-Na_2O-P_2O_5$ ( )
(g) Graphite   7. Substrate material of heart valve disk ( )

(a) PMMA   1. Polyolefin ( )
(b) Polyamide   2. Styrene/butadiene copolymer ( )
(c) PolyHEMA   3. Natural rubber ( )
(d) Polyacrylamide   4. Teflon® ( )
(e) PTFE   5. Nylon ( )
(f) SBR   6. Soft contact lens ( )
(g) Polyisoprene   7. Bone cement ( )
(h) Polypropylene   8. Hydrogel ( )

11-3. A bioengineer is trying to understand the biomechanics of a hole created in the skin for a transcutaneous implant. He made a hole using a circular biopsy drill in the dorsal skin of a dog. The diameter of the drill is 5 mm. If the hole became an ellipse with a minor and a major axis of 3 and 7 mm, answer the following questions.

a. In which direction is the internal stress greater?
b. In which direction are the collagen fibers more oriented?
c. How can the bioengineer obtain a circular rather than elliptical hole for the implant?
d. Assuming the implant is nondeformable compared to the skin, what problems will arise between skin and implant when a load or force is applied to the skin or implant by handling accidentally?

11-4. Answer the following:

a. Calculate the maximum tension developed for a 0.5-cm-diameter artery. Assume the maximum pressure will be 250 mm Hg and the artery is uniform in length.
b. If one wishes to replace a 5-cm section of the artery, what is the maximum force exerted on the wall?
c. Can one use silicone rubber for the replacement material if the wall thickness is 1 mm and the safety factor is 10?

11-5. Design a heart valve and give specific materials selected for each part and your reasons for selecting each material.

11-6. Design a blood access device for kidney dialysis or other long-term use and give specific materials selected for each part and your reasons for selecting each material.

REFERENCES

1. L. A. G. Rutter, Natural materials, in: *Modern Trends in Surgical Materials*, L. Gills (ed.), p. 208, Butterworths, London, 1958.
2. J. G. Thacker, G. Rodeheaver, J. W. Moore, J. J. Kauzlarich, L. Kurtz, M. T. Edgerton, and R. F. Edlich, Mechanical performance of surgical sutures, *Am. J. Surg. 130*, 374–380, 1975.
3. I. L. Rosenberg, T. G. Brennan, and R. G. Giles, How tight should tension sutures be tied? A controlled clinical trial, *Br. J. Surg. 62*, 950–951, 1975.

4. R. W. Postlethwait, J. F. Schaube, M. L. Dillon, and J. Morgan, Wound healing. II. An evaluation of surgical suture material, *Surg. Gynecol. Obstet. 108*, 555–566, 1959.

5. S. D. Elek and P. E. Conen, The virulence of *Staphylococcus pyogenes* for man: A study of the problems of wound infection, *Br. J. Exp. Pathol. 38*, 573–586, 1957.

6. R. W. Postlethwait, Long term comparative study of nonabsorbable sutures, *Ann. Surg. 171*, 892–898, 1970.

7. W. B. Conolly, T. K. Hunt, B. Zederfeldt, H. T. Cafferata, and J. E. Dunphy, Clinical comparison of surgical wounds closed by suture and adhesive tapes, *Am. J. Surg. 117*, 318–322, 1969.

8. L. J. Ordman and T. Gillman, Studies in the healing of cutaneous wounds. III. A critical comparison in the pig of the healing of surgical incisions closed with sutures or adhesive tape based on tensile strength and clinical and histological criteria, *Arch. Surg. 93*, 911–928, 1966.

9. S. N. Levine, Polymers and tissue adhesives, *Ann. N.Y. Acad. Sci. 146*, 143, 1968.

10. G. W. Hastings, Adhesives and tissues, in: *Biocompatibility of Implant Materials*, D. F. Williams (ed.), Chapter 17, Sector, London, 1976.

11. S. Houston, J. W. Hodge, Jr., D. K. Ousterhout, and F. Leonard, The effect of alpha-cyanoacrylate on wound healing, *J. Biomed. Mater. Res. 3*, 281–289, 1969.

12. J. E. Salvatore and M. P. Mandarino, Polyurethan polymer, its use in osseous lesions: An experimental study, *Ann. Surg. 149*, 107–109, 1959.

13. P. Y. Wang, Performance of tissue adhesives in vivo, in: *Biocompatibility of Implant Materials*, D. F. Williams (ed.), Chapter 18, Sector, London, 1976.

14. D. R. Beech, Adhesion to tooth structure by covalent or ionic bonds, in: *Biocompatibility of Implant Materials*, D. F. Williams, (ed.), Chapter 19, Sector, London, 1976.

15. F. R. Eirich, Bioadhesion as an interphase phenomenon, in: *Biocompatibility of Implant Materials*, D. F. Williams (ed.), Chapter 17, Sector, London, 1978.

16. R. Nowotny, A. Chalupka, C. Nowotny, and P. Bosch, Mechanical properties of fibrinogen-adhesive material, *Transactions, First World Biomaterials Congress*, p. 4.4.3, Baden, Austria, 1980.

17. H. Matras, Fibrin clot sealants in maxillofacial surgery, *Transactions, First World Biomaterials Congress*, p. 4.4.1, Baden, Austria, 1980.

18. A. F. von Recum and J. B. Park, Percutaneous devices, *CRC Crit. Rev. Bioeng. 5*, 37–77, 1979.

19. C. Grosse-Siestrup, *Entwicklung und Klinische Erprobung von Hautdurchleitungen Veterinaermedizin*, Dissertation, Free University, Berlin, 1978.

20. F. W. Cooke, W. B. Tarpley, S. J. Scheifele, and R. A. Erb, Study of a System for the Permanent Implantation of Cannulas, Contract NIH-69-83, The Franklin Institute Research Laboratory, Philadelphia, 1971.

21. B. D. T. Daly, M. Szycher, M. B. Lewis, M. Worthington, F. W. Quimby, and C. C. Haudenschild, Development of percutaneous energy transmission systems, in: *Devices and Technology Branch Contractors Meeting Proceedings*, pp. 81–82, NIH Publication 81-2022, 1979.

22. T. C. Robinson, P. Y. Schoenfeld, and B. D. Folkman, Clinical evaluation of a microporous tissue interface A-V shunt, in: *Proceedings, 6th Annual Contractors Conference*, Artificial Kidney Chronic Uremia Program, National Institute of Arthritic, Metabolic and Digestive Diseases, Washington, D.C., 1976.

23. C. Kablitz, T. Kessler, P. A. Dew, R. L. Stephen, and W. J. Kolff, Subcutaneous peritoneal catheter: 2 1/2 years experience, *Artif. Organs 3*, 210–217, 1979.

24. I. V. Yannas and J. F. Burke, Design of an artificial skin. I. Basic design principles, *J. Biomed. Mater. Res. 14*, 65–81, 1980.

25. I. V. Yannas, J. F. Burke, P. L. Gordon, G. Huang, and R. H. Rubenstein, Design of an artificial skin. II. Control of chemical composition, *J. Biomed. Mater. Res. 14*, 107–131, 1980.

26. N. Dagalakis, J. Flink, P. Stasikelis, J. F. Burke, and I. V. Yannas, Design of an artificial skin. III. Control of pore structure, *J. Biomed. Mater. Res. 14*, 511–528, 1980.

27. G. B. Park, Burn wound coverings: A review, *Biomater. Med. Devices Artif. Organs 6*, 1–35, 1978.

28. M. Chvapil, Considerations on manufacturing principles of a synthetic burn dressing: A review, *J. Biomed. Mater. Res. 16*, 245–263, 1982.

29. A. H. Bulbulian, *Facial Prosthetics*, Thomas, Springfield, Ill., 1973.

30. V. A. Chalian and R. W. Phillips, Materials in maxillofacial prosthetics, *J. Biomed. Mater. Res. Symp. 5*, 349–361, 1973.

31. J. N. Kent and R. B. James, Alloplasts in maxillofacial surgery, *Transactions, 4th Annual Meeting of the Society for Biomaterials*, San Antonio, 1978.

32. J. J. Shea, F. Sanabria, and G. D. L. Smyth, Teflon piston operation for otosclerosis, *Arch. Oto-laryngol. 76*, 516–521, 1962.

33. J. L. Sheehy, Stapes surgery when the incus is missing, in: *Hearing Loss-Problems in Diagnosis and Treatment*, L. R. Boies, (ed.), p. 141, Saunders, Philadelphia, 1969.

34. J. R. Tabor, Reconstruction of the ossicular chain, *Arch. Otolaryngol. 92*, 141–146, 1970.

35. Richard Manufacturing Co., *Plasti-Pore Material Technical Information*, Technical Publication 4240, Memphis, Tenn., 1980.

36. J. J. Shea, Alloplastic materials for otologic reconstruction, *Transactions, 4th Annual Meeting of the Society for Biomaterials*, San Antonio, 1978.

37. D. G. McPherson and J. M. Anderson, Keratoplasty with acrylic implant, *Br. Med. J. 1*, 330, 1953.

38. H. Cardona, Keratoprosthesis, *Am. J. Ophthalmol. 54*, 284, 1962.

39. W. H. Dobelle, M. G. Mladejovsky, and J. P. Girvin, Artificial vision for the blind: Electrical stimulation of visual cortex offers hope for a functional prosthesis, *Science 183*, 440–444, 1974.

40. Dow Corning Co., Silastic, Hospital-Surgical Products, Bulletin 51-051A, Midland, Mich., 1972.

41. F. E. Nulsen and E. B. Spitz, Treatment of hydrocephalus by direct shunt from ventricle to jugular vein, *Surg. Forum. 2*, 399–403, 1951.

42. R. H. Ames, Ventriculo-peritoneal shunts in the management of hydrocephalus, *J. Neurosurg. 27*, 525–529, 1967.

43. J. R. Little, A. L. Rhoton, and J. F. Mellinger, Comparison of ventriculoperitoneal and ventriculo-atrial shunts for hydrocephalus in children, *Proc. Staff Meet. Mayo Clin. 47*, 396, 1972.

44. J. J. Kaufman, Ureteral replacements, in: *The Ureter*, H. Bergman (ed.), Harper & Row, New York, 1967.

45. H. Lee and K. Neville, *Handbook of Biomedical Plastics*, Chapters 4 and 13, Pasadena Technology Press, Pasadena, Calif., 1971.

46. *Guidelines for Blood–Material Interactions*, Report of the National Heart, Lung, and Blood Institute Working Group, NIH Publication 80-2185, 1980.

47. W. S. Edwards, *Plastic Arterial Grafts*, Thomas, Springfield, Ill., 1965.

48. S. A. Wesolowski, C. C. Fries, A. Martinez, and J. D. McMahon, Arterial prosthetic materials, *Ann. N.Y. Acad. Sci. 146*, 325–344, 1968.

49. A. D. Haubold, H. S. Shim, and J. C. Bokros, Carbon cardiovascular devices, in: *Assisted Circulation*, F. Unger (ed.), pp. 520–532, Academic Press, New York, 1979.

50. K. Botzko, R. Snyder, J. Larkin, and W. S. Edwards, *In vivo/in vitro* life testing of

vascular prostheses, in: *Corrosion and Degradation of Implant Materials*, ASTM STP 684, B. C. Syrett and A. Acharya (ed.), pp. 76–88, ASTM, Philadelphia, 1979.
51. L. B. Housman, W. A. Pitt, J. H. Mazur, B. Litchford, and S. A. Gross, Mechanical failure (leaflet disruption) of a porcine aortic heterograft: Rare cause of acute aortic insufficiency, *Thorac. Cardiovasc. Surg. 76*, 212–213, 1978.
52. J. D. Hardy (ed.), *Human Organ Support and Replacement: Transplantation and Artificial Prostheses*, Thomas, Springfield, Ill., 1971.
53. D. O. Cooney, *Biomedical Engineering Principles*, Dekker, New York, 1976.
54. W. J. Kolff, *Artificial Organs*, Wiley, New York, 1976.
55. W. J. Kolff, Artificial organs and their impact, in: *Polymers in Medicine and Surgery*, R. L. Kronenthal, Z. Oser, and E. Martin (ed.), pp. 1–28, Plenum Press, New York, 1975.
56. W. Greatbatch, Metal electrodes in bioengineering, *CRC Crit. Rev. Bioeng. 5*, 1–36, 1981.
57. R. J. Solar, Materials for cardiac pacemakers encapsulation, in: *Biocompatibility in Clinical Practice*, Volume II, D. F. Williams (ed.), Chapter 13, CRC Press, Boca Raton, Fla., 1982.
58. D. Amundsen, W. McArthur, and M. Mosharrafa, A new porous electrode for endocardial stimulation, *Pace 2*, 40, 1979.
59. S. R. Ash, R. G. Barile, P. G. Wilcox, D. L. Wright, J. A. Thornhill, C. R. Dhein, D. P. Kessler, and N. H. L. Wang, The sorbent suspension reciprocating dialyzer: A device with mineral sorbent saturation, *ASAIO J. 4*, 28–40, 1981.
60. A. Gordon, O. S. Better, M. A. Greenbaum, L. B. Marantz, T. Gral, and M. H. Maxwell, Clinical maintenance hemodialysis with a sorbent-based low-volume dialysate regeneration system, *Trans. Am. Soc. Artif. Intern. Organs 17*, 253–256, 1971.
61. R. A. Ward, Investigation of the risk and hazards with devices associated with peritoneal dialysis and sorbent regenerated dialysate delivery systems, FDA Contract 223-81-5001, revised draft report, 1982.

BIBLIOGRAPHY

J. Black, *Biological Performance of Materials*, Dekker, New York, 1981.
S. D. Bruck, *Blood Compatible Synthetic Polymers: An Introduction*, Thomas, Springfield, Ill., 1974.
A. H. Bulbulian, *Facial Prosthetics*, Thomas, Springfield, Ill., 1973.
M. Chvapil, Considerations on manufacturing principles of a synthetic burn dressing: A review, *J. Biomed. Mater. Res. 16*, 245–263, 1982.
W. S. Edwards, *Plastic Arterial Grafts*, Thomas, Springfield, Ill., 1965.
W. Greatbatch, Metal electrodes in bioengineering, *CRC Crit. Rev. Bioeng. 5*, 1–36, 1981.
C. A. Homsy and C. D. Armeniades (ed.), *Biomaterials for Skeletal and Cardiovascular Applications*, J. Biomed. Mater. Symp. 3, 1972.
S. F. Hulbert, S. N. Levine, and D. D. Moyle (ed.), *Prosthesis and Tissue: The Interfacial Problem*, J. Biomed. Mater. Symp. 5, 1974.
H. Lee and K. Neville, *Handbook of Biomedical Plastics*, Chapters 3–5 and 13, Pasadena Technology Press, Pasadena, Calif., 1971.
S. N. Levine, Polymers and tissue adhesives, *Ann. N.Y. Acad. Sci. 146*, 136, 1968.
G. H. Gyers and V. Parsonnet, *Engineering in the Heart and Blood Vessels*, Wiley, New York, 1969.
J. B. Park, *Biomaterials: An Introduction*, Plenum Press, New York, 1979.
R. W. Postlethwait, Long term comparative study of nonabsorbable sutures, *Ann. Surg. 171*, 892–898, 1970.

Transactions, American Society for Artificial Internal Organs, published yearly and contains studies related to this chapter.

A. F. von Recum and J. B. Park, Percutaneous devices, *CRC Crit. Rev. Bioeng. 5*, 37–77, 1979.

L. Vroman and F. Leonard (ed.), *The Behavior of Blood and Its Components at Interfaces, Ann. N.Y. Acad. Sci. 283*, 1976.

D. F. Williams (ed.), *Biocompatibility in Clinical Practice*, Volumes I and II, CRC Press, Boca Raton, Fla., 1982.

D. F. Williams (ed.), *Systemic Aspects of Blood Compatibility*, CRC Press, Boca Raton, Fla., 1981.

D. F. Williams (ed.), *Fundamental Aspects of Biocompatibility*, Volumes I and II, CRC Press, Boca Raton, Fla., 1981.

G. D. Winter, Epidermal regeneration studied in the domestic pig, in: *Epidermal Wound Healing*, H. I. Maibach and D. T. Rovee (ed.), Chapter 4, Year Book Medical Publishers, Chicago, 1972.

HARD TISSUE REPLACEMENT IMPLANTS

When we try to replace a joint or help heal a fractured bone, it is logical that bone repairs should be made according to the best repair course that the tissues follow. Therefore, if they are healed faster when a compressive force or strain is exerted, then we should provide compression through an appropriate implant design.[1] Likewise, if compression is detrimental to healing of the wound, the opposite approach should be taken. Unfortunately, the effects of compressive or tensile forces on the repair of long bones are not fully understood. Worse yet, experimental results provide a completely opposite conclusion.

It is believed that the secret of osteogenic and osteoclastic activity is related to the normal activities of the bone *in vivo*. Thus, the equilibrium between osteogenic and osteoclastic activity can be balanced according to the static and dynamic force applied *in vivo*, that is, if more load is applied the equilibrium tilts toward more osteogenic activity to counteract the load and vice versa (Wolff's law). Of course, this should be done without excessive load, which can damage the cells rather than enhance their viabilities.

The Wolff's law may also be related to the piezoelectric phenomenon of bone and other tissues in which the strain-generated electric potentials may trigger the tilting of the equilibrium.[2] This is the basis of the electrically stimulated fracture repair of clinical nonunions.

The design principles and manufacturing criteria for orthopedic implants are the same as for any other engineering applications requiring a dynamic load-bearing member. Although it is tempting to duplicate the natural tissues with materials having the same strength and shape, this has

357

not been practical nor desirable since natural tissues have one major advantage over man-made materials, that is, their ability to adjust to a new set of circumstances by remodeling their micro- and macrostructure. Consequently, the fatigue of tissues is minimal unless a disease hinders the natural cell turnover processes.

Although the exact mechanism of bone fracture repair is not known at this time, stability of the implant with respect to the wound surfaces is an important factor to be considered. Whether the fixation is accomplished by compressive or tensile force, the reduction should be anatomical and be firmly fixed so that the healing processes cannot be disturbed by unnecessary micro- and macromovement.[3]

12.1. INTERNAL FRACTURE FIXATION DEVICES

As mentioned in Chapter 1, fracture fixation devices were the first to be developed as implants. These devices range from simple pins to complicated hip nails. Almost all of these devices are made from metal alloys, as discussed in Chapter 8.

12.1.1. Wires, Pins, and Screws

The simplest but most versatile implants are the various metal wires [Kirschner wires for a diameter less than 3/32 in. (2.38 mm) and Steinman pins for a larger diameter] which can be used to hold fragments of bones together. Wires are also used to reattach the greater trochanter in hip joint replacements or for long oblique or spiral fracture of long bones. The common problems of fatigue combined with corrosion of metals may weaken the wires *in vivo*. The added necessity of twisting and knotting the wires for fastening aggravates the problem since strength can be reduced by 25% or more due to the stress concentration effect. The deformed region of the wire will be more prone to corrosion than the undeformed region due to the higher strain energy (see Section 8.5). The wires are classified as given in Table 12-1.

The Steinman pin is also a versatile implant and is often used for internal fixation in cases when a plate is difficult to apply or when adequate stability cannot be obtained by any other means. The tip of the pin is designed to penetrate bone easily and flute angles of the pin are opposite of the screws to push the bone chips forward due to the lack of space between the hole created and the pin. Three types of tip designs are shown in Figure 12-1. The trochar tip is the most efficient in cutting and hence is used often for cortical bone insertion.

Table 12-1. Nomenclature and Specifications of Surgical Wires

Suture size	Wire gauge No.	American wire gauge in.	mm	Standard wire gauge in.	mm	Diameter (mm) Min.	Max.	Knot-pull tensile strength (kg) class 3
10-0						0.013	0.025	0.05
9-0						0.025	0.038	0.06
8-0						0.038	0.051	0.11
7-0						0.051	0.076	0.16
6-0	40	0.0031	0.079	0.0048	0.122			
						0.076	0.102	0.27
6-0	38	0.0040	0.102	0.0060	0.152			
5-0	35	0.0056	0.142	0.0084	0.213	0.102	0.152	0.54
4-0	34	0.0063	0.160	0.0092	0.234			
						0.152	0.203	0.82
4-0	32	0.0080	0.203	0.0108	0.274			
000	30	0.0100	0.254	0.0124	0.315	0.203	0.254	1.36
00	28	0.0126	0.320	0.0148	0.376	0.254	0.330	1.80
0	26	0.0159	0.404	0.0180	0.457	0.330	0.406	3.40
1	25	0.0179	0.455	0.0200	0.508	0.406	0.483	4.76
2	24	0.0201	0.511	0.0220	0.559	0.483	0.559	5.90
3	23	0.0226	0.574	0.0240	0.610	0.559	0.635	7.26
4	22	0.0254	0.643	0.0280	0.712	0.635	0.711	9.11
5	20	0.0320	0.813	0.0360	0.915	0.711	0.813	11.40
6	19	0.0359	0.912	0.0400	1.020	0.814	0.914	13.60
7	18	0.0403	1.061	0.0480	1.220	0.914	1.016	15.9
	10	0.0109	2.590	0.1280	3.260			
	1	0.2893	7.340	0.3000	7.630			
	0	0.3249	8.250	0.3240	8.230			
	6-0	0.5800	14.700	0.4640	11.800			
	7-0			0.5000	12.700			

Figure 12-1. Types of Steinman pin tips.

Diamond   Trochar   Cove

Screws are one of the most widely used implants for fixation of bones to themselves or in conjunction with fracture plates. Figure 12-2 illustrates the various parts of the screw with various head designs. There are basically two types of screws: self-tapping and non-self-tapping (Figure 12-3). As the name indicates, the self-tapping screw does not require tapping or a predrilled hole which is slightly larger than the minor diameter of the screw. The extra step of tapping required for the non-self-tapping screws makes them less desirable although the holding power (or pull-out strength) of the two types of screws is about the same.[5] The variations of thread design (Figure 12-4) do not influence the holding power. However, the radial stress transfer between the screw thread and the bone is slightly less for the

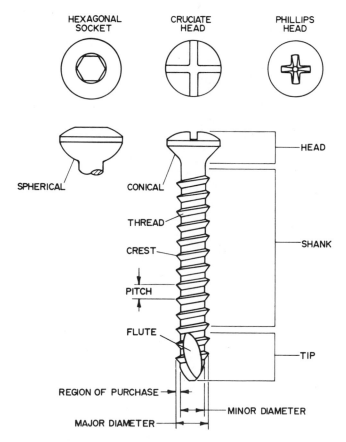

Figure 12-2. Illustration showing the various parts of a self-tapping bone screw, including various head designs. (From Ref. 4.)

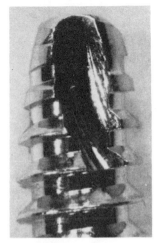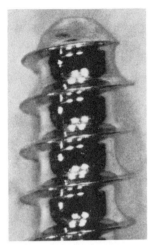

Self-tapping   **Nonself-tapping**

Figure 12-3. Photographs of the points of a self-tapping and a non-self-tapping screw.

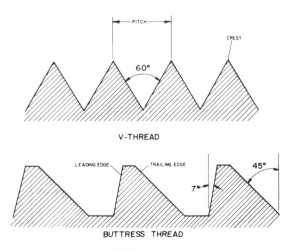

Figure 12-4. Types of screw threads.

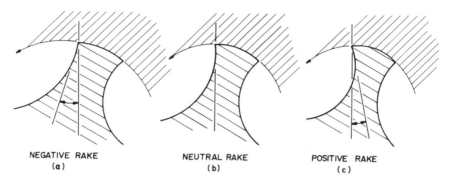

NEGATIVE RAKE   NEUTRAL RAKE   POSITIVE RAKE
(a)   (b)   (c)

Figure 12-5. Illustration showing the relationship of various rake angles to the outer edge of the cross-section of a cutting flute.

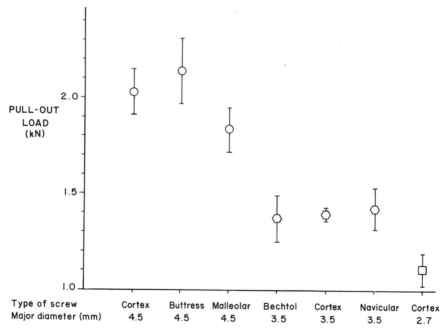

Figure 12-6. Average pull-out strength with 95% confidence limits for various Richards and Richards Osteo bone screws. (From Ref. 4.)

V-shaped thread than the buttress thread, indicating the latter can withstand the longitudinal load slightly better.[6,7]

The rake angle of the cutting edge is also an important factor in screw
design (Figure 12-5). The positive rake angle requires higher cutting force
yet results in lower cutting temperature due to less drag; the opposite effect
is obtained by the negative rake angle.[6] Almost all bone screws are made
with positive rake angles; hard metal cutting drills are made with negative
rake angles which can withstand larger cutting loads.

The pull-out strength or holding strength of the screws is an important
factor in the selection of a particular screw design. However, regardless of
the differences in design, the pull-out strength depends only on the size
(diameter) of the screw, as illustrated in Figure 12-6. The larger screw, of
course, has a higher pull-out strength.

The tissues immediately adjacent to the screw often necrose and are
reabsorbed initially, but if the screws are firmly fixed the dead tissues are
replaced by living tissues. When micro- or macromovement exists, a collagenous fibrous tissue encapsulates the screw.[8] This is the reason why the

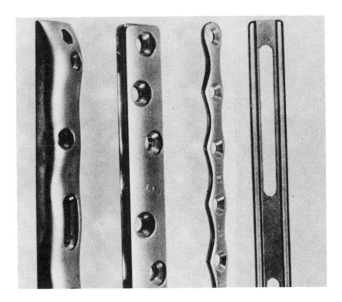

Figure 12-7. Bone plates. From left: Richards–Hirschorn plate (stress is evenly distributed throughout the length of the plate); AO compression plate; Sherman plate (stress is
evenly distributed throughout the length of the plate, but the plate is much weaker than
the two plates on the left); Egger's plate. (From Ref. 10.)

loading of the repaired bone should be delayed until firm fixation takes place between the screw and the bone.[9]

12.1.2. Fracture Plates

There are many different types and sizes of fracture plates, as shown in Figure 12-7. Since the forces generated by the muscles in the limbs are very large (see Table 12-2), the plates must be strong. This is especially true for the femoral and tibial plates. The bending moment and angle of various devices are plotted in Figure 12-8. In comparison with the bending moment at the proximal end of the femur (cf. Table 12-2), one can see that the plates cannot withstand the maximum bending moment applied. Therefore, some type of restriction on movement is essential in the early stage of healing.

Equally important is adequate fixation of the plate to the bone via the screws, as mentioned previously. However, overtightening may result in necrosed bone as well as deformed screws which may fail later due to the strain or stress energy corrosion process (see Section 8.5).

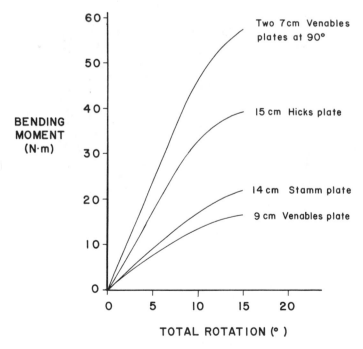

Figure 12-8. Bending moment versus total rotation of various bone plates. (From Ref. 11.)

There have been some efforts to compress the fractured bones together as shown in Figure 12-9. Most companies manufacture these devices with some variations in design. The added complexity of the devices and the controversy as to whether the compressive force or strain is beneficial or not[13] had hindered early acceptance of the devices. It is interesting to note that traditionally the large amount of callus formation had been considered to be a favorable sign and even essential for good healing. However, with the use of the compression plate, the opposite is thought to be a more favorable sign of healing. In this situation the amount of callus formed is proportional to the amount of motion between the plate and the bone.[14] One major drawback of the healing by rigid plate fixation is the weakening of the underlying bone such that refracture may occur following removal of the plate. This is largely due to the stress-shield effect.[15,16]

Woo's group[17] has experimented with a rigid Co–Cr alloy and a flexible bone plate. The flexible plate was made of a composite graphite, carbon fibers, and polymethylmethacrylate resin (GFMM), which has an elastic modulus approximately 1/10th (10–40 GPa) that of the Co–Cr alloy (250 GPa), as shown in Figure 12-10. Each plate was onlayed in the intact right (Co–Cr alloy) and left (GFMM) femurs of dogs using fixation screws. After 1 year, the animals were sacrificed and the femurs were immediately excised followed by removal of plates. The central diaphysis of each femur was cut and bone strips 5.5 cm long and 3 mm wide were obtained from four quadrants, as shown in Figure 12-11. Four-point bending tests were performed for each 3-mm-wide specimen on a special jig and load–deformation curves were obtained as shown in Figure 12-12. From these curves the maximum load (P_{max}) and the maximum energy stored at failure (A_{max}) similar to fracture toughness were calculated.

Table 12-2. Greatest Resistable Bending Moment at Proximal End of Femur[a]

Muscle group	No. of subjects tested		Bending moment (N-m)	
	Men	Women	Range	Mean
Hamstring	11		54–93	72
		17	26–54	35
Quadriceps	6		42–60	51
Hip abductors	6		38–108	63
		3	24–48	39
Hip adductors	6		60–126	81
		3	32–40	30

[a] From Ref. 11.

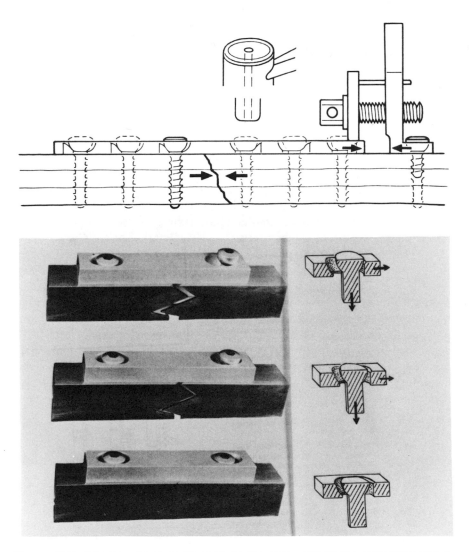

Figure 12-9. Principle of a dynamic compression plate: method of compression with a device (upper) and the principle (lower). (From Ref. 12.)

Figure 12-10. Photograph of the GFMM composite and Co–Cr alloy internal fixation plates used to plate the intact canine femurs. (From Ref. 17.)

Figure 12-13 shows that the P_{max} is much greater for the less rigid GFMM-plated bone than the Co–Cr-plated bone and the A_{max} of the less rigid GFMM-plated side was over twice that of the Co–Cr-plated side. However, Table 12-3 shows that there were no statistically significant differences in the modulus E and ultimate tensile strength, σ_{yt} for the normal and plated femurs, indicating the mechanical properties of the bone were not altered significantly by plating.

It is also interesting that the cortical thickness and total count (amount) of bone of the plated anterior region (combined A, AL, and AM) and unplated posterior region (combined P, PL, and PM) showed a statistically significant difference between GFMM- and Co–Cr-plated bones, as shown in Figure 12-14.

These results, along with histomorphological studies not mentioned here, indicate that long bone *homeostasis* is dependent on the level of the applied stress following Wolff's law. Therefore, the stress reduction in bone with the less rigid GFMM plate was considerably less, and consequently, atrophy of the cortical bone was not nearly as pronounced as for Co–Cr-plated bone. The decrease in the strength of the plated bone was mainly due to the thinning or resorption of cortical bone, rather than changes in the intrinsic properties of the bone per se,[17] which can be reversed by removal of the plates.[18]

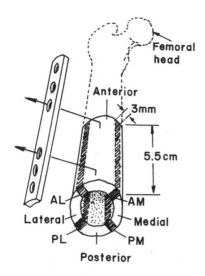

Figure 12-11. Schematic diagram showing the anatomical locations where the 3-mm-wide bone strips were obtained after the plate was removed. (From Ref. 17.)

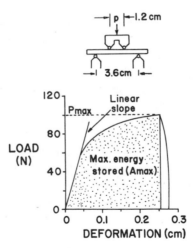

Figure 12-12. A typical load–deformation diagram of a 3-mm-wide bone strip taken from the canine femur subjected to four-point bending test to failure. Similar curves were obtained for swine femurs. In the latter case, the bone strips tested were 4 mm wide and the outer supports of the four-point bending test jig were set at 3.0 cm apart. (From Ref. 17.)

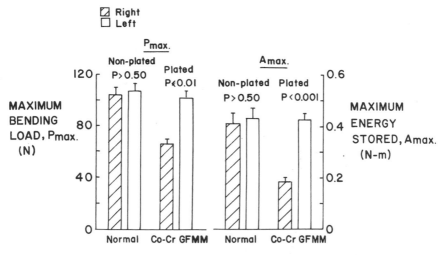

Figure 12-13. A comparison of the structural strength properties of bone strips from normal and plated canine femurs. (From Ref. 17.)

A considerable amount of care must be exercised when fixing cancellous bone since there are far fewer spiculae to support the load than found in cortical bone. An example of the fixation of the ends of a long bone is shown in Figure 12-15, in which the fractured bones are fixed with a combination of screws, plates, bolts, and nuts. The bulk necessary for adequate stabilization of the fracture increases the chance of infection near the site of implants.

Table 12-3. Comparisons of the Mechanical Properties
of Bone Strips of Normal and Plated
Canine Femurs[a]

	Flexural modulus of elasticity (GPa)	Ultimate tensile yield strength (MPa)
Normal femurs		
Left	19 ± 1	181 ± 20
Right	19 ± 1	185 ± 8
Plated femurs		
Left (GFMM)	18 ± 2	182 ± 10
Right (Co–Cr alloy)	21 ± 3	184 ± 12

[a] From Ref. 16.

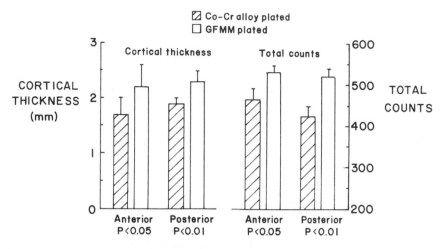

Figure 12-14. A comparison of cortical thickness and the total counts of bone from the anterior and posterior portion of the plated canine femurs. (From Ref. 17.)

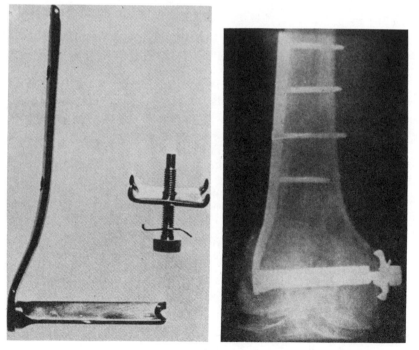

Figure 12-15. Devices to fix a cancellous bone of a supracondylar fracture of the femur. (From Ref. 19.)

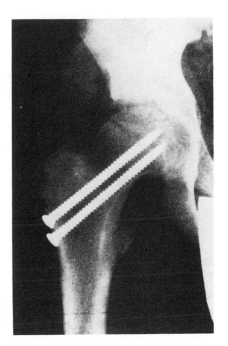

Figure 12-16. An example of a simple fracture fixation of a cancellous bone. (From Ref. 20.)

Sometimes one can fix a cancellous bone fracture by using a simple nail as shown in Figure 12-16. This is a special case since the patient was a young child (who incidentally has two to three times the trabecular bone mass normally present in the cancellous femoral head and neck region) and since the epiphyseal plate lies close to the hip joint the loading is essentially normal to the fracture surface.[20] Also the freely mobile hip joint relieves stress except for the compression cycle. Obviously, a wide range of choices is available. The choice is largely determined by the surgeon(s), not by the patients or bioengineers.

12.1.3. Intramedullary Devices

Intramedullary devices are used to fix fractures of the long bones by snugly inserting them into the intramedullary cavity (Figure 12-17). This type of implant should have some spring in it to exert some elastic force inside the bone cavity to prevent rotation and to fix the fracture firmly.

Compared to plate fixation, the intramedullary device is better positioned to resist bending since it is located in the center of the bone. However, its torsional resistance is much less than that of plate fixation.[11]

Figure 12-17. Illustration of intramedullary device used in the femur.

It is also believed that the intramedullary device destroys the intramedullary blood supply although it does not disturb the periosteal blood supply for the case of plate fixation. Another advantage of the intramedullary device is that it does not require opening of the wound, and the device can be nailed through a small inicision.

Many studies have examined the medullary blood supply and fracture healing in view of the extensive damage to the medullary canal caused by the insertion of the device. The long bone blood supply comes from three sources: the nutrient arteries and their intramedullary branches, the metaphyseal arteries, and the periosteal arteries. If fracture occurs, the extraosseous circulation from the surrounding soft tissues becomes active and forms the fourth source of blood supply. When the intramedullary canal was reamed and a device inserted, the nutrient artery and its branches were destroyed, but did not significantly damage the viability of the bone. Thus, a solid reunion can be achieved with this method of treatment.[21] Others have demonstrated that the tight-fitting nail delays healing due to the time necessary to reestablish an intracortical (or intramedullary) circulation.[22,23]

There are many different types of intramedullary devices varying to a large extent only in their cross-sectional shapes. For a given size of device, the resistance to bending and torque is different for different devices, as shown in Figures 12-18 and 12-19. The closed (solid) designs of the four-flanged nail (Schneider) and the diamond nail showed higher resistance to torsion than the open-cloverleaf nail.[21] However, in bending, the cloverleaf nail showed the highest resistance due to its larger bending moment of inertia.

The intramedullary nail and a plate are sometimes used in a simultaneous array for the fixation of a femoral neck and intertrochanteric bone fracture as shown in Figure 12-20. There are many different types of cross-sectional area designs to prevent rotation after fixation, as shown in Figure 12-21. Note that all of them have a guide hole in the center except the V-shaped device. The femoral fixation is usually made to compress the broken together which also helps to stabilize the fracture.

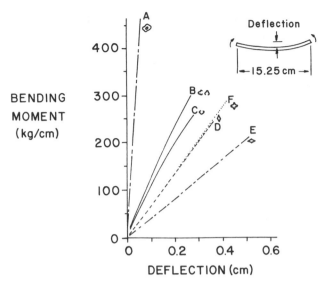

Figure 12-18. Bending deflections of the femur and of three intramedullary nails (9-mm cloverleaf, Schneider, and diamond-shaped) are compared as they would appear under identical loading. The length of each structure is taken as 15.25 cm. Curve A for the femur shows the bone to be more rigid than any of the nails. The cloverleaf nail is stiffer with the slot in tension (curve B) then in compression (curve C). The diamond-shaped nail is 50% more rigid when bent in its major plane (curve D) than in its minor plane (curve E). The Schneider nail has the same rigidity as the diamond-shaped nail (curve F), but even in its most unfavorable orientation the Schneider nail has a higher ultimate bending strength (not shown). (From Ref. 24.)

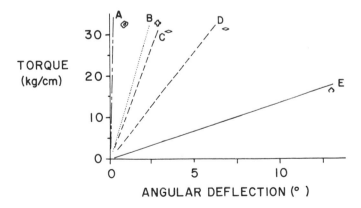

Figure 12-19. Deflections of the femur of three intramedullary nails. The femur is more rigid than any of the nails (curve A). The Schneider nail (curve B) is about 1/10th as rigid as the femur. As the length of the diamond-shaped is doubled, the rigidity is halved (curves C and D). The cloverleaf nail is the least rigid (curve E). The length of each nail tested to 20.25 cm except that for curve D. (From Ref. 25.)

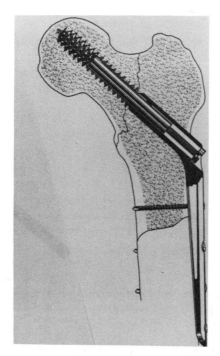

Figure 12-20. An illustration of hip nail fixation of a fractured femoral head. (Courtesy of DePuy, Division of Bio-Dynamics, Inc., Warsaw, Ind.)

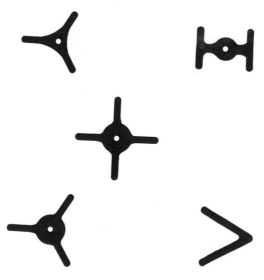

Figure 12-21. Cross-section of various hip nails. (From Ref. 26.)

12.1.4. Spinal Fixation Devices

When the spinous elements are deformed in such a manner that the length of the anterior elements is longer than that of the posterior ones, the resulting structure is bent backward, termed *lordosis*. The opposite condition is *kyphosis*. In severe cases of such spinal deformities, an internal and external fixation is called upon to correct the situation.[27,28] There are several designs to stabilize or straighten the curvatures, one of which is shown in Figure 12-22. Other designs include plates which are attached to the spinous processes by using bolts and expanding spinal jacks which are hooked to the spine through the articular process so that the distraction can be adjusted during implantation.

The main problems with these devices involve the adjustment or extension of the device as the spine is being straightened and the necrosis of the bones where the fixation device is attached due to the concentrated load. This necrosis results from the tremendous moment (> 100 N-m) exerted by the trunk muscles. As the spine is straightened, it is harder to distract without hooks since the leverage on the spine becomes smaller. Thus, multiple hooks are sometimes attached to obviate this problem.[29]

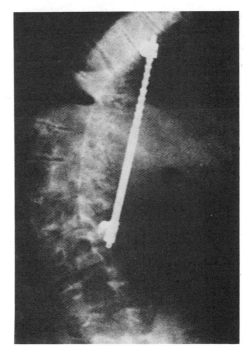

Figure 12-22. Harrington spinal distraction rod. (From Ref. 26.)

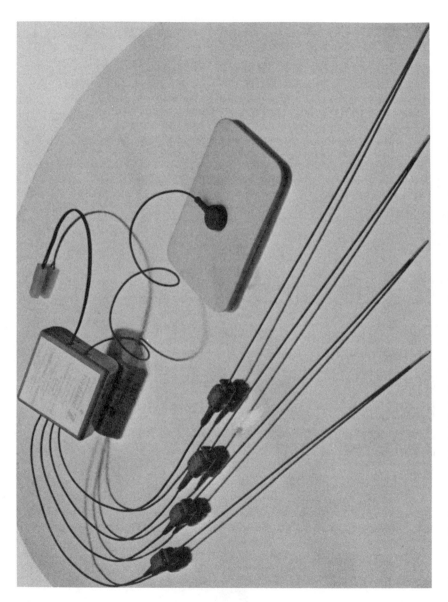

Figure 12-23. A commercial electrical stimulator for bone fracture repair. (Courtesy of Zimmer USA, Warsaw, Ind.)

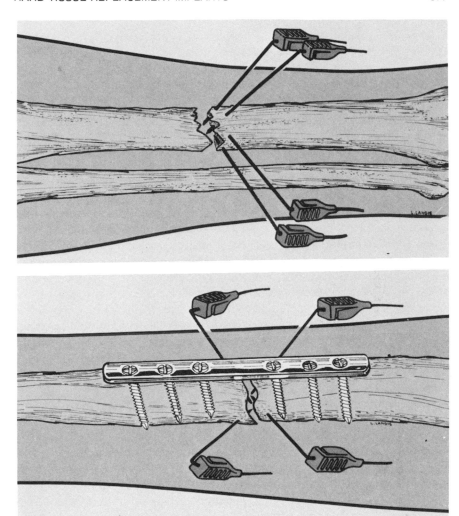

Figure 12-24. Schematic illustration of the use of an electrical stimulator with or without fracture fixation devices.

12.1.5. Fracture Healing by Electrical and Electromagnetic Stimulation

It has been recognized that some of the most challenging problems in orthopedic surgery are the clinical nonunions or delayed fractures and congenital and acquired pseudoarthroses. Until recently, the usual methods of treatments were grafting, plating, and nailing. When one or a combination of any of the methods failed to repair, especially in the lower extremi-

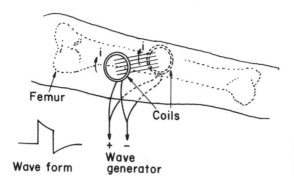

Femur

Coils

Wave form

Wave generator

+ −

Figure 12-25. Representation of pulsed electromagnetic field bone stimulation.

ties, amputation was the only alternative.[30] Recently, many investigators have demonstrated that the application of electrical energy by means of direct (or alternating) current or electromagnetic field can enhance and stimulate the osteogenic activities.[30–33]. Although the exact mechanism of the stimulation is not yet fully elucidated, the electrical stimulation may be related to the highly electronegative nature of the fracture site due to the increases in ionic and metabolic activities (see Figure 7-6). The extra electrical energy input into the wound area seems to signal more osteogenic activities. Of course, the tissue can only respond to the right amount of energy input (10–40 μW, 5–20 μA) without excessive electrical potential (<1 V).[34,35] The stimulation is also closely related to the nature of electrode material, surface area, and location.[34,36]

One commercially available electrical stimulator is shown in Figure 12-23. The four electrodes (cathodes or negative poles) are inserted into the fracture site transcutaneously after drilling the bone and the positive electrode (anode) is placed over the skin using a conducting pad. The electrode can be used with or without internal fixation devices, as shown in Figure 12-24.

The magnetic stimulators use a pair of Helmholtz coils, which are aligned across the wound site and a magnetic field with a monophasic, 150-ms phase, with a repetition rate of 75 Hz, is applied as shown in Figure 12-25. This pulse amplitude induces 1–2 mV/cm of potential in the bone.[37] Magnetic stimulation has one big advantage over direct-current stimulation, viz. it is a noninvasive technique. The efficacy of both types of stimulation is about the same, over 70% success rate.[33,37]

12.2. JOINT REPLACEMENTS

The articulation of joints poses some additional problems as compared with long bone fracture repairs. These include wear and corrosion and their products, as well as complicated load transfer dynamics. In addition, the

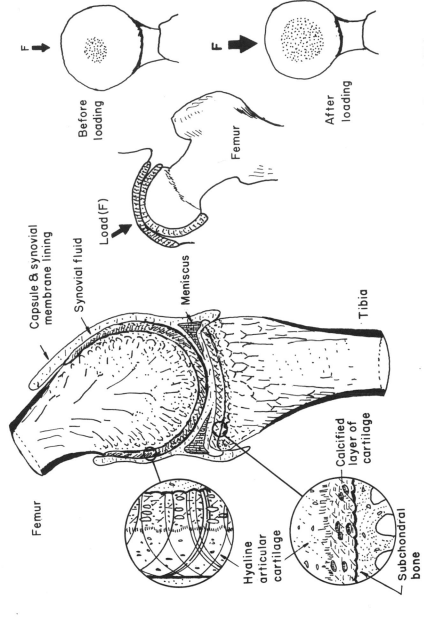

Figure 12-26. Structural arrangement of a knee joint and a hip joint before and after loading. See Figure 6-27 for more details on the articular cartilage surface. (From Ref. 20.)

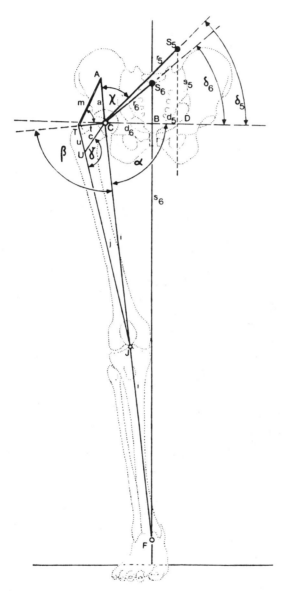

Figure 12-27. Some distances and angles in the human pelvis and leg skeleton for biomechanical analysis of the joints. (From Ref. 38.)

massive nature of (total) joint replacements such as the knee and the elbow near the skin also renders a greater possibility of infection. More importantly, if the replacement fails for any reason, it is much more difficult to replace the joint a second time since a large portion of the natural tissue has already been destroyed.

For these reasons orthopedic surgeons try to salvage the existing joint whenever possible and only use implants as a last resort. However, the hip prosthesis has shown favorable acceptance in recent years by older patients.

The hip and shoulder joints have a ball-and-socket articulation whereas other joints such as the knee and the elbow have a hinge-type articulation. However, they all possess two opposing smooth cartilaginous articular surfaces which are lubricated by viscous synovial fluid. This fluid is made of polysacchardies (Section 6.1.2) which adhere to the cartilage and upon loading can be permeated out onto the surface to reduce friction. The cartilage is not vascularized, and the repair and nutrition of the tissues appear to be diffusional processes.

Nature provided the surface of the joint with a large area to minimize the load concentration effect, as shown in Figure 12-26, for the hip and the knee joints. The shock of loading can be absorbed further by the trabecular subchondral bone underlying the cartilaginous tissue which also transfers the load gradually due to its viscoelastic properties.

The actual articulation of the joint is performed by the ligaments, tendons, and muscles. The analysis of the forces acting on the various tendons and ligaments is very complicated (Figure 12-27). Even the center of rotation of the knee joint cannot be determined with any great precision; in fact, it shifts position with each movement. The eccentric joint movement helps to distribute the load throughout the surfaces of the entire joint.

Table 12-4. Average Maximum Values of Forces
at Hip and Tibiofemoral Joints
during a Range of Activities[a]

Activity	Maximum joint force (multiples of body weight)	
	Hip	Knee
Level walking		
Slow	4.9	2.7
Normal	4.9	2.8
Fast	7.6	4.3
Up stairs	7.2	4.4
Down stairs	7.1	4.9
Up ramp	5.9	3.7
Down ramp	5.1	4.4

[a] From Ref. 39.

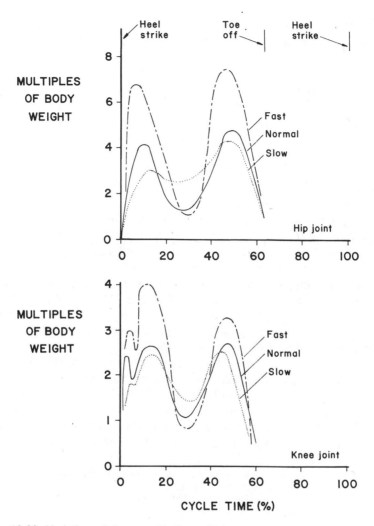

Figure 12-28. Variation of forces with time of hip and knee joint in walking. (From Ref. 39.)

Some joints such as the knee have fibrous, cartilaginous menisci shaped like wedges located between the sliding surfaces (Figure 12-26). It is believed that the main function of the meniscus is to transfer the load over a larger area than is possible without it.

The joint forces applied during a range of activities are given in Table 12-4. Of course, the forces applied during walking vary considerably with each motion, as shown in Figure 12-28. It should not be surprising that the forces are up to 8 times the body weight since they act upon the joints in a

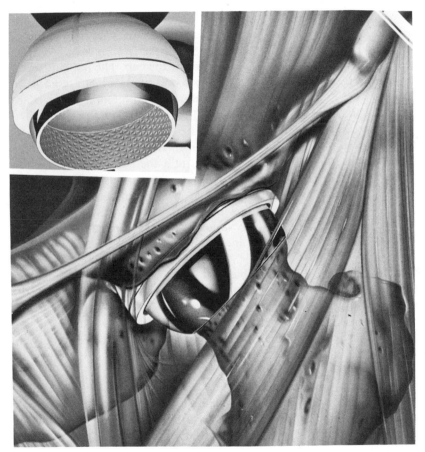

Figure 12-29. An example of modern double cup arthroplasty. In early days, only the femoral cup was placed (mold arthroplasty) in order to obtain a movable joint. (Courtesy of Howmedica, Inc., Rutherford, N.J.)

dynamic rather than a static manner. This type of biomechanical analysis can help design a better implant although we will not study this subject any further.

12.2.1. Hip Joint Replacements

The early methods of correcting the diseased or fractured hip joints only involved the acetabular cup or femoral head. One technique of restoring hip joint function is to place a cup over the femoral head while the surface of the acetabulum is also resected to fit the cup. The implant serves as a *mold* interposing the two surfaces which eventually adapt according to the function of the joint. Today, both acetabulum and femoral head surface arthroplasty can be performed as shown in Figure 12-29.

Some have tried to replace the femoral head after resection with various designs, as shown in Figure 12-30. The wide variety of implants reflects the limited knowledge of the function of joints and the ability of the joint to accommodate any insult imposed upon it by the various implants. Most femoral head replacements are performed with the installation of an acetabular cup. This is the so-called total hip joint replacement, which is frequently performed bilaterally. The various types of hip implants can be grouped into ball and socket, retained ball and socket, trunnion bearing, and floating acetabulum and double cup, as illustrated in Figure 12-31.

The single most difficult problem of hip joint as well as other joint replacements is fixation of the implants.[41] This is due to the fact that the implant lies on the cancellous bone, which has few trabeculae to support the large load imposed. Also, the stress concentration of the implant at points of sharp contact, such as the calcar region and the end of the femoral stem, makes the already weakened bone more necrotic. In fact, the first wide acceptance of the total hip replacement was achieved by providing an acceptable fixation using an acrylic bone cement (see Section 10.6.2). The cement is inserted when it is in the dough state, after mixing the polymer powder and monomer liquid throughly, and the prosthesis is press-fitted into the drilled hole. The cement is often used in the fixation of the acetabular cup.

The cement serves not only the initial attachment of the implant with bone, but it also acts as a shock absorber since it is a viscoelastic polymer. The bone cement also helps to spread the load more evenly over a large area and reduces the stress concentration on the bone by the prosthesis. However, the stress on the bone in the distal stem is much higher than in the proximal region (calcar) when the stem is inserted by using bone cement, as shown in Figure 12-32. This causes bone resorption of the calcar region due to the reduced stress transfer, which, in turn, will lead to either loosening or fracture of the stem.[42]

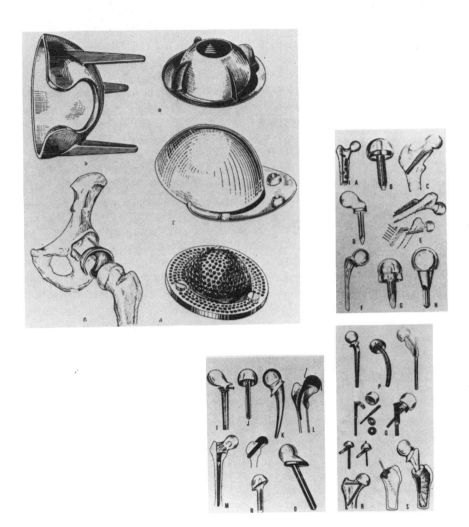

Figure 12-30. Various designs of acetabular and femoral head components of hip prostheses. (From Ref. 40.)

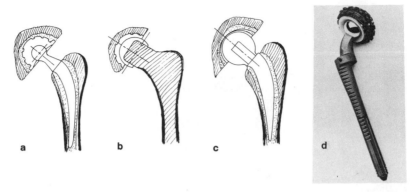

Figure 12-31. Different types of total hip implants: (a) ball and socket, (b) double cup, (c) trunnion, and (d) retained ball and socket. (Sivash design, courtesy of United States Surgical Corp., New York, N.Y.)

To obviate this problem, a more desirable higher loading condition in the proximal region is obtained by making the neck portion of the stem longer. However, this arrangement increases the moment applied in the midstem, causing fracture more readily. Actually, the new trend in stem design and insertion technique is to make it straighter, thus decreasing the moment.

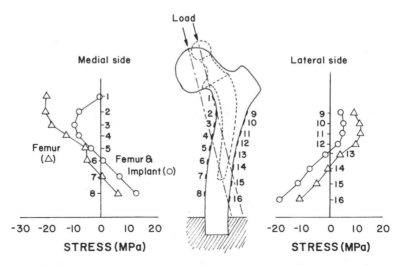

Figure 12-32. The stresses on the surface of the femoral stem by a load of 4000 N. The numbers indicate the location of strain gauges to measure the deformations. Note that there is no stress in position 1 (calcar region) after insertion of the implant. (From Ref. 41.)

The cement itself sometimes causes problems such as monomer vapors interfering with systemic function, thereby decreasing the blood pressure. The highly exothermic polymerization reaction can cause a local temperature rise which can result in cell necrosis as mentioned in Section 10.6.2. Also, the extensive intramedullary cavity preparation for the cement space can block the bone sinusoids, enhancing tissue necrosis and fat embolism.

Another added problem is the difficulty of removal and the extent of tissues destroyed if the implant has to be removed for any reason. Thus, the replaceability of the implant is an important aspect of the design. In this regard, the original Sivash implant has an inherent weakness due to its design, that is, the whole prosthesis has to be replaced even if only one component has failed. It is now designed so that one part can be replaced without removing the other implant. Also, the fixation of the device is accomplished by direct apposition to the bone, which results in stress concentration at the end of the stem and sharp edges of the acetabular cup in contact with bone.

The friction between the ball and cup of the hip joint is significant when it creates a rotational torque. Especially at high loading rates, the frictional torque becomes very significant for the Co–Cr alloy hip joint, as shown in Figure 12-33. The stainless steel–polyethylene and Co–Cr alloy–polyethylene combinations are better for reducing the frictional torque and wear than the all-metal system. The high frictional torque of the all-metal system may also be due to the larger surface contact area since the

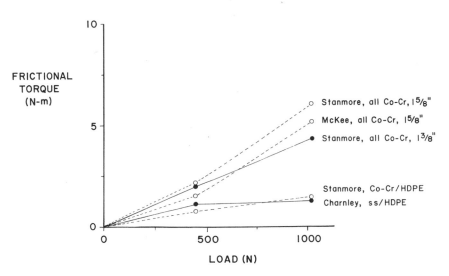

Figure 12-33. Frictional torque versus applied load for various hip prostheses. (From Ref. 43.)

femoral head is much larger than in the metal–polymer prosthesis. In actual use, the all-metal system works well without exerting high frictional torque. This is mainly due to the lubrication of the surfaces by tissue fluids.[44]

The failure of hip joint replacements is usually due to the loosening of the acetabular and femoral components. Loosening can be largely divided into mechanical and radiological loosening which may not be related to any clinical symptoms. In one study, the incidence of radiological loosening was 19.5% (76/389 hips), of which 10.3% (40 hips) were stem/cement loosening and 11.1% (43 hips) were bone/cement loosening.[45] Inadequate surgical and cementing techniques were thought to be the main factors responsible for loosening. Another study attributed the loosening to blood clots inter-

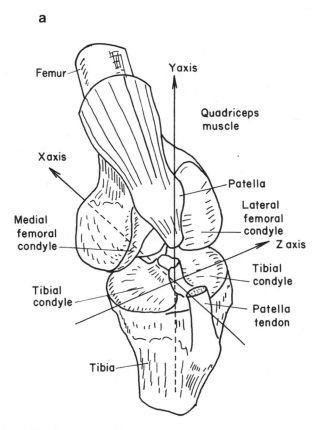

Figure 12-34. (a) The knee joint with the surrounding capsule removed and the femur and tibia separated, drawn in about 30° of flexion. (b,c) The knee joint as a closed kinetic chain. In addition, a lateral motion is also encountered. [(a) from Ref. 47; (b,c) from Ref. 38.]

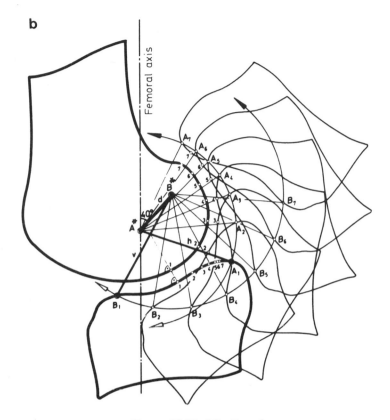

Figure 12-34. (*Continued*)

posed at the time of surgery and shrinkage of bone cement during poly-merization.[46]

12.2.2. Knee Joint Replacements

The development and acceptance of knee joint prostheses have been slower than for the hip joint due to its more complicated geometry and biomechanics of movements (Figure 12-34). The incidence of knee joint degeneration is higher than for any other joints, as shown in Figure 12-35. The four primary indicators for any joint replacement are pain, instability, stiffness, and deformity.[49]

The knee joint implants can be classified into hinged and nonhinged types. The latter is further divided into uni- and bicompartmental, accord-ing to Mears.[50] Table 12-5 gives a brief summary of various knee joints shown in Figure 12-36. The selection of a particular implant depends on the

c

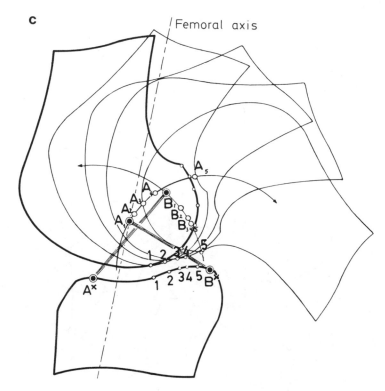

Figure 12-34. (*Continued*)

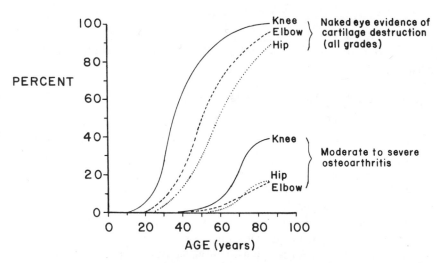

Figure 12-35. Incidence of joint degeneration. (From Ref. 48.)

health of the knee, the type of disease, and the range of activities required. As in the case of hip joint replacement, the major problems are loosening and infection (deep infection was reported to be 1.5% to over 10%).[65] The sinking of the tibial plateaus can be corrected by enlarging the prosthesis, which can be placed over the cortical bone, and a metallic backing is made under the polymer [ultrahigh-molecular-weight polyethylene (UHMWPE)], as shown in Figure 12-37.

The porous-coated implant is designed to induce bony tissue ingrowth for making a dynamic interface of bone and implant (Figure 12-38). Of course, this implant does not require the use of bone cement as most other

Table 12-5. Types of Knee Joint Replacements[a]

Type	Region of replacement	Remarks	Example
Unicompartmental	Only the surfaces of load-bearing arcs of the femur and tibia	High stresses on the implant, reduced stability polycentric motion	Marmor,[51] Figure 12-36a Charnley[52]
Bicompartmental	Resurfacing with retention of anatomical geometry of the femoral condylar and tibial surfaces	Large flexion, greater functional replication of anatomical motion	UCI[53] Duocondylar[54] Freeman–Swanson (ICLH),[55] Figure 12-36b
Surface replacements with enhanced stability	Severely compromised knee with moderate to severe deformities	Stabilization attained by ball-and-socket and hinge-stabilizing stem in femoral component. Reduced resistance to extension and flexion	Deane[56] Spherocentric,[57] Figure 12-36c Attenborough,[58] Figure 12-36c
Hinged-type prosthesis	Severely deformed and painful knee with no intrinsic stability	Unnatural stress patterns at the bone ends in diaphysis result in osteoporosis with failure of fixation	Walldius,[59] Figure 12-36d Shiers[60] Buchholz[61] Guepar[62] Stanmore[63]
Patellofemoral arthroplasty	After patellectomy to relieve pain	Can be used with other total replacements	Bechtol,[64] Figure 12-36e

[a] From Ref. 79.

implants do. Knee surgery with a bone cement requires a complete cleanup
of the cement and bone chip debris, which can cause severe damage to the
articulating surfaces, especially the tibial plateaus. The porous-coated implants should be used for a relatively healthy knee since their stability is
entirely dependent on the ingrown tissues. It is also expected that the
ambulation time will be much longer than in the cement-fixed case since it
will take some time for the tissues to grow into the pores and premature
loading may be detrimental for the ingrowth process.[67]

The wear of the surface of the tibial plateaus can be significant, as
shown in Figure 12-39, which is an *in vitro* measurement. Note the unusu-

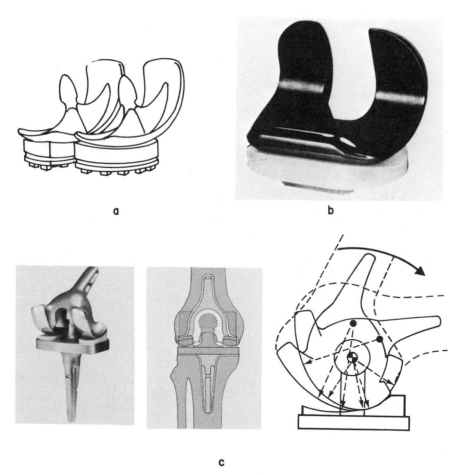

Figure 12-36. Examples of various types of knee replacements. (a) Marmor[51]; (b)
Freeman–Swanson ICLH[55]; (c) spherocentric[57]; (d) Walldius[59]; (e) Bechtol.[64]

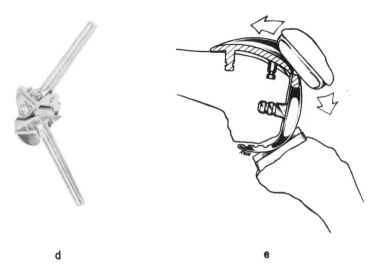

d   e

Figure 12-36. (Continued)

ally high wear rate of the UCI-type knee. The *in vivo* performance may be quite different.

12.2.3. Ankle Joint Replacements

The ankle joint consists of three articulating surfaces: the distal tibia and the superior surface of the talus, the medial malleolus and the medial side of the talus, and the lateral malleolus or fibula and the lateral side of the talus as shown in Figure 12-40. The ankle joint has movements of

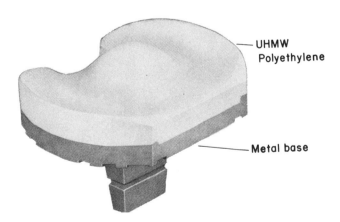

Figure 12-37. Metal-base tibial implant designed to prevent sinking of the component. (From Ref. 65.)

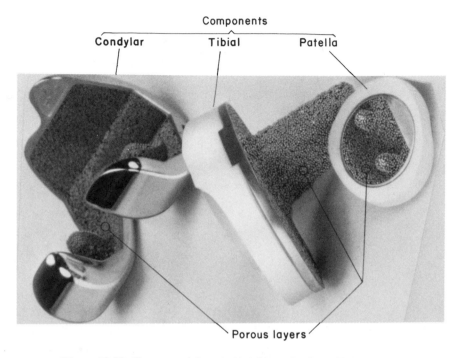

Figure 12-38. Porous metal-coated total knee implant. (From Ref. 66.)

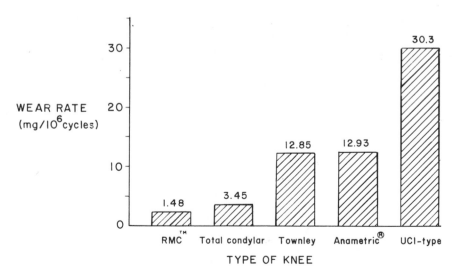

Figure 12-39. Bar graph of the wear rate results. (From Ref. 68.)

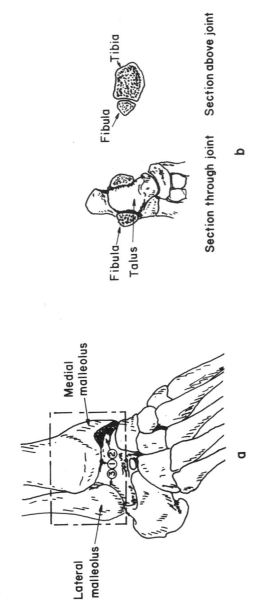

Figure 12-40. (a) Anatomical features of the ankle joint articulating surfaces: (1) the distal tibia and the superior surface of the talus; (2) the medial malleolus and the medial side of the talus; and (3) the lateral malleolus or fibula and the lateral side of the talus. (b) Sectional views through and above the ankle joint. (From Ref. 50.)

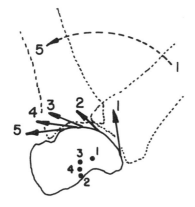

Figure 12-41. In this schematic view of the lateral ankle joint of a normal non-weight-bearing male, all of the instant centers are located within the talar body. Surface velocity shows the distraction at the beginning of motion but sliding thereafter. (From Ref. 71.)

dorsiflexion and plantar flexion.[69] The joint, however, works as a universal joint and can rotate up to 14° in normal walking.[70] The motion of the joint is not entirely the hinge and slight rotating type of motion as in the knee but has a gliding motion as shown in Figure 12-41 which makes it more difficult to duplicate in an implant.

Ankle joint prostheses are of two types: congruent and incongruent.[72] Figure 12-42 illustrates three types of incongruent ankle prostheses: trochlear, concave/convex, and convex/convex. These types of implants may suffer a high stress concentration effect and intrinsic instability.

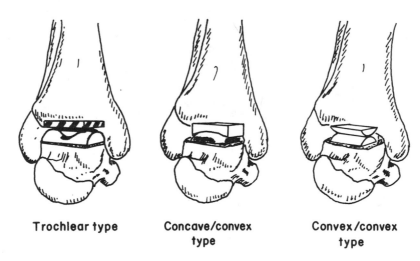

Trochlear type   Concave/convex   Convex/convex
type   type

Figure 12-42. Incongruent surface types of total ankle joint replacements. (From Ref. 72.)

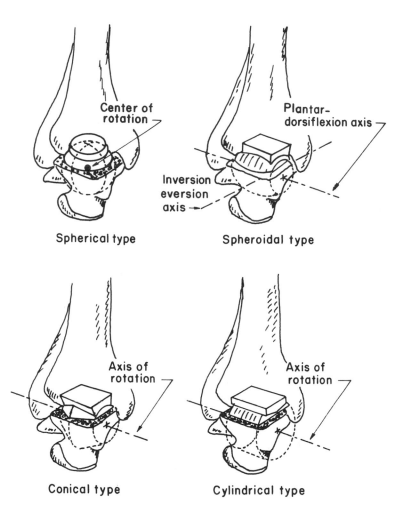

Figure 12-43. Congruent surface types of total ankle joint replacements. (From Ref. 72.)

There are four basic variations of the congruent type of implant (Figure 12-43): spherical (ball-and-socket), spheroidal (barrel-shaped), conical, and cylindrical. These designs may give greater stability of the joint and lessen the stress concentration more than the incongruent type, due to the larger contact surfaces.

The materials used to construct ankle joints are usually Co–Cr alloy and UHMWPE. Recently, a carbon fiber-reinforced UHMWPE was used to fabricate the tibial component to provide higher strength and creep resistance. Although these implants have been used for a relatively short period of time and clinical experience is lacking, the usual problems are prosthetic loosening, limited motion, and mechanical pain at the ankle joint.[50]

12.2.4. Shoulder Joint Replacements

The major shoulder joint motion originates from the ball-and-socket articulation of the gleno-humeral joint as shown in Figure 12-44. The hemispherical, incongruent joint provides the largest motion in the body. As

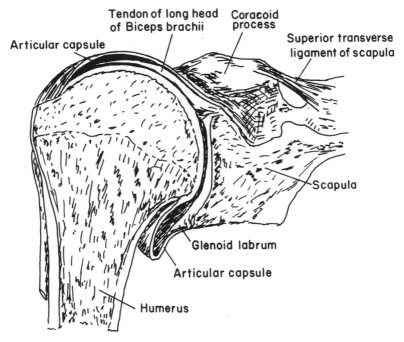

Figure 12-44. The anatomy of the glenohumeral joint. (From Ref. 73.)

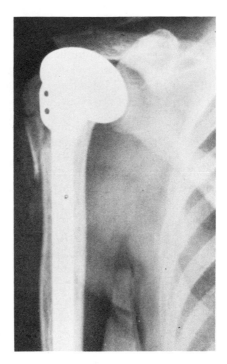

Figure 12-45. Neer's humeral implant. (Courtesy of Richards Manufacturing Co., Memphis, Tenn.)

in the hip joint replacements, the first shoulder joint replacement by Neer was attempted by merely replacing the humeral head, as shown in Figure 12-45. Note a large articulating surface and two holes to provide better fixation and resistance to rotation. The implant can be fixed with or without bone cement. Figure 12-46 shows some shoulder implant designs, most of which are a cup-and-head type as in hip joint prostheses.

A more challenging task in shoulder replacements is confronted when the rotator cup does not function normally. The Neer prosthesis is designed to function with the rotator cuff while others can be used without it. One of the latest designed glenohumeral prostheses has claimed to be more anatomical, but the real problem was again the fixation in the glenoid, not in the humeral part, due to the small bone stock and greater loading imposed on it.[76]

12.2.5. Elbow Joint Replacements

The elbow joint is a hinge-type joint allowing mostly flexion and extension, but also polycentric motion.[77] The distal end of humerus has two articulating surfaces: the trochlea and the capitulum, as shown in Figure 12-47.

Most elbow joint implants are either hinge or surface replacement types, as given in Table 12-6. Major problems are loosening of the implant and limited soft tissue coverage of the implant, making it vulnerable to infection.[80] An *in situ* elbow prosthesis is shown in Figure 12-48.

12.2.6. Wrist Joint Replacements

The wrist joint allows flexion, extension, adduction, and abduction primarily through a radiocarpal joint, as shown in Figure 12-49. Since the wrist joint arthroplasty for prosthetic replacements includes the removal of capitate, where the anatomical instant center or axis of motion of a radiocarpal complex is located, it is very difficult to place the prosthesis in the correct position. The unnatural position of the implant will constrain its movement, causing an excessive generation of bending moments in adduction or abduction.[50] This is also a major cause of complications of total wrist replacements.[81] A more advanced treatment of biomechanics of the wrist can be found elsewhere.[82]

Figure 12-50 illustrates some wrist implants. Meuli's and Volz's implants are ball-in-socket while Swanson's is the space-filler type made of silicone rubber, as for his finger joint prosthesis. The clinical results are poorer than other joint replacements due to the more complex motion of the wrist and surgical inconsistency due to lack of landmarks to rely on.

12.2.7. Finger Joint Replacements

A finger is made of distal, middle, and proximal joints and is provided with adequate controls by ligaments and tendons so as not to collapse under a compressive load as shown in Figure 12-51. The exact mechanics of joint movements are rather complex.[86] Burton[87] has divided the various types of resectional and implant arthroplasties into five major categories as shown in Figure 12-52. The traditional treatment is resection of the joint (Figure 12-52a), which usually relieves pain and corrects deformity but lacks stability and strength. The alternate approaches are: hinge type (Figure 12-52b), polycentric type (Figure 12-52c), space filler type (Figure 12-52d), and a combination of space-filler and hemiresection arthroplasty (Figure 12-52e). These are summarized in Table 12-7, and some actual implants are shown in Figure 12-53.

The concept of fixing the implant with an encapsulated collagenous membrane, which is created around the implant, is believed to be a clinically

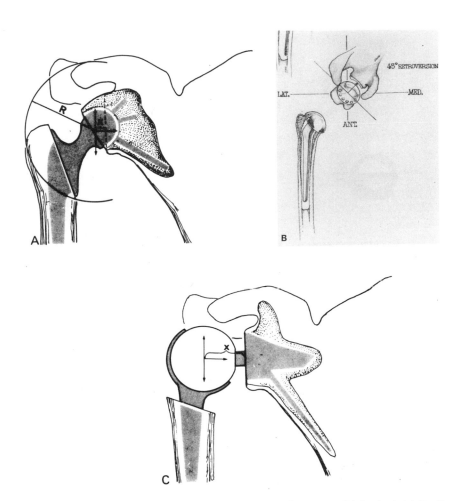

Figure 12-46. Types of shoulder joint prostheses. (a) Stanmore; (b) Bechtol; (c) Fenlin.

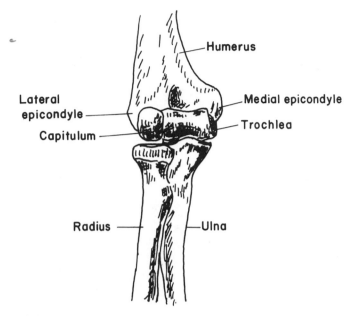

Figure 12-47. A detailed anatomical drawing of the elbow joint. (From Ref. 78.)

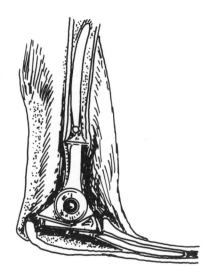

Figure 12-48. A schematic lateral view of a Dee elbow prosthesis. Apart from a portion of the humeral intramedullary stem, the remainder of the implant is buried within the bone. (From Ref. 80.)

Table 12-6. Various Types of Elbow Prostheses[a,b]

Author	Type	Humeral component	Fixing	Ulnar component	Fixing	Comments
Attenborough	Loose hinge	Co-Cr	Stem	HDPE	Stem	Stabilizing rod: handed; two HDPE snap-in bushes
Cavendish	Surface replacement	SS double cone, tapered ends	Wedge + cement	HDPE	Wedge + cement	Not handed
Dee	Loose hinge	Metal	IMS	Metal + HDPE hook	IMS	Snap on: handed
Engelbrecht	Loose hinge	Metal, HDPE bobbin	Two stems in medullary canal	Metal hook	IMS	Possible radius retainer, snap on
Ewald	Surface replacement	Metal	IMS	HDPE	IM peg	Radial head replacement: handed (4 stem angles)
Gschwend	Loose hinge	Metal + HDPE ring	IMS + cap	Metal	IMS	New design
Lowe	Surface replacement	HDPE	Cement	Ti alloy	Cement	Not handed
Pritchard	Narrow loose hinge	HDPE	Long IMS	Co-Cr-Mo handed	IMS	Snap-in axle: HDPE + metal core, handed
Roper	Surface replacement	Metal	Cement	Plastic	Cement	Cement keys
Scales	Narrow rigid hinge	Co-Cr-Mo + HDPE bush	IMS	Co-Cr-Mo	IMS	Snap-in HDPE bush: handed
Stevens	Surface replacement	Metal (Ti alloy)	Friction	—	—	Single component: not handed

[a] From Ref. 79.
[b] SS, stainless steel; IMS, intramedullary stem; HDPE, high-density polyethylene.

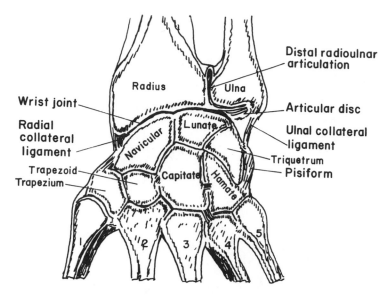

Figure 12-49. Anatomical features of the wrist joint. (From Ref. 73.)

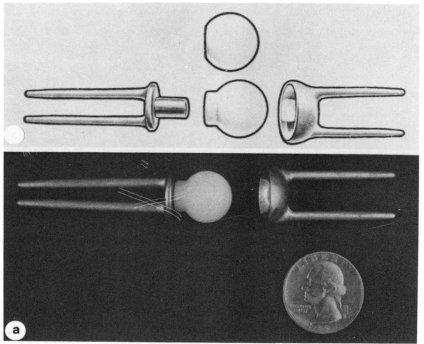

Figure 12-50. Various types of wrist prostheses. (a) Meuli[83]; (b) Volz[81]; (c) Swanson.[84]

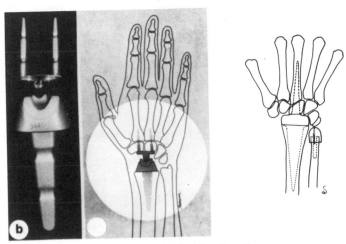

Figure 12-50. (*Continued*)

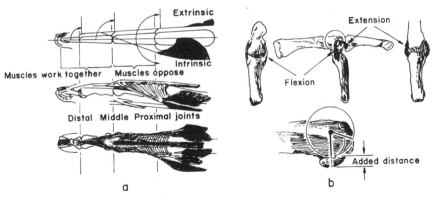

a

b

Figure 12-51. (a) This schematic diagram of a finger indicates the principal musculo-tendinous structures. Note the arrangement of muscles and tendons. (b) A schematic diagram of the metacarpophalangeal joint in extension and flexion. When the joint is in extension, the collateral ligaments are slack to permit abduction and adduction. The joint is stabilized in flexion because the ligaments are tightened in both longitudinal and transverse planes. (From Ref. 85.)

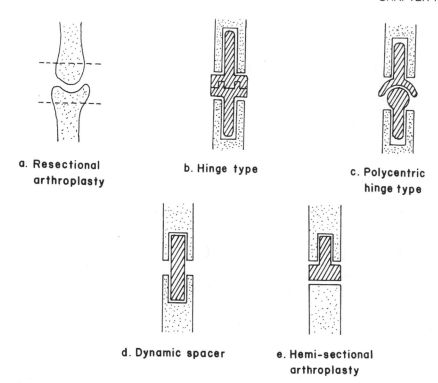

a. Resectional   b. Hinge type   c. Polycentric
   arthroplasty   hinge type

d. Dynamic spacer   e. Hemi-sectional
                            arthroplasty

Figure 12-52. Schematic diagrams indicating the various types of resectional and implant arthroplasties available for treatment of joints in the hand. (From Ref. 87.)

Table 12-7. Types of Finger Joint Replacements

Type	Remarks	Examples
Hinge	Loosening, excessive wear, and breakage, subsidence of implants UHMWPE/Co–Cr alloy	Schultz (Figure 12-53a) St. Georg (Figure 12-53b)
Polycentric	Good duplication of anatomical motion, unstable joint UHMWPE/Co–Cr alloy	Steffee (Figure 12-53c)
Space-filler	Good clinical success, tear sensitive, strong grip not possible due to weakness of implant Silicone rubber (d), silicone rubber–Dacron® composite (f), polypropylene (e)	Swanson (Figure 12-53d) Calnan-Nicolle (Figure 12-53e) Niebauer-Cutter (Figure 12-53f)

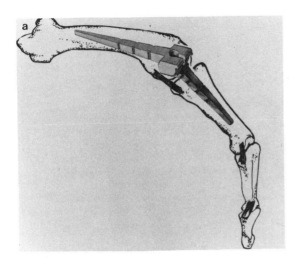

Figure 12-53. Various types of finger joint prostheses. (a) Schultz; (b) St. Georg; (c) Steffee; (d) Swanson; (e) Calnan-Nicolle; (f) Niebauer-Cutter; (g) Lord's Bonded Bion elastomy titanium joint.

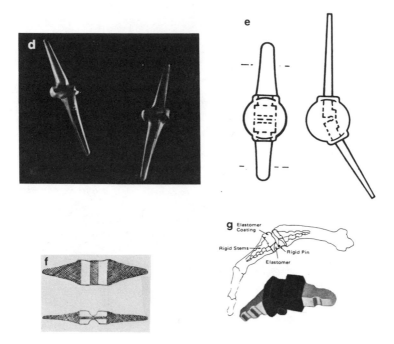

Figure 12-53. (*Continued*)

sound one. This is primarily taking advantage of nature's response to any implant in the body, that is, encapsulation, especially a moving implant. The finger joint is best suited for this means of fixation since the cortical bones are often too thin to withstand a concentrated load if the implant is fixed either by bone cement or by tissue ingrowth.[88] This encapsulation phenomenon occurs in all implant fixations, and it is strongly advocated that it be taken advantage of.[89]

12.3. DENTAL IMPLANTS

We are all familiar with some form of dental implant, such as amalgam for cavities. However, this type of implant is outside of our interest. Instead, we are going to study the total tooth or alveolar replacements with man-made materials.

Tooth replacement has been a challenging problem due to its transcutaneous (or percutaneous) nature in the hostile oral environment, which continually changes its chemical composition, pH, temperature, etc. Teeth undergo the most severe compressive stresses in the body (up to 850 N), and a satisfactory material or technique has not yet been found which can withstand not only the compressive stress but also the added torque and shear stress during mastication.

In a recent comprehensive conference on the benefits and risks of dental implants,[90] dental implants were classified largely into two categories: subperiosteal/staple/transosteal implants and endosseous tooth implants; the former serves to support denture and the latter to restore the function of teeth with or without supporting the bridge network.

12.3.1. Endosseous Tooth Implants

The endosseous implant is inserted into the site of missing or extracted teeth to restore the original function. The ideal implant is the tooth itself, pulled from the same socket, if it can be replanted, although this is not a solution to the problems of tooth replacements. There are many different types of designs for endosseous implants as shown in Figure 12-54. The main idea behind the various root portions of self-tapping screws, spiral, screw-vent, and blade-vent implants, is to achieve immediate stabilization as well as long-term viable fixation. The post will be covered with an appropriate crown after the implant has been fixed firmly for about 1–4 months. Some implants use a more complicated system of fixation as shown in Figure 12-55, in which the implant root is first implanted in the tooth extraction site (preferrably after complete healing of the site) and completely buried, then the post is installed through a punctured hole on the mucosa using a cement, and finally the crown is made. Despite the more elaborate work and design of the implants, the success rate of this implant system did not exceed that of other implants such as blade-vents (Linkow types). The major problems were (1) initially the quality control of the vitreous carbon (Vitrident®) implants was not stringent (almost 20% of one batch was deemed defective by one investigator),[92] (2) the brittle nature of the vitreous carbon makes the implant bulky, consequently removing a greater amount of alveolar bone, and (3) the implants were distributed indiscriminately to general dentists lacking the qualifications attending a rather complicated implant, and thus the high level of failure.[93]

The survival rates of the popular blade-vent implants vary for various investigators due to many factors: surgical techniques, patient selection, evaluation criteria, etc. Figure 12-56 shows the unadjusted and adjusted survival rates for free-end blade implants by various investigators.[94] The variations are largely due to differences in the evaluation criteria used by different investigators. Due to these large differences in reported results, the

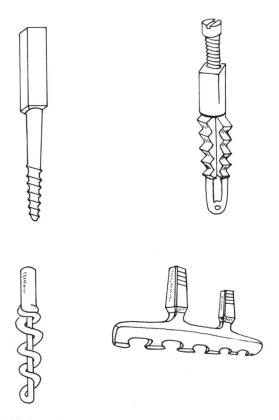

Figure 12-54. Various designs of self-tapping dental implants. (From Ref. 91.)

NIH–Harvard Consensus Development Conference participants recommended "no specific survival estimates can be quoted because of either insufficient sample size or conflicting results" for the endosseous dental implants including vitreous carbon implants.[90]

Most of the blade-vent endosseous implants are made of stainless steel, Co–Cr alloy, Ti, and Ti6Al4V alloy.[91] There have been efforts to coat the surfaces of the implants with ceramics (alumina and zirconium) and polytetrafluoroethylene composite (Proplast®) with little significant improvement in their performance.[95,96] Others used pyrolytic carbon,[97,98] polycrystalline alumina,[99,100] and single-crystal alumina.[101] Recently, surface-textured implants[102] and porous implants with electrical stimulation[103–105] have been tested. Some have even tried to use anorganic

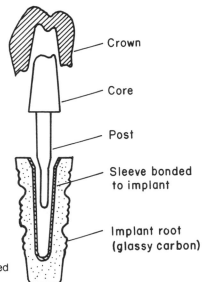

Figure 12-55. The tooth root implant fabricated from glassy carbon. (From Ref. 91.)

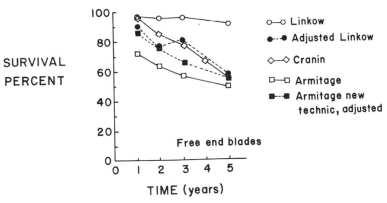

Figure 12-56. Unadjusted and adjusted survival rates for all free-end blade implants. The midpoint between grades 2 and 3 of radiographic bony defect was used as the criterion for adjusting the survival rates in the Linkow and Armitage data. (From Ref. 94.)

bone/acrylic polymer (PMMA) composite material to induce bony tissue replacements in place of anorganic bone without much success.[106]

12.3.2. Subperiosteal and Staple/Transosteal Implants

Implants have been successfully used to provide a framework for dentures for the edentulous alveolar ridge, as shown in Figure 12-57. Although similar functions can be duplicated by implanting osseous dental implants, the periosteal and/or staple/transosteal implants are required due to the thin alveolar ridge for many edentulous patients to provide better support for dentures or sometimes whole arches of bridgework.

Closely related problems encountered with dental tooth implants have been met with these implants: inflammation and tissue reaction around the post and exposure of the framework. Also, similar success rates seen in dental implants were reported,[108] while others have claimed much better results.[109] This prompted the same restrictions on the use of the data as for the dental implants as mentioned in the previous section.

Materials used for these implants are primarily metals, stainless steel, Co–Cr alloy, and Ti alloy, for their ease of fabrication in the conventional dental laboratory. Some have advocated use of a coating of metals with other inert materials, such as carbon and ceramics. It is suspected that these coatings will result in a marginal improvement, as in endosseous dental implants.

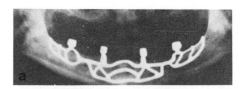

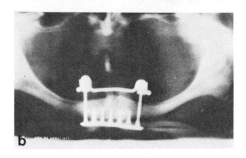

Figure 12-57. (a) A mandibular subperiosteal implant. (b) Seven-pin mandibular staple. [(a) from Ref. 90; (b) from Ref. 107.]

PROBLEMS

12-1. A bioengineer is trying to determine the amount of gap developed between bone and cement when a femoral hip is placed. He assumed the system as concentric cylinders. Calculate the gap developed between bone and cement if the temperatures of cement, implant, and bone reached 55, 50, and 45°C, respectively, throughout each component uniformly [assume the thermal expansion coefficient (α) of the implant is $17 \times 10^{-6}/°C$].

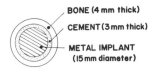
BONE (4 mm thick)
CEMENT (3 mm thick)
METAL IMPLANT
(15 mm diameter)

Figure P12-1.

12-2. A bioengineer is asked to construct a bone fracture plate from the following materials: 316L SS, Ti6Al4V, cast CoCrMo, wrought CoNiCrMo, Al_2O_3, carbon fiber–carbon composite.

a. Which two would be chosen for their biocompatibility?
b. Which one would be chosen for its strength?
c. Which one would be chosen for its specific strength (σ/ρ)?
d. Which ones have been approved by the FDA for implants?
e. Discuss the advantages and disadvantages of Ti6Al4V in the fabrication of a fracture plate.

12-3. Explain possible degradation mechanisms of polymers *in vitro* and *in vivo*.

12-4. Describe the sequence of events by which the production of wear debris in an artificial joint can lead to joint pain and loosening of the prosthesis.

12-5. A simple way to prevent the stem of a hip prosthesis from breaking through the cortex of a weak femur would be to use a metal band which circumscribed the weak area and distributed the load around the cortex. Describe the advantages and disadvantages of this approach.

12-6. Assuming that you wished to attach a femoral head prosthesis by using a porous coating on a metallic stem, would you choose a ceramic, metallic, or plastic coating material? Discuss all aspects of the problem including fabrication, compatibility, stress distribution, etc.

12-7. What are some of the advantages and disadvantages of porous materials (for the attachment of joint replacements) as compared to PMMA?

12-8. What is the Oppenheimer effect? (Read B. S. Oppenheimer *et al.*, *Cancer Res.*, *21*, 137, 1961 before answering.)

REFERENCES

1. S. Olenoid and G. Danckwardt-Lilliestrom, Fracture healing in compression osteosynthesis: An experimental study in dogs with an avascular, diaphyseal, intermediate fragment, *Acta Orthop. Scand. Suppl. 137*, 1–44, 1971.
2. J. Black, S. P. Richardson, R. U. Mattson, and S. R. Pollack, Haversian osteons: Longitudinal variation of internal structure, *J. Biomed. Mater. Res. 14*, 41–53, 1980.
3. K. Pierkarski, A. M. Wiley, and J. E. Bartels, The effect of delayed internal fixation on fracture healing: An experimental study, *Acta Orthop. Scand. 40*, 543–551, 1970.

4. Richards Manufacturing Co., *Bone Screw Technical Information*, Technical Publication 4167, Memphis, Tenn., 1980.

5. J. Schatzker, R. Sanderson, and J. P. Murnaghan, The holding power of orthopedic screws *in vivo*, *Clin. Orthop. Relat. Res. 108*, 115–126, 1975.

6. E. Koranyi, C. E. Bowman, C. D. Knecht, and M. Janssen, Holding power of orthopedic screws in bone, *Clin. Orthop. Relat. Res. 72*, 283–286, 1970.

7. J. Schatzker, J. G. Horne, and G. Summer-Smith, The effect of movement on the holding power of screw in bone, *Clin. Orthop. Relat. Res. 111*, 257–262, 1975.

8. J. Schatzker, J. G. Horne, and G. Summer-Smith, The reaction of cortical bone to compression by screw threads, *Clin. Orthop. Relat. Res. 111*, 263–265, 1975.

9. H. K. Uhthoff, Mechanical factors influencing the holding power of screws in compact bone, *J. Bone Jt. Surg. 55B*, 633–639, 1973.

10. J. A. Albright, T. R. Johnson, and S. Saha, Principles of internal fixation, in: *Orthopedic Mechanics: Procedures and Devices*, D. N. Ghista and R. Roaf (ed.), pp. 123–229, Academic Press, New York, 1978.

11. M. Laurence, M. A. R. Freeman, and S. A. V. Swanson, Engineering considerations in the internal fixation of fractures of the tibial shaft, *J. Bone Jt. Surg. 51B*, 754–768, 1969.

12. M. Allgower, P. Matter, S. M. Perren, and T. Ruedi, *The Dynamic Compression Plate, DCP*, p. 18, Springer-Verlag, Berlin, 1973.

13. A. A. White, Fracture treatment: The still unsolved problem, *Clin. Orthop. Relat. Res. 106*, 279–284, 1975.

14. R. V. Lindholm, T. S. Lindholm, S. Toikkanen, and R. Leino, Effect of forced interfragmental movements on the healing of tibial fractures in rats, *Acta Orthop. Scand. 40*, 721–728, 1970.

15. H. K. Uhthoff and F. L. Dubuc, Bone structure changes in the dog under rigid internal fixation, *Clin. Orthop. Relat. Res. 81*, 165–170, 1971.

16. S. L. Y. Woo, W. H. Akeson, R. D. Coutts, L. Rutherford, D. Doty, G. F. Gemmott, and D. Amiel, A comparison of cortical bone atrophy secondary to fixation with plates and large differences in bending stiffness, *J. Bone Jt. Surg. 58A*, 190–195, 1976.

17. S. L. Y. Woo, The relationships of changes in stress levels on long bone remodeling, in: *Mechanical Properties of Bone*, S. C. Cowin (ed.), pp. 107–129, American Society of Mechanical Engineers, New York, 1981.

18. Z. F. G. Jaworski, M. Liskova-Kiar, and H. K. Uhthoff, Regional disuse osteoporosis and factors influencing its reversal, in: *Current Concepts of Internal Fixation of Fractures*, H. K. Uhthoff (ed.), pp. 17–26, Springer-Verlag, Berlin, 1980.

19. A. Brown and J. C. D'Arcy, Internal fixation for supra-condylar fractures of the femur in the elderly patients, *J. Bone Jt. Surg. 53B*, 420–424, 1971.

20. H. M. Frost, *Orthopedic Biomechanics*, p. 444, Thomas, Springfield, Ill., 1973.

21. G. Kuntscher, *Practice of Intramedullary Nailing*, Thomas, Springfield, Ill., 1967.

22. F. W. Rhinelander, Effects of medullary nailing on the normal blood supply of the diaphysial cortex, in *A. A. O. S. Instructional Course Lectures*, Volume 12, p. 161, Mosby, St. Louis, 1973.

23. F. W. Rhinelander, Circulation in bone, in: *The Biochemistry and Physiology of Bone*, G. H. Bourne (ed.), 2nd ed., Volume 2, pp. 1–77, Academic Press, New York, 1972.

24. R. Soto-Hall and N. P. McCloy, Cause and treatment of angulation of femoral intramedullary nails, *Clin. Orthop. Relat. Res. 2*, 66–74, 1953.

25. W. C. Allen, G. Piotrowski, A. H. Burstein, and V. H. Frankel, Biomechanical principles of intramedullary fixation, *Clin. Orthop. Relat. Res. 60*, 13–20, 1968.

26. D. F. Williams and R. Roaf, *Implants in Surgery*, Saunders, Philadelphia, 1973.

27. W. Kuehnegger, The systematic development of a cervical-thoracic-lumbo-sacral orthesis and its clinical applications, in: *Orthopedic Mechanics: Procedures and Devices*, D. N. Ghista and R. Roaf (ed.), pp. 231–286, Academic Press, New York, 1978.

28. P. R. Harrington, The treatment of scoliosis, correction and internal spine instrumentation, *J. Bone Jt. Surg. 44A*, 591–610, 1962.

29. R. Roaf, A new plate for correcting scoliosis, *Proc. R. Soc. Med. 62*, 272–273, 1969.

30. L. S. Lavine, I. Lutrin, and M. H. Shamos, Treatment of congenital pseudoarthrosis of the tibia with direct current, *Clin. Orthop. Relat. Res. 124*, 69–74, 1977.

31. J. A. Spadaro, Electrically stimulated bone growth in animals and man, *Clin. Orthop. Relat. Res. 122*, 325–332, 1977.

32. C. A. L. Bassett, R. J. Pawluk, and A. A. Pilla, Acceleration of fracture repair by electromagnetic fields—a surgically non-invasive method, *Ann. N.Y. Acad. Sci. 238*, 242–263, 1974.

33. C. T. Brighton, Z. B. Friedenberg, E. I. Mitchell, and R. E. Booth, Treatment of nonunion with constant direct current, *Clin. Orthop. Relat. Res. 124*, 106–123, 1977.

34. J. Black and C. T. Brighton, Mechanisms of stimulation of osteogenesis by D. C. currect, in: *Electrical Properties of Bone and Cartilage*, C. T. Brighton, J. Black, and S. R. Pollack (ed.), pp. 215–224, Grune & Stratton, New York, 1979.

35. C. R. Hassler, E. F. Rybicki, R. B. Diegle, and L. C. Clark, Studies of enhanced bone healing via electrical stimuli, *Clin. Orthop. Relat. Res. 124*, 9–19, 1977.

36. J. A. Spadaro, S. E. Chapin, and R. O. Becker, Cathode composition and electrical osteogenesis, 25th Annu. Orthop. Res. Soc. Meet., p. 85, San Francisco, 1979.

37. C. A. L. Bassett, A. A. Pilla, and R. J. Pawluk, A non-operative salvage of surgically resistant pseudoarthroses and nonunions by pulsing electromagnetic fields, *Clin. Orthop. Relat. Res. 124*, 128–143, 1977.

38. B. Kummer, Biomechanics of the hip and knee joint, in: *Advances in Hip and Knee Joint Technology*, M. Schaldach and D. Hohmann (ed.), pp. 24–52, Springer-Verlag, Berlin, 1976.

39. J. P. Paul, Loading on normal hip and knee joints and joint replacements, in: *Advances in Hip and Knee Joint Technology*, M. Schaldach and D. Hohmann (ed.), pp. 53–70, Springer-Verlag, Berlin, 1976.

40. K. M. Sivash, *Alloplasty of the Hip Joint: A Laboratory and Clinical Study*, Medical Press, Moscow, 1967.

41. S. A. V. Swanson and M. A. R. Freeman (ed.), *The Scientific Basis of Joint Replacement*, Wiley, New York, 1977.

42. I. D. Oh, J. D'Errico, and W. H. Harris, Studies of strain in the proximal femur in simulated one-legged stance: The role of collar–calcar contact of THR in protection of the femoral stem, 24th Annu. Orthop. Res. Soc. Meet., p. 276, Dallas, 1978.

43. J. N. Wilson and J. T. Scales, Loosening of the total hip replacements with cement fixation, *Clin. Orthop. Relat. Res. 72*, 145–160, 1970.

44. J. H. Dumbleton, *Tribology of Natural and Artificial Joints*, Elsevier, Amsterdam, 1981.

45. H. C. Amstutz, K. L. Markolf, G. M. McNeices, and T. A. W. Gruen, Loosening of total hip components: Cause and prevention, in: *The Hip*, pp. 102–116, Mosby, St. Louis, 1976.

46. J. Miller, D. L. Burke, J. W. Stachiewics, A. N. Ahmed, and L. C. Kelebay, Pathophysiology of loosening of femoral components in total hip arthroplasty, in: *The Hip*, pp. 64–86, Mosby, St. Louis, 1978.

47. P. S. Walker, Engineering principles of knee prostheses, in: *Disorders of the Knee*, A. J. Helfet (ed.), p. 262, Lippincott, Philadelphia, 1974.

48. A. S. Greenwald and M. B. Matejczyk, Knee joint mechanics and implant evaluation, in: *Total Knee Replacement*, A. A. Savastano (ed.), pp. 11–30, Appleton–Century–Crofts, New York, 1980.

49. A. A. Savastano, Indications for knee joint replacement, in: *Total Knee Replacement*, A. A. Savastano (ed.), pp. 31–39, Appleton–Century–Crofts, New York, 1980.

50. D. C. Mears, *Materials and Orthopedic Surgery*, Williams & Wilkins, Baltimore, 1979.

51. L. Marmor, The Marmor type of knee replacement, in: *Total Knee Replacement*, A. A. Savastano (ed.), pp. 107–123, Appleton–Century–Crofts, New York, 1980.

52. J. Charnley, *The Charnley Load-Angle Inlay Arthroplasty of the Knee*, Leeds, Thackey, 1975.

53. T. R. Waugh and P. M. Evanski, University of California, Irvine (UCI) knee replacement —Design, operative technique and results, in: *Total Knee Replacement*, A. A. Savastano (ed.), pp. 217–232, Appleton–Century–Crofts, New York, 1980.

54. J. N. Insall, The total condylar prosthesis, in: *Total Knee Replacement*, A. A. Savastano (ed.), pp. 83–105, Appleton–Century–Crofts, New York, 1980.

55. M. A. R. Freeman, The ICLH arthroplasty of the knee joint, in: *Total Knee Replacement*, A. A. Savastano (ed.), pp. 59–82, Appleton–Century–Crofts, New York, 1980.

56. G. Deane, *The Deane Knee*, p. 1, Institution of Mechanical Engineers, London, 1975.

57. D. A. Sonstegard, H. Kaufer, and L. S. Matthews, The spherocentric knee: Biomechanical testing and clinical trial, *J. Bone Jt. Surg. 59A*, 602–616, 1977.

58. C. G. Attenborough, The Attenborough total knee replacement, *J. Bone Jt. Surg. 60B*, 302–326, 1978.

59. B. Walldius, Arthroplasty of the knee—27 years experience, in: *Total Knee Replacement*, A. A. Savastano (ed.), pp. 195–216, Appleton–Century–Crofts, New York, 1980.

60. L. G. Shiers, Arthroplasty of the knee: Interim report of a new method, *J. Bone Jt. Surg. 42B*, 31–39, 1960.

61. E. Englebrecht, A. Siegel, J. Rottger, and H. W. Buchholz, Statistics of total knee replacement: Partial and total knee replacement, design St. Georg. A review of a 4 year observation, *Clin. Orthop. Relat. Res. 120*, 54–64, 1976.

62. A. Deburge, Guepar, "Guepar hinge prosthesis: Complications and results with two years' follow-up, *Clin. Orthop. Relat. Res. 120*, 47–53, 1976.

63. *Stanmore Total Knee Replacement*, Booklet No. 169, Orthopedic Equipment Co., Bourbon, Ind., 1978.

64. C. O. Bechtol, Bechtol patello-femoral joint replacement system, Richards Manufacturing Co., Memphis, Tenn.

65. E. A. Salvati and J. N. Insall, The management of sepsis in total knee replacement, in: *Total Knee Replacement*, A. A. Savastano (ed.), pp. 49–58, Appleton–Century–Crofts, New York, 1980.

66. *The Porous Coated Anatomic (PCA) Total Knee System*, Orthopaedic Division, Howmedica, Inc., Rutherford, N.J., 1981.

67. P. Ducheyne, M. Martens, P. DeMeester, E. Aernoudt, and J. C. Mulier, Influence of a functional dynamic loading on bone ingrowth into surface pores of orthopaedic implants, *J. Biomed. Mater. Res. 11*, 811–838, 1977.

68. Richards Manufacturing Co., *In vitro* testing of the RMC total knee, R & D Technical Monograph 3468, Memphis, Tenn., 1978.

69. H. J. Hicks, The mechanics of foot, *J. Anat. 87*, 345–357, 1953.

70. D. G. Wright, S. M. Desai, and W. H. Henderson, Action of the subtalar and ankle-joint complex during the stance phase of walking, *J. Bone Jt. Surg. 46A*, 361–382, 1964.

71. G. J. Sammarco, A. H. Burstein, and V. H. Frankel, Biomechanics of the ankle: A kinematic study, *Orthop. Clin. N. Am. 4*, 75–96, 1973.

72. M. Pappas, F. F. Buechel, and A. F. DePalma, Cylindrical total ankle joint replacement: Surgical and biomechanical rationale, *Clin. Orthop. Relat. Res. 118*, 82–92, 1976.

73. C. M. Goss (ed.), *Gray's Anatomy*, p. 316, Lea & Febiger, Philadelphia, 1975.

74. C. S. Neer, Replacement arthroplasty for glenohumeral osteoarthritis, *J. Bone Jt. Surg. 56A*, 1–13, 1974.

75. J. M. Fenlin, Jr., Total glenohumeral joint replacement, *Orthop. Clin. N. Am. 6*, 565–583, 1975.

76. I. C. Clarke, T. A. W. Gruen, A. Sewhoy, D. Hirschowitz, S. Maki, and H. C. Amstutz, Problems in glenohumeral surface replacements—real or imagined?, *Eng. Med. 8*, 161–175, 1979.

77. B. F. Morrey and E. Y. S. Chao, Passive motion of the elbow joint, *J. Bone Jt. Surg. 58A*, 501–508, 1976.

78. I. A. Kapandji, *Physiology of the Joints*, p. 81, Livingstone, Edinburgh, 1970.

79. Institution of Mechanical Engineers Report, Joint replacement in the upper limb, *Eng. Med. 6*, 90–93, 1977.

80. N. Gschwend, Design criteria, present indication, and implantation techniques for artificial knee joints, in: *Advances in Artificial Hip and Knee Joint Technology*, M. Schaldach and D. Hohmann (ed.), pp. 90–114, Springer-Verlag, Berlin, 1976.

81. R. G. Volz, Total wrist arthroplasty: A new approach to wrist disability, *Clin. Orthop. Relat. Res. 128*, 180–189, 1977.

82. Y. Youm, R. Y. McMurty, A. E. Flatt, and T. E. Gillespie, Kinematics of the wrist. I. An experimental study of radial ulnar deviation and flexion extension, *J. Bone Jt. Surg. 64A*, 423–431, 1978.

83. H. C. Z. Meuli, Alloarthropstik des Handgelenks, *Z. Orthop. Ihre Grenzgeb. 113*, 476–478, 1975.

84. A. B. Swanson, *Flexible Implant Resection Arthroplasty in the Hand and Extremities*, Mosby, St. Louis, 1973.

85. A. E. Flatt (ed.), *The Care of Minor Hand Injuries*, Mosby, St. Louis, 1972.

86. R. L. Linshend and E. Y. S. Chao, Biomechanical assessment of finger function in prosthetic joint design, *Orthop. Clin. N. Am. 4*, 317–320, 1973.

87. R. I. Burton, Implant arthroplasty in the hand: An introduction, *Orthop. Clin. N. Am. 4*, 313–316, 1973.

88. J. B. Park and K. Margraf, Interfacial shear stress strength of implant/intramedullary bone in geese, in: *Biocompatible Polymers, Metals, and Composites*, M. Szycher (ed.), Chapter 28, Technomic, Lancaster, Pa., 1982.

89. E. E. Frisch, Functional considerations in implant design, Proceedings of the First Southern Biomedical Engineering Conference, S. Saha (ed.), pp. 299–304, Pergamon Press, Elmsford, N.Y., 1982.

90. P. A. Schnitman and L. B. Schulman (ed.), *Dental Implants: Benefits and Risk*, NIH–Harvard Consensus Development Conference, NIH Publication 81-1531, 1980.

91. D. E. Grenoble and D. Voss, Materials and designs for implant dentistry, *Biomater. Med. Devices Artif. Organs 4*, 133–169, 1976.

92. J. E. Lemons, Biomaterials science protocols for clinical investigations on porous alumina ceramic and vitreous carbon implants, *J. Biomed. Mater. Res. Symp. 4*, 9–16, 1975.

93. E. D. McCoy, Risk of vitreous carbon implants, in: *Dental Implants: Benefits and Risk*, P. A. Schnitman and L. B. Schulman (ed.), NIH–Harvard Consensus Development Conference, pp. 211–221, NIH Publication 81-1531, 1980.

94. K. K. Kapur, Benefit and risk of blade implants: A critique, in: *Dental Implants: Benefits and Risk*, P. A. Schnitman and L. B. Schulman (ed.), NIH–Harvard Consensus Development Conference, pp. 305–314, NIH Publication 81-1531, 1980.

95. A. N. Cranin, P. A. Schnitman, M. F. Rabkin, and T. Dennison, Alumina and zirconium coated Vitallium oral endosteal implants in beagles, *J. Biomed. Mater. Res. Symp. 9*, 257–262, 1975.
96. J. N. Kent, C. A. Homsy, B. D. Gross, and E. C. Hinds, Pilot studies of a porous implant in dentistry and oral surgery, *J. Oral Surg. 30*, 608–615, 1972.
97. H. S. Shim, The strength of LTI carbon dental implants, *J. Biomed. Mater. Res. 11*, 435–445, 1977.
98. S. F. Hulbert, J. N. Kent, J. C. Bokros, H. S. Shim, and O. M. Reed, Design and evaluation of LTI-Si carbon endosteal implants, *Oral Implant. 6*, 79–94, 1975.
99. T. D. Driskell and A. L. Heller, Clinical use of aluminum oxide endosseous implants, *Oral Implant. 7*, 53–76, 1977.
100. W. Schulte, C. M. Busing, B. D'Hoedt, and G. Heinke, Endosseous implants of aluminum oxide ceramics: A 5 year study in humans, in: *Implantology and Biomaterials in Stomatology*, H. Kawahara (ed.), pp. 157–167, Ishiyakn, Tokyo, 1980.
101. A. Sawa, A. Fujisawa, A. Yamagami, U. Tsunosue, K. Hoshino, H. Agariguchi, F. Ozawa, I. Kuroyama, S. Shimodaira, M. Chin, and T. Wada, Statistical study of the clinical cases using ceramic implants, in: *Implantology and Biomaterials in Stomatology*, H. Kawahara (ed.), pp. 141–150, Ishiyakn, Tokyo, 1980.
102. A. M. Weinstein, S. D. Cook, J. J. Klawitter, L. A. Weinberg, and M. Zide, An evaluation of ion-textured aluminum oxide dental implants, *J. Biomed. Mater. Res. 15*, 749–756, 1981.
103. S. O. Young, J. B. Park, G. H. Kenner, R. R. Moore, B. R. Meyers, and B. W. Sauer, Dental implant fixation by electrically mediated process. I. Interfacial strength, *Biomater. Med. Devices Artif. Organs 5*, 111–126, 1978.
104. J. B. Park, S. O. Young, G. H. Kenner, A. F. von Recum, B. R. Meyers, and R. R. Moore, Dental implant fixation by electrically mediated process. II. Tissue ingrowth, *Biomater. Med. Devices Artif. Organs 5*, 291–301, 1978.
105. W. J. Whatley, J. B. Park, G. H. Kenner, and A. F. von Recum, The effects of load on electrically stimulated porous dental implants, in: *Implantology and Biomaterials in Stomatology*, H. Kawahara (ed.), pp. 173–179, Ishiyakn, Tokyo, 1980.
106. L. Gettleman, D. Nathanson, R. L. Myerson, and M. Hodosh, Porous heat cured poly(methyl methacrylate) for dental implants, *J. Biomed. Mater. Res. Symp. 6*, 243–249, 1975.
107. I. A. Small, Benefit and risk of mandibular staple bone plates, in: *Dental Implants: Benefits and Risk*, P. A. Schnitman and L. B. Schulman (ed.), NIH–Harvard Consensus Development Conference, pp. 139–151, NIH Publication 81-1531, 1980.
108. N. I. Goldberg, Risk of subperiosteal implants, in: *Dental Implants: Benefits and Risk*, P. A. Schnitman and L. B. Schulman (ed.), NIH–Harvard Consensus Development Conference, pp. 89–95, NIH Publication 81-1531, 1980.
109. R. L. Bodine and R. T. Yanase, Benefit of subperiosteal implants, in: *Dental Implants: Benefits and Risk*, P. A. Schnitman and L. B. Schulman (ed.), NIH–Harvard Consensus Development Conference, pp. 75–85, NIH Publication 81-1531, 1980.

BIBLIOGRAPHY

M. Allgower, P. Matter, S. M. Perren, and T. Ruedi, *The Dynamic Compression Plate, DCP*, Springer-Verlag, Berlin, 1973.
C. O. Bechtol, A. B. Ferguson, and P. G. Laing, *Metals and Engineering in Bone and Joint Surgery*, Ballière, Tindall & Cox, London, 1979.

APPENDIX B

SI UNITS

The International System of Units or SI (Système International d'Unitès) units define the *base units* as follows:

Base Units

Quantity	Unit	Symbol
Length	Meter	m
Mass	Kilogram	kg
Time	Second	s
Electric current	Ampere	A
Temperature	Kelvin	K
Amount of substance	Mole	mol

The *derived units* for our interest arc as follows:

Derived Units

Quantity	Unit	Symbol	Formula
Frequency	Hertz	Hz	$1/s$
Force	Newton	N	$kg\text{-}m/s^2$
Pressure, stress	Pascal	Pa	N/m^2
Energy, work, quantity of heat	Joule	J	$N\text{-}m$
Power	Watt	W	J/s
Absorbed dose	Gray	Gy	J/kg

The common prefixes used in this book are:

	Prefix	Symbol
10^9	Giga	G
10^6	Mega	M
10^3	Kilo	k
10^{-2}	Centi	c
10^{-3}	Milli	m
10^{-6}	Micro	μ
10^{-9}	Nano	n

The conversion factors between the various units and SI units are:

To convert from...	to...	multiply by...
Angstrom	m	10^{-10}
Inch	m	0.0254
Free fall, standard (g)	m/s^2	9.80665
Calorie	J	4.1868
Erg	J	10^{-7}
Dyne	N	10^{-5}
kg force	N	9.80665
Pound force	N	4.448
Pound (mass)	kg	0.4535924
Atmosphere (standard)	Pa	0.1
Inch of Hg (60°F)	Pa	3.37685×10^3
Dyne/cm^2	Pa	0.1
kg/mm^2	Pa	9.807×10^6
lb/in.^{2} (psi)	Pa	6.894757×10^3
MPa	psi	145
Poise	Pa-s	0.1
Rad	Gy	0.01

ANSWERS TO PROBLEMS

1-1. 1 year: 8% failure; 10 years: 13% failure rate.

1-2. Advantages of transplants: complete freedom of movement; more complete restoration of kidney function especially secretion of enzymes, hormones, etc; lesser damage to the blood and no problem of infection, trauma, etc. due to the catheterization; psychologically better solution; lower cost in the long run; some patients cannot be machine-dialyzed.

Disadvantages: hard to obtain donors; problems related to rejection and the chemical and radiation treatment; the operation itself is major surgery and thus involves a certain danger; once rejected, it is harder to have a second operation.

1-3. Your own.

2-1. a. See figure A2-1.
   b. $E = 198$ GPa, 0.2% offset yield strength $= 560$ MPa, $\sigma = 1354$ MPa.
   c. $\sigma = 3488$ MPa.

2-2. a. $\lambda = 1.443$ h.
   b. $\epsilon = 0.484$ (48.4% which is 96.9% recovery of the strain since the original strain was 50%).

2-3. a. See Figure A2-3.
   b. $M_c = 7500$ g/mol.

2-4. Your own.

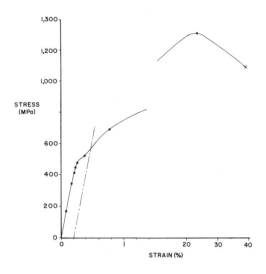

Figure A2-1.

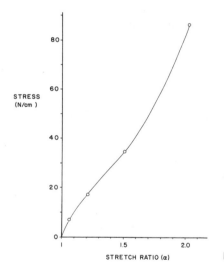

Figure A2-3.

2-5. a. E' and E'' vary according to the frequency, ω:

	ω (rads/s)						
	0	0.1	1	10	100	1000	10,000
E' (MPa)	0	0.01	1	50	100	100	100
E'' (MPa)	0	1	10	50	10	1	0.1

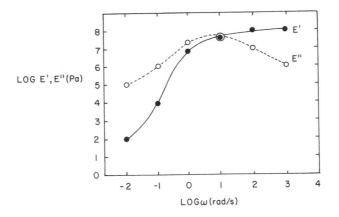

Figure A2-5.

b. See Figure A2-5.

2-6. Your own.

3-1. 7.9 g/ml.

3-2. 0.732.

3-3. 32 w/o S.

3-4. Your own.

3-5. $M_n = 3646$ g/mol, $M_w = 4550$ g/mol, $M_w/M_n = 1.248$.

3-6. 1.633.

3-7. 0.08 v/o (increase in volume when hcp → bcc).

3-8.

	fcc	bcc	sc
Side of unit cell	$4R/2^{1/2}$	$4R/3^{1/2}$	$2R$
Body diagonal	$4R$	$4R(2/3)^{1/2}$	$2R2^{1/2}$
Face diagonal	$4R(3/2)^{1/2}$	$4R$	$2R3^{1/2}$

3-9. 0.68 and 0.85×10^{23} Fe atoms/ml.

4-1. $n/N = 1 \times 10^{-5}$ (1 in 10 sites are vacancies).

4-2. $A = 6/\rho d$.

Diameter (cm)	A (cm^2/g)
0.1	26.1
0.01	261
0.001	2,610
0.0001	26,100
0.00001	261,000

4-3. 61.3 g of austenite (γ) will transform into pearlite at temperatures below 723°C. At 723°C 92.85 g of ferrite (α) and 7.15 g of carbide; at room temperature 92.5 g of α and 7.5 g of carbide.

4-4.

T (°C)	Phases	Composition	Rel. amount
1600	L	0.8%C + 99.2%(Fe + Mn)	All
1400	L	1.8%C + 98.2%(Fe + Mn)	5%
	γ	0.75%C + 99.25%(Fe + Mn)	95%
1000	γ	0.8%C + 99.2%(Fe + Mn)	All
724	γ	0.8%C + 99.2%(Fe + Mn)	All
722	α	0.025%C + 99.975%(Fe + Mn)	89%
	Fe$_3$C	6.67%C + 93.33%(Fe + Mn)	11%

4-5. See Figure A4-5.

4-6. a.

Liquid	Contact angle	γ_{GL}	γ_{LS}
A	103°	63	34.17
B	80°	50	11.3
C	45°	28	0.2

b. $\gamma_A > \gamma_B > \gamma_C$ for LS surface and there is a linear correlation between γ_{GL} and contact angle. See Figure A4-6.

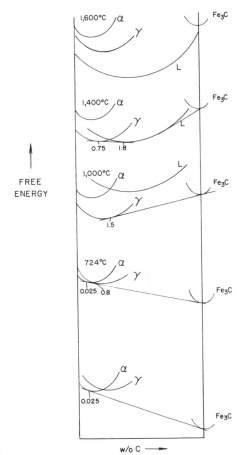

FREE
ENERGY

Figure A4-5.

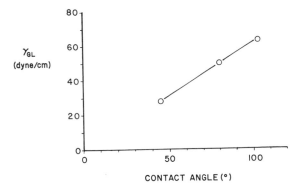

Figure A4-6.

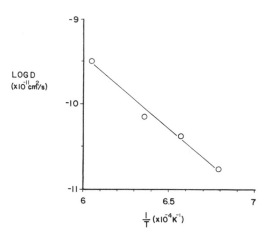

Figure A4-7.

4-7. a. See Figure A4-7.
   b. $D_0 = 5$ cm^2/s.
   c. $D(800°C) = 9.4 \times 10^{-16}$ cm^2/s.

4-8. Your own.

4-9. a. The C atom is too small to dissolve substitutionally in Fe according to Hume–Rothery rules (the difference in atomic radii <15%).
   b. γFe, which has larger voids although it is small. Note that the radius ratio C/Fe is about 0.6, which is larger than the radius ratio of fcc octahedral voids.
   c. It is more likely the C atoms will occupy the octahedral sites (very limited dissolution takes place).

5-1. $a_0 = 0.12$ nm (the atomic radius of Fe is 0.124 nm).

5-2. $2c = 2.6$ nm.

5-3. 383 and 100 MPa depending on the different theories.

5-4. $G/G_1 = 1.06$.

5-5. $1/E_c = (V_m/E_m) + (V_f/E_f)$

5-6. Corollary: The strength increases as the mean free path between disruptions decreases.

   a. The solute atoms interfere with the movement of dislocations which in turn will increase the strength.

   b. Cold-working introduces more dislocations which lead to the entanglement of the dislocations.

5-7. $c_p/c_e = (\gamma + p)/\gamma = 1001$.

5-8. Fatigue failures of laboratory specimens can be completely eliminated by periodically etching away the surface of the specimens. This is not a practical way of eliminating fatigue but it does prove unequivocally that the surface cracks are the culprits. Therefore, hardening surfaces (by shot-peening or treating with nitrogen to induce compression strains in the surface) is an effective treatment. This concept is also used to make glass less susceptible to fracture by putting the surface in compression (tempering potassium ion bombardment).

6-1. $x = 0.9922$ (99.22%; most of the collagen fibers are arranged in the longitudinal direction).

6-2. a. $\sigma = 10.8$ MPa.
   b. $E_i = 0.34$ MPa, $E_s = 38$ MPa.
   c. 1.83 MPa/m.

6-3. a. $P_c/P_t = 0.9989$ (99.89% of load borne by collagen).
   b. Collagen contributes to the total load at high strain ($> 55\%$) but elastin contributes a larger proportion at lower strain.

6-4. 76.73 Gpa (this value is closer to tooth enamel).

6-5. a. Model A:

$$\frac{dS}{dt} + \frac{G_1 + G_2}{\eta_2} S = G_1 \frac{d\gamma}{dt} + \frac{G_1 G_2}{\eta_2} \gamma$$

Model B:

$$\frac{dS}{dt} + \frac{g_2}{v_2} S = (g_1 + g_2) \frac{d\gamma}{dt} + \frac{g_1 g_2}{v_2} \gamma$$

b. Model A:

$$\gamma = -\frac{S}{G_2}\exp\left(-\frac{G_2 t}{\eta_2} + \frac{G_1 + G_2}{G_1 G_2}\right)S$$

Model B:

$$\gamma = \frac{-S_2 g_2}{g_1(g_1 + g_2)}\exp\left[-\frac{g_1 g_2}{(g_1 + g_2)v_2}t\right] + \frac{S}{g_1}$$

c.

$$g_1 = \frac{G_1 G_2}{G_1 + G_2}$$

$$G_2 = \frac{g_1(g_1 + g_2)}{g_2}$$

d. Shortcomings of the models: (1) the models describe only one-dimensional behavior; (2) the assumptions are not realistic, i.e., the bone is highly anisotropic and inhomogeneous; (3) they cannot predict the fracture point.

7-1. The most difficult problem is that the surface of the implant is covered with proteins and other biomolecules as soon as it is in contact with tissues and blood. This absorbed layer will change the surface properties and mask the effect of intrinsic material properties. Another difficulty is that the evaluation of biocompatibility is to a large extent subjective rather than objective; hence, it is difficult to quantitate which in turn makes it difficult to evaluate and compare with others.

7-2. a. 0.536 (53.6%).
    b. 0.27 MPa.
    c. 3.4 and 5.2%.
    d. About the same recovery.
    e. 0.71 μm/year.

7-3. 42.65 weeks (or 299 days).

7-4. a. 2 mg/min.
    b. 1000 min (or 16 h 40 min).

7-5. The increased surface area of the powder form increases the surface energy a great deal to which the tissue has to counteract. This will sharply increase the cellular activities and the body tries to reject individual particles. It is also possible that in powder form, covalent and ionically bonded materials (polymers and ceramics) may contain residual free radicals.

7-6. a. The thickness of the collagenous tissue depends largely on the weight of the implant and its relative movement with respect to the surrounding tissues, and not the intrinsic properties. Therefore, thickness measurement is not a reliable method.

b. It is generally accepted that thinner encapsulation is better.

7-7.

	Muscle				Kidney			
	6 weeks		16 weeks		6 weeks		16 weeks	
Element	Av.	Norm.	Av.	Norm.	Av.	Norm.	Av.	Norm.
Cr	38	122.6	5	16	3.3	10.6	6.3	20
Co	30	48.5	7.5	12	35	56.5	28.3	45.7
Ni	28	1842	7.5	493	13.3	875	13.3	875
Mo	3.3	70	31.7	670	23.3	492	38.3	810
Fe	43	7049	267	43,770	90	14,754	106.7	17,491

This indicates that the amount of metal ions retained by various organs is extremely variable with time and type of organ and not necessarily in proportion to the relative amount of metal elements present in the alloy.

7-8. The higher energy state of metals tends to be more reactive with tissues than ceramics which are already oxidized and polymers which have giant molecules (chains).

7-9. a. Since the maximum temperature reached is well over 55°C at which proteins will begin to denature, the adjacent tissues will be damaged.

b. The temperature will decrease since (1) the metal acts as a heat sink and (2) the blood carries away (transfers) heat continuously.

c. A gap will develop between the bone and the cement.

7-10. The tissue response to an "inert" material is very like normal wound healing. No foreign body giant cells appear and a thin fibrous capsule is formed. The tissue in this capsule differs very little from normal scar tissue. In response to an irritant material, foreign body giant cells appear and an inflammatory response ensues. There is an abundance of leukocytes and macrophages and granulation tissue forms. This tissue is a precursor to tissue, serving the functions of phagocytosis and organization, and appears only under circumstances of irritation or infection. Healing is slow and a thick capsule forms. If the material is chemically reactive or mechanically irritating, necrosis of surrounding tissue may result. It has been established that the size and shape of an implant have an effect on the type of tissue reaction seen.

8-1. $\Delta V = 0.7$ (70% volume increase by oxidation).

8-2. 0.0073 cm^3/g.

8-3. 0.01 J/C (or V).

8-4. a. Aluminum.
  b. $F_c = 859.5$ N.
  c. $E_c = 85.21$ GPa.
  d. $\alpha_c = 19.3 \times 10^{-6}/°C$.
  e. $\rho_c = 3.27$ g/ml.
  f. Aluminum; however, if it is passivated, then the steel will be anode.

8-5. a. 8.93 g/ml.
  b. 0.74.
  c. Grain size would tend to increase since there is more time for grain growth after nucleation of the grains.
  d. Grain size would continue to increase due to the thermal energy input.

8-6. See Figure A8-6.

8-7. a. See Figure A8-7.
  b. 10.38 g/ml.
  c. From Figure P8-7, only solid phase (α) exists which has the same compositions as the original sterling silver.
  d. Yes, since the second phase (β) can be precipitated out after quenching.

8-8. Pearlite steels are susceptible to corrosion due to their inhomogeneity. This naturally leads to the formation of galvanic cells where the α phase is

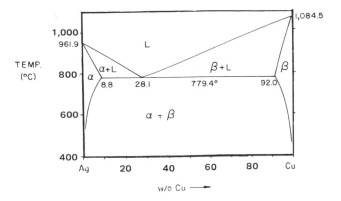

Figure A8-6.

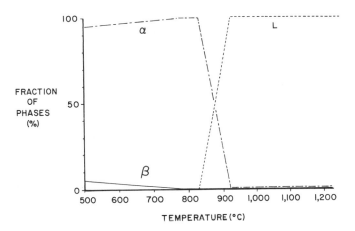

Figure A8-7.

preferentially attacked (cf. Table 8-14). The type of protection depend to a great extent on the environment but, in general, the fine-grained pearlite is less susceptible than a coarse-grained pearlite so that fine grain size is a desirable factor.

8-9. a. The corrosion occurs just at the water line (in the "splash area") where the O_2 concentration is highest. (It is also high above the water line but no electrolyte is present there!) The oxidation reaction $Fe \rightarrow Fe^{2+} + 2e^-$ occurs deeper in the water where the O_2 concentration is low. Therefore, the pattern is pitting in the deep water and rust deposition near the water line.

b. Sacrificial coatings are the best solution initially and sacrificial anodes may be used later. Paint is not easily applied underwater.

8-10. Welding of 316 SS produces an area rich in intergranular chromium carbides near the weld. These areas contain chromium-depleted grains next to the carbides (the chromium entered the carbides) and these areas are particularly susceptible to corrosion. By selecting 316L SS for welding, we take advantage of reduced (by 2/3) carbon content and therefore reduced carbide formation. Even so, the weld areas are more susceptible than unwelded areas so that welding is performed only when no other methods will suffice. The reduced carbon content should alter the mechanical properties only slightly, resulting in a slightly more ductile, less hard and strong steel. Young's modulus is unchanged.

8-11. The stressed sheet becomes anode. emf $= 15$ mV.

9-1. a. 7×10^{42} years in water, 1.59×10^{30} years in blood.

b. The dynamic load exceeds the strength of alumina and failure may occur immediately.

9-2. 140 and 179 MPa in air and in Ringer's solution, respectively, from Figure 9-6. This shows that Ringer's solution has a greater detrimental effect on the strength of the alumina so that it has to undergo a higher proof stress testing than in air.

9-3. $S_0 = 550 d^{-1/3}$ (MPa).

9-4. Unlike bone, carbon has large variations in its properties, that is, it is hard to predict its ultimate properties. Hence, we have to make a prosthesis much larger than the bone in order to eliminate this uncertainty. Also, the bone can be regenerated, repairing microfailures or cracks, and thus its long-term fatigue properties are much better than the carbon.

10-1. a. Tensile stresses: for control, 3.03 MPa; for implanted, 2.81 MPa. Moduli: for control, 165 MPa; for implanted, 55 MPa.

b. The implanted sample has ingrown tissues which are hard to remove from the porous surface and when subjected to testing they exhibit a random tearing which results in the uneven curve. The porous polymer deteriorates upon implantation, resulting in lower strength than the control.

10-2.

$$
\begin{array}{cc}
\text{H} & \text{H} \\
| & | \\
-\text{C}-\text{C}- \\
| & | \\
\text{H} & \text{C}=\text{O} \\
& | \\
& \text{O} \\
& | \\
& \text{CH}_3 \\
& | \\
& \text{O} \\
& | \\
& \text{CH}_3
\end{array}
$$

M.W. = 86,000 g/mol.

10-3. a. 60 days.
      b. 2.26 MPa.

10-4. a. $\eta = 4 \times 10$ Pa-s.
      b. $E = 100$ MPa.

10-5. The hydrolyzable bonds are generally more susceptible to attack by body fluids and biomolecules such as enzymes.

10-6. a, greases; b, soft waxes; c, brittle waxes; d, tough waxes; e, hard and soft plastics.

10-7. 3.08 nm (very short!). The average end-to-end distance of a chain for amorphous polymers is quite short although their total (stretched) distance is quite long (~ 60 nm in this example). In crystalline polymers the chains are not randomly distributed, and thus this calculation does not apply.

10-8. a, polyisoprene; b, polysiloxane (silicone rubber); c, polyethylene; d, polypropylene; e, polyvinyl chloride; f, nylon 6; g, polyvinyl alcohol; h, polyoxymethylene; i, polystyrene; j, polymethylmethacrylate; k, polytetrafluoroethylene; l, polycarbonate; m, polysulfone; n, polyethylene terephthalate; o, polyhydroxyethylmethacrylate.

11-1. After 5 h, $C_t = 39$ mg%; after 10 h, $C_t = 15$ mg%. Note that the extra 5 h of dialysis removed only 24 mg% of urea nitrogen compared to 61 mg% removed in the first 5 h.

11-2. 1-(d), 2-(f), 3-(b), 4-(c), 5-(e), 6-(a)
    1-(c), 2-(d), 3-(a), 4-(b), 5-(f), 6-(e), 7-(g)
    1-(h), 2-(f), 3-(g), 4-(e), 5-(b), 6-(d), 7-(a), 8-(c)

11-3. a. In the major axis.
    b. In the major axis.
    c. By making an elliptic hole with a major and minor axis in the opposite direction made by the circular hole.
    d. All the load will be transferred through the implant and the deformation will be stopped at the interface. It is likely that the interface will fail if it cannot absorb the deformation.

11-4. a. $T = 83$ N/m.
    b. $F = 4.2$ N.
    c. $\sigma = 0.084$ MPa. Since the safety factor is 10, the maximum wall stress should be 0.84×10^6 Pa or 0.84 MPa. From Table 10-14 the tensile strength of silicone rubber is 6 MPa which is adequate for this purpose; however, it is not used due to the inadequate cyclic fatigue property, absorption of small lipid molecules *in vivo*, and above all thrombus formation.

11-5. Your own.

11-6. Your own.

12-1. Gap between bone and cement is 1.4 μm.

12-2. a. Al_2O_3 and carbon fiber–carbon composite.
    b. Wrought CoNiCrMo.
    c. Ti_6Al_4V (cf. Figure 8-15).
    d. 316L SS, cast CoCrMo, wrought CoNiCrMo.
    e. Your own.

12-3. *In vivo*: OH^-, ionic and enzymatic attack, lipid adsorption. *In vitro*: hydrolysis, crystallization, UV, O_2, O_3, radicals, mechanical fatigue, wear, crazing.

12-4. Read pp. 112–122 of *The Scientific Basis of Joint Replacement*, S. A. V. Swanson and M. A. R. Freeman (ed.), Wiley, New York, 1977.

12-5. Some advantages are: a firmer fixation of oblique long bone fractures; easier to apply and remove, shortening surgical time and cost. Some disadvantages are: cutting off of the soft tissue blood supply and subsequent necrosis of tissues; many fractures are not suitable for banding.

12-6.

Coating material	Advantages	Disadvantages
Ceramics	Excellent biocompatibility	Hard to fabricate (no good adhesion metals), uneven stress distribution
Polymers	Easy to fabricate, good stress distribution	Pores can be collapsed, adhesion to the metal is not strong
Metals	Easy to fabricate, good stress distribution	Corrosion possibility, reduced strength of stem

12-7. Major advantage of porous implants is that once fixation is established, a continuous viable fixation can be maintained. However, it takes long for the tissues to grow into the pores and the growth and resorption are dictated by Wolff's law. Thus, even if initial tissue ingrowth has taken place, stress-shielding may occur if the tissues are not subjected to microstrain or microstress, resulting in loosening of implants.

The major disadvantages of PMMA cement fixation are toxicity of monomer, exothermic heat generation during polymerization, no real bonding between bone cement and bone or prosthesis (eventual failure is predictable), and once failed it is hard to remove the cement from the intramedullary cavity. The advantages are fast fixation and subsequent ambulation, easier to use during surgery.

12-8. Briefly, neoplasia can be induced by changing solid materials into powders or films, indicating that the shape and size of implants have a significant role in inducing the type of tissue response in addition to the chemical composition (read B. S. Oppenheimer *et al.*, *Cancer Res. 21*, 137, 1961).

NAME INDEX

SUBJECT INDEX

Acetabular cup, 384
Acrylamide, 189
   N, N-methylene-bis-acrylamide, 281
Acrylic, *see* Polymethylmethacrylate
Acrylonitrile (CH_2=CHCN), 270
Activation energy, E^*, 62
   diffusion, 79
Adenosine diphosphate (ADP), 187
Adhesive
   strength of tissues (Figure 11-2), 309
   tensile strength (Table 11-2), 310
   tissue, 305
Adventitia, 155, 187
AgCl, *see* Silver chloride
AISI (American Iron and Steel Institute),
   74
Albumin, 189
Alkali silicate glass, 104
Allotropy, 65
Alpha S-2, 308
Alumina or aluminum oxide (Al_2O_3), 4,
   88
   crystal structure, 47, 235
   density, 257
   entropy, 61
   fiber properties, 104, 112
   mechanical properties (Table 9-2), 257
   probability of failure, 241
   theoretical strength, 88
Aluminum, 180, 212
   oxide, *see* Alumina
   slip system, 90
Alveolar bone, 400
Ames shunt, 323
Amide (R′ CONHR), 105

Amine (R-NH_2), 105
Amino acid, 120
   acid polar, 123
   basic polar, 123
   chemical formula, 120
   naturally occurring (Table 6-2), 122
   ω, 278
Amorphous, 51
Ankle joint, 393
   anatomy (Figure 12-40), 395
   implants (Figures 12-42, 12-43), 396, 397
Anodic
   back emf, 216
   reaction, 218
Antioxidant, 291
Aortic valve, 333
Arrhenius equation, 62
Arterial wall elastin variation (Figure 6-31),
   158
Asbestos, properties, 112
ASTM (American Society for Testing and
   Materials), 195
   F75, 201
   F86, 201
   F90, 201
   F562, 201
   F563, 201
   F648, 276
Atactic, 53
Atrio-ventricular (AV) node, 341
Austenite, γ-Fe, 74, 90
Autoclave, *see* Sterilization, steam

Bakelite, *see* Phenolformaldehyde
   Bainite, 75

Corrosion (*cont.*)
  Galvanic, 198, 205
  grain boundary, 198
  pitting, 195
  products, 227
  stress, 222
Cotton, 306
Covalent bonding, *see* Bonding
Crack (c), 101
Creep
  properties, 20
  recovery, 29
  testing, 140
Critical nucleus size (r*), 72
Critical resolved shear stress (CRSS), 91
Cross-link, 55, 274, 290, 306, 316
Crystal system, 45
  illustrations, (Figure 3-8), 45
  imperfections, 48
Cuprophane, 345
Curie temperature, 252
Cyanoacrylate, 308, 316

Dacron, *see* Polyethylene terephthalate
Dashpot, 26
Deciduous teeth, 134
Defects
  grain boundaries, 50
  line, 49
  point, 48
Degree of polymerization (DP), 269
Dentin, 134
Dentinal tubules, 134
Denture, 412
Derlin, 293
Desmosine, structure (Figure 6-4), 126
Deterioration
  metals, 218
  polymers, 295
Diamond (C), 88
  entropy, 61
  heat of vaporization, 44
  theoretical strength, 88
Dibenzoyl peroxide, *see* Peroxide
Diffusion, 76
  constant, 76
  vacancy, 79
Diffusivity, *see* Diffusion, constant
Dimethylchlorosilane, 291
Dimethylsiloxane, 184; *see also* Silicone rubber

Dislocations
  climb, 94
  cross-slip, 94
  density, 97
  edge, 50, 92
  energetics, 91
  line defects, 49
  movement, 93
  screw, 50, 92

Ear
  anatomy (Figure 11-12), 323
  implant types (Figure 11-13), 329
  implants, 321
Eastman 910, 308
Elastic
  constant, 13
  energy, 49, 102
  limit, 32
  modulus of materials (Table 2-1), 13
  region, 14
Elasticity, 11
  of non-Hookean materials, 32
Elastin, 125
  composition (Table 6-4), 126
  variation in arterial wall (Figure 6-31), 158
Elastomer, 32, 54
Elbow
  anatomy (Figure 12-47), 402
  implant types (Table 12-6), 403
  joint replacement, 399
Electrets, 189
Electric potential, 178
  of fractured tibia (Figure 7-6), 178
Electrical stimulation, 337
Electrode
  porous, 343
  potential (Table 8-13), 221
  standard hydrogen, 220
Electromagnetic stimulation, 337
Electron paramagnetic resonance (EPR), 108, 284
Embolus, 186
Enamel, 134
Endosteum, 132
Endothelial cell, 172
Endurance limit, 24
Energy
  activation, 62